Reliability, Maintenance and Logistic Support
- A Life Cycle Approach

Reliability, Maintenance and Logistic Support
- A Life Cycle Approach

by

U Dinesh Kumar
Indian Institute of Management, Calcutta

John Crocker
J Knezevic
M El-Haram
Mirce Akademy, UK

Kluwer Academic Publishers
Boston/Dordrecht/London

Distributors for North, Central and South America:
Kluwer Academic Publishers
101 Philip Drive
Assinippi Park
Norwell, Massachusetts 02061 USA
Telephone (781) 871-6600
Fax (781) 871-6528
E-Mail < kluwer@wkap.com >

Distributors for all other countries:
Kluwer Academic Publishers Group
Distribution Centre
Post Office Box 322
3300 AH Dordrecht, THE NETHERLANDS
Telephone 31 78 6392 392
Fax 31 78 6546 474
E-Mail < orderdept@wkap.nl >

 Electronic Services < http://www.wkap.nl >

Library of Congress Cataloging-in-Publication Data
Reliability maintenance and logistic support: a life cycle approach / by U. Dinesh
Kumar ... [et al.].
     p.cm.
  Includes bibliographical references and index.
  ISBN 0-412-84240-8
   1. Reliability (Engineering) 2. Maintainability (Engineering) 3. Accelerated life
testing. 4. Industrial equipment--Reliability. 5. Plant maintenance. I. Kumar, Dinesh.

TS173.R46 2000
620'.0045--dc21   00-056140

Printed on acid-free paper.

Printed in the United States of America

To my Wonderful wife Haritha and Gorgeous daughter Prathusha

U Dinesh Kumar

To Frances, Elizabeth and Louise for their love, support and encouragement.

John Crocker

To the memory of Centre M.I.R.C.E, University of Exeter

J Knezevic

To my family

M El-Haram

Contents

Preface

There has to be a beginning to every great undertaking

Sir Francis Drake

It happens that we are writing this preface, just days after the start of the 2000 *Formula 1* motor racing season. As usual, at this time in the season, many of the cars failed to finish. It has often been observed that, *"to finish first you must first finish"*. Formula 1, and their American cousins, *Indy 500* cars represent the pinnacle of automotive engineering design: they are aerodynamic, fast, light, manoeuvrable and, by and large, safe. These cars are, however, not intended to last decades. If they survive to the end of the race (between 200 and 500 miles) without the need for any maintenance, that takes more than 10 seconds, then they can be considered to have met one of their prime requirements. In much the same way, a combat aircraft is not expected to cruise at high altitudes, at sub-sonic speeds for 18 hours a day nor is a commercial airliner expected to fly at Mach 2, at altitudes below 250 feet (80 metres) pulling between –4.5 and +9 G. Having said that, however, the owners of both types of aircraft want and expect maximum availability at minimum (through-life) cost. Both also want to be able to carry a maximum payload, over maximum range with minimum fuel burn, minimum maintenance, minimum support and maximum safety and reliability.

Reliability, Maintenance and Logistic Support play a crucial role in achieving a competitive product. While manufacturing and equipment cost are important for the success of a product, they are not the sole domains in realising its competitive edge. Improved manufacturing and operating quality and performance coupled with reduced acquisition cost and in-service cost of ownership are important in achieving business success. For example, Airlines need equipment with Reliability, Maintenance and Supportability "designed in" so that their aircraft can have high levels of dispatch reliability and availability with affordable maintenance and support costs. The early phase of design offers the best opportunity to address reliability, maintenance and logistic support and thus the life cycle effectiveness. Life cycle cost analysis provides a meaningful way of integrating reliability, maintenance and supportability to enhance the product performance and sales opportunities.

The main objective of the book is to provide an integrated approach to reliability, maintainability, maintenance and logistic support analysis. We not only look at ways we can improve the design process to ensure the product offers *value for money, more for less, more bangs per buck, better cheaper faster* but we also consider how owners can get the most from these products once they have been entered service.

The additional objectives of the book are:

1. Introduce the concept of reliability, maintenance and logistic support and their role in system life cycle and effectiveness.
2. Introduce the basic probability and statistical techniques that are essential for modelling reliability, maintenance and supportability problems.
3. Introduce reliability measures: how to predict them; how to determine them from in-service data; how to use them.
4. Analysis of advanced models in Reliability.
5. Discuss basic and advanced concepts in maintenance including preventive, corrective and condition based maintenance.
6. Discuss maintenance management and optimisation concepts, such as reliability-centred maintenance and age-related maintenance.
7. Provide basic concepts in supportability and integrated logistic support.
8. Discuss techniques for design for reliability, maintenance and supportability.
9. Analysis of simple and advanced models in spares forecasting and optimisation.
10. Discuss data analysis, data management and data mining techniques.

In the first chapter we introduce the concept of reliability, maintenance and supportability (RMS) and their role in product success and life cycle effectiveness. A case study on *Apollo 13* is used to illustrate some of the problems during design and operation stage that can affect the reliability of a product. In Chapter 2, we introduce the concept of probability, random variables and probability distributions. In particular, we will look at ways of describing time-to-failure, time-to-repair and time-to-support using such distributions as the exponential, normal, Weibull, Gamma, and lognormal and how these can be used to provide better RMS models.

Chapter 3 introduces number of useful reliability measures including the failure function, reliability function, hazard function, mean time between failure (MTBF) and maintenance free operating period (MFOP). The characteristics and applications of these measures are discussed. We will look at how these might be used to provide more meaningful measures of reliability. The concept of life exchange rate matrix (LERM) is introduced to recognise that not all components in a system have same duty cycle. LERM provides a normalised unit for measuring the age of a system.

Chapter 4 deals with the basic mathematical tools required for predicting reliability of series, parallel, series-parallel, complex and network systems. Many real life examples will be used for illustration and a case study based on an aircraft engine is presented.

Basic maintainability and maintenance concepts are introduced in Chapter 5. Maintainability measures, level of maintenance and maintenance classification are discussed in this chapter. In addition, Maintenance policies such as corrective, preventive and condition-based maintenance are discussed. Maintenance models and optimisation procedures such as reliability centred maintenance, age related maintenance with case study on aircraft engine is discussed in Chapter 6.

Chapters 7, 8 and 9 are dedicated to supportability issues. Chapter 7 introduces the important role of supportability in product life cycle and supportability measures. Chapter 8 discusses several models for forecasting spares requirements. In particular, we investigate the advantages and disadvantages, when to use and when not to use Poisson models, Renewal theory, marginal analysis and Monte Carlo simulations. Chapter 9 is dedicated to Integrated Logistics Support (ILS). Case studies from British Airways and Aircraft engine are used to illustrate some of the concepts discussed in these chapters.

In chapter 10, we discuss availability concepts like inherent, operational and achieved availability. The chapter also analyses the effect of reliability, maintenance and logistic support on availability as well as looking at how different maintenance and support policies can help to realise the full potential of the system. Chapter 11 considers some of the more traditional approaches such as reliability, maintainability and supportability allocation, Failure Mode Effects and Criticality Analysis (FMECA), Fault Tree Analysis (FTA), Fault Tolerant Software (FTS) and Life Cycle Costing (LCC).

We conclude the book with a chapter on various methods for estimating the type and parameters of time-to-failure, time-to-repair and time-to-support distributions. This explains some of the standard methods including mean and median rank regression and maximum likelihood estimation. We will also look at how to get the maximum information from in-service data and how to use it to demonstrate reliability and maintenance requirements.

Most of the materials presented in this book are tested from time to time with different types of students, starting with technicians up to top level management from different multi-national companies around the world. The book is intended for Under-graduate and Post-graduate students from all engineering disciplines. Above all, the book is designed to be a useful reference by reliability, maintenance and supportability engineers from all types of industry and those people encumbered with tasks of operating, maintaining and supporting the complex systems produced by these industries safely and cost-effectively.

As Alan Mulally, in the capacity of the General Manager of Boeing 777 division use to say frequently: *we are where we are,* certainly this book can be improved, and we are looking forward to receiving critical reviews of the book from students, teachers, and practitioners. We hope you will all gain as much knowledge, understanding and pleasure from reading this book as we have from writing it.

U Dinesh Kumar   John Crocker
Indian Institute of Management Calcutta   Mirce Akademy, UK

J Knezevic   M El-Haram
Mirce Akademy, UK   Mirce Akademy, UK

Acknowledgements

This book is closely based on lectures given by us at several Universities in UK and around the World, and at Mirce Akademy, UK. We thank all our past and present students for providing a stimulating discussion during the lecture, which ultimately gave energy and enthusiasm to write this book.

It gives us a great pleasure in thanking Dr. Dinesh Verma of Lockheed Martin Federal Systems, Robert Knotts of Mirce Akademy, Clive Nicholas of University of Exeter, and Prof. Malcom Horner of Dundee University for their help with few sections of this book. Special thanks to Indian Institute of Management Calcutta and Mirce Akademy, UK for providing the resources required for completing this book.

We would like to thank our families, partners, friends and colleagues without whose support and patience we would never have completed our task.

Finally we would like to thank Gary Folven and Carolyn Ford of Kluwer Academic Publishers for their continuous help and encouragement throughout the preparation of this book.

Chapter 1

Reliability, Maintenance, and Logistic Support – Introduction

All the business of war, and indeed all the business of life, is to endeavour to find out what you don't know from what you do.

Duke of Wellington

1.1.   INTRODUCTION

Ever since the Industrial Revolution began some 2½ centuries ago, customers have demanded *better, cheaper, faster, more for less,* through greater reliability, maintainability and supportability (RMS). As soon as people set themselves up in business to provide products for others and not just for themselves, their customers have always wanted to make sure they were not being exploited and that they were getting *value for money* and products that would be *fit for purpose.*

Today's customers are no different. All that has changed is that the companies have grown bigger, the products have become more sophisticated, complex and expensive and, the customers have become more demanding and even less trusting. As in all forms of evolution, the *Red Queen Syndrome* (Carroll, L. 1871, Ridley, R. 1993) is forever present – in business, as in all things, you simply have to keep running faster to stand still. No matter how good you make something, it will never remain good enough for long

Operators want infinite performance, at zero life-cycle cost, with 100% availability from the day they take to delivery to the day they dispose of it. It is the task of the designer/manufacturer/supplier/producer to get as near as possible to these extremes, or, at the very least, nearer than their competitors. In many cases, however, it is not simply sufficient to tell the (potential) customer how well they have met these requirements, rather, they will be required to produce demonstrable evidence to substantiate these claims. In the following pages, we hope to provide you with the techniques and methodologies that will enable you to do this and, through practical examples, explain how they can be used.

The success of any business depends on the effectiveness of the process and the product that business produces. Every product in this world is made to perform a function and every customer/user would like her product to maintain its functionality until has fulfilled its purpose or, failing that, for as long as possible. If this can be done with the minimum of maintenance but, when there is a need for maintenance, that this can be done in the minimum time, with the minimum of disruption to the operation requiring the minimum of support and expenditure then so much the better. As the consumer's awareness of, and demand for, quality, reliability and, availability increases, so too does the pressure on industry to produce products, which meet these demands. Industries, over the years, have placed great importance on engineering excellence, although some might prefer to use the word "hubris". Many of those, which have survived, however, have done so by manufacturing highly reliable products, driven by the market and the expectations of their customers.

The operational phase of complex equipment like aircraft, rockets, nuclear submarines, trains, buses, cars and computers is like an orchestra, many individuals, in many departments doing a set of interconnected activities to achieve maximum effectiveness. Behind all of these operations are certain inherent characteristics (design parameters) of the product that plays a crucial role in the overall success of the product. Three such characteristics are reliability, maintainability and supportability, together we call them RMS. All these three characteristics are crucial for any operation. Billions of dollars are spent by commercial and military operators every year as a direct consequence of the unreliability, lack of maintainability and poor supportability of the systems they are expected to operate.

Modern industrial systems consist of complex and highly sophisticated elements, but at the same time, users' expectations regarding trouble free operation is ever present and even increasing. A Boeing 777 has over 300,000 unique parts within a total of around 6 million parts (half of them are nuts, bolts and rivets). Successfully operating, maintaining and supporting such a complex system demands integrated tools, procedures and techniques. Failure to meet high reliability, maintainability and supportability can have costly and far-reaching effects. Losing the services of airliners, such as the Boeing 747, can cost as high as $ 300,000 per day in forfeited revenue alone. Failure to dispatch a commercial flight on time or its cancellation is not only connected to the cost of correcting the failure, but also to the extra crew costs, additional passenger handling and loss of passenger revenue. Consequently, this will have an impact on the competitiveness, profitability and market share of the airline concerned. 'Aircraft on Ground' is probably the most dreaded phrase in the commercial airlines' vocabulary. And, although the costs and implications may be different, it is no more popular with military operators.

Figure 1.1 shows the factors contributing to delays suffered by Boeing 747s in service with a long haul airline (Knotts 1996); technical delay and cancellations account for about 20% of the total. Costs per minute delay for different aircraft type are shown in Figure 1.2. Here the delay costs are attributable to labour charges, airport fees, air traffic control costs, rescheduling costs, passenger costs (food, accommodation, transport and payoffs).

Boeing 747 - Delay Causes

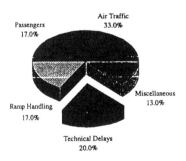

Figure 1.1 Boeing 747 Delay Causes

Industries have learned from past experience and through cutting edge research how to make their products safe and reliable. NASA, Boeing, Airbus, Lockheed Martin, Rolls-Royce, General Electric, Pratt and Whitney, and many, many more, are producing extremely reliable products. For example, over 25% of the jetliners in US have been in service for over 20 years and more than 500 over 25 years, nearing or exceeding their original design life (Lam, M., 1995). The important message is that these aircraft are still capable of maintaining their airworthiness; they are still safe and reliable. But, we cannot be complacent, even the best of organisations can have their bad days. The losses of the Challenger Space Shuttle in 1986, and Apollo 13 are still very fresh in many of our memories.

Customers' requirements generally exceed the capabilities of the producers. Occasionally, these go beyond what is practically, and sometimes even theoretically, possible. An example of this could be the new reliability requirement, *maintenance and failure free operating period*, (Hockley *et al* 1996, Dinesh Kumar et al, 1999, 2000). High reliability is certainly a desirable function, but so to is maintainability and excellent logistic support. It is only through all three that the life-cycle cost can be driven down whilst the level of availability is driven up.

Combat aircraft are expensive and so are their crews, so no operator wants to lose either. At the same time, deploying large ground forces to maintain and support them is also expensive and, potentially hazardous. It is therefore not surprising that the operators are looking to the manufacturers to produce aircraft so reliable that they can go for weeks without any maintenance. The question is, however, can we achieve the necessary level of reliability, with sufficient confidence, at an affordable price, to meet this requirement?

Recent projects such as the Ultra Reliable Aircraft (URA) and Future Offensive Air Systems (FOAS) place a new dimension to the reliability requirement. The operators/users would like to have Maintenance Free Operating Periods (MFOP), during which the probability that the system will need restorative maintenance is very low. Between each of these periods, sufficient maintenance will done to ensure the system will survive the next MFOP with the same probability. Only time will tell whether this policy becomes adopted but there is no doubt that the days of the MTBF (mean time between failures) and its inverse, the [constant] failure rates are surely numbered. Science, mathematics and probability theory are slowly finding their way into the after-market business and with them will come the need for better educated people who understand these new concepts, techniques and methodologies. And, it will not just affect military aircraft, buyers of all manufactured products will demand greater value for money, at the time of purchase, of course, but more than that they will expect it throughout its life. Manufacturers who have relied on unreliability will need to re-think their policies, processes and finances.

1.2.   THE LIFE CYCLE OF A SYSTEM

Fundamental to any engineering design practice is an understanding of the cycle, which the product goes through during its life. The life cycle begins at the moment when an idea of a new system is born and finishes when the system is safely disposed. In other words, the life cycle begins with the initial identification of the needs and requirements and extends through planning, research, design, production, evaluation, operation, maintenance, support and its ultimate phase out (Figure 1.3).

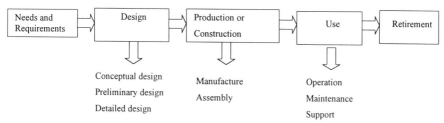

Figure 1.3 Life cycle of the system.

Manufacturers who specialise in military hardware will often be approached, either directly or through an advertised "invitation to tender" to discuss the latest defence requirement. For most other manufacturers, it is generally up to them to identify a (potential) market need and decide whether they can meet that need in a profitable way. The UK MoD approached BAE Systems to bring together a consortium (including representatives of the MoD and RAF) for an air system that would out-perform all existing offensive systems, both friend and foe, and that would include all of the concepts identified as practical in the URA research project. Airbus Industries, on the other hand, decided, based on their extensive market research, that there was a sufficient market need for a very large aircraft that could carry well in excess of 500 passengers, at least across the Pacific from Tokyo to Los Angeles and possibly even non-stop between London and Sydney. It will be many years before we will know whether either of these aircraft will get off the ground and very much longer to see if they prove a business success for their manufacturers.

The first process then is a set of tasks performed to identify the needs and requirements for a new system and transform them into its technically meaningful definition. The main reason for the need of a new system could be a new function to be performed (that is there is a new market demand for a product with the specified function) or a deficiency of the present system. The deficiencies could be in the form of: 1. Functional deficiencies, 2.

Inadequate performance, 3. Inadequate attributes. 4. Poor reliability, 5. High maintenance and support costs, 5. Low sales figures and hence low profits.

The first step in the conceptual design phase is to analyse the functional need or deficiency and translate it into a more specific set of qualitative and quantitative requirements. This analysis would then lead to conceptual system design alternatives. The flow of the conceptual system design process is illustrated in Figure 1.4 (D Verma and J Knezevic, 1995). The output from this stage is fed to the preliminary design stage. The conceptual design stage is the best time for incorporating reliability, maintainability and supportability considerations. In the case of FOAS, for example, various integrated project teams with representatives of the users, suppliers and even academia will drawn together to come up with new ideas and set targets, however, impractical. It was largely a result of this activity that the concepts of the MFOP and the uninhabited combat air vehicle (UCAV) were born.

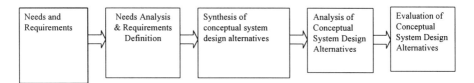

Figure 1.4 Conceptual system design process

The main tasks during the preliminary design stage are system functional analysis such as operational functions, maintenance functions, allocations of performance and effectiveness factors and the allocation of system support requirement (Blanchard, 1991). It is at this time that the concepts are brought down to earth out of the "blue sky". Groups will be required to put these ideals into reality possibly via technical development programs or abandon them until the next time.

The main tasks performed during the detailed design stage 1. Development of system/product design, 2. Development of system prototype, and 3. System prototype test and evaluation. Design is the most important and crucial stage in the product life cycle. Reliability, maintainability and supportability depend on the design and are the main drivers of the operational availability and costs. It is during this stage that safety, reliability and maintainability demonstrations can be performed and, from these, maintenance and support plans can be decided.

The production/construction process is a set of tasks performed in order to transform the full technical definition of the new system into its physical existence. The main tasks performed during this process are 1. Manufacture/Production/Test of prime system elements, 2. System assessment, 3. Quality Assurance, and 4. System Modification. During the production/construction process the system is physically created in accordance with the design definition. The input characteristics of the production process are the raw material, energy, equipment, facilities and other ingredients needed for the production/construction of the new system. The output characteristics are the full physical existence of the functional system.

1.3.   CONCEPT OF FAILURE

As with so many words in the English language, *failure* has come to mean many things to many people. Essentially, a *failure of a system* is any event or collection of events that causes the system to lose its *functionability* where *functionability is the inherent characteristic of a product related to its ability to perform a specified function according to the specified requirements under the specified operating conditions.* (Knezevic 1993) Thus a system, or indeed, any component within it, can only be in one of two states: *state of functioning* or; *state of failure*.

In many cases, the transition between these states is effectively instantaneous; a windscreen shatters, a tyre punctures, a blade breaks, a transistor blows. There is insufficient time to detect the onset or prevent the consequences. However, in many other cases, the transition is gradual; a tyre or bearing wears, a crack propagates across a disc, a blade "creeps" or the performance starts to drop off. In these circumstances, some form of *health monitoring* may allow the user to take preventative measures. Inspecting the amount of tread on the tyres at regular intervals, scanning the lubricating oil for excessive debris, boroscope inspection to look for cracks or using some form trending (e.g. Kalman Filtering) on the specific fuel consumption can alert the user to imminent onset of failure. Similarly, any one of the many forms of non-destructive testing may be used (as appropriate) on components that have been exposed during the recovery of their parent component to check for damage, deterioration, erosion, corrosion or any of the other visible or physically detectable signs that might cause the component to become non-functionable.

With many highly complex systems, whose failure may have serious or catastrophic consequences, measures are taken, wherever possible, to mitigate against such events. Cars are fitted with dual braking systems, aircraft with (at least) triple hydraulic systems and numerous other instances of *redundancy*. In these cases, it is possible to have a failure of a component without a failure of the system. The recovery of the failed item, via a maintenance action, may be deferred to a time which is more convenient to the operator, safe in the knowledge that there is an acceptably high probability that the system will continue operating safely for a certain length of time. If one of the flight control computers on an aircraft fails, its functions will instantly and automatically be taken over by one of the other computers. The flight will generally be allowed to continue, uninterrupted to its next scheduled destination. Depending on the level of redundancy and regulations/certification, further flights may be permitted, either until another computer fails or, the aircraft is put in for scheduled maintenance.

Most commercial airliners are fitted with two, or more, engines. Part of the certification process requires a practical demonstration that a fully loaded aircraft can take-off safely even if one of those engines fails at the most critical time; "rotation" or "weight-off-wheels". However, even though the aircraft can fly with one engine out of service, once it has landed, it would not then be permitted to take-off again until that engine has been returned to a state of functioning (except under very exceptional circumstances). With the latest large twins (e.g. Airbus 330 and Boeing 777), a change in the airworthiness rules has allowed them to fly for extended periods following the in-flight shutdown of one of the engines, generally referred to ETOPS (which officially stands for *extended twin operations over sea* or, unofficially, *engines turn or passengers swim*). This defines the maximum distance (usually expressed in minutes of flying time) the aircraft can be from a suitable landing site at any time during the flight. It also requires an aircraft that has "lost" an engine to fly to immediately divert to a landing site that is within this flying time. Again, having landed, that aircraft would not be permitted to take off until it was fitted with two functionable engines. In this case, neither engine is truly redundant but, the system (aircraft) has a limited level of fault/failure tolerance.

Most personal computers (PC) come complete with a "hard disc". During the life of the PC, it is not uncommon for small sectors of these discs to become unusable. Provided the sector did not hold the file access table (FAT) or key system's files, the computer is not only able to detect these sectors but it will mark them as unusable and avoid writing any data to them. Unfortunately, if there was already data on these sectors before they become unusable, this will no longer be accessible, although with special software, it may be possible to recover some of it. Thus, the built-in test software of the computer is able to provide a level of fault tolerance which is often totally invisible to the user, at least until the whole disc crashes or the fault affects a critical part of a program or data. Even under these circumstances, if that program or data has been backed up to another disc/storage medium, it should be possible to restore the full capability of the system usually with a level of manual intervention. So there is both fault tolerance and redundancy although the latter is usually at the discretion of the user.

1.4. APOLLO 13 – CASE STUDY

On March 14, 1968, a NASA subcontractor shipped a cryogenic oxygen tank to a California assembly plant where it was installed into a spacecraft for an upcoming mission element, which was containing a thermostat, designed for only 28 volts. Later the impact of this defect was exacerbated by ground procedure that inadvertently overheated the tank and destroyed insulation protecting its internal wiring. On April 13, 1970, at 55 hours, 54 minutes and 53 seconds into the mission, some 200 000 miles from Earth, when astronaut Jack Swigert, responding to a routine daily request from the ground, switched on the cryogenic fan to stir up the contents of the oxygen tanks, a spark was generated that ignited insulation material, raising the temperature and pressure to the point where the tank exploded. The oxygen inside the tank flashed instantly into gas and filled bay four of the service module, blowing out the ship's external panel which collided with the orbiter's high-gain antenna and causing the failure that caused a loss of breathable oxygen and power in the command-service module. The three-man crew was forced to abandon the spacecraft and survive in the Lunar Excursion Module, for over 80 hours, until just a few hours before splashdown.

The Apollo spacecraft's electrical system was designed to operate on 28 volts of current. Consequently, When North American first awarded tank contract to Beech Aircraft, they were told that the thermostat switches, like other switches and systems aboard the ship, should be made compatible with the spacecraft's 28-volts power grid. However, this voltage was not the only current the spacecraft would ever be required to accept. During the weeks and months preceding the launch, the ship spent much of its time connected to launch-pad generators at Cape Canaveral, so that pre-flight equipment could be run. The generators used there were dynamos, which charge out current of 65 volts. Having learned this, North American became concerned that high voltage would cook delicate heating system in the cryogenic tanks before the craft ever left the pad, and decided to change the specification. The subcontractor was informed that it should change the original heater plans and replace the entire heating system. Inexplicably, the engineers neglected to change the specifications on the thermostat switches, leaving the originally designed 28-volt switches in the new 65-volt heaters. Despite rigorous control by Beech, North American and NASA technicians, the discrepancies were not discovered.

The tanks that flew aboard Apollo 13 were shipped to North American plant in Downey California. There, they were attached to a metal frame, or shelf, and installed in service module 106, which was scheduled to fly during 1969's Apollo 10 mission. As additional technical improvements were made in the design of oxygen tanks the engineer decided to remove the existing tanks from the Apollo 10, service module and replace them with newer ones. The tanks that had been installed on the ship would be upgraded and placed in another service module, for use on another flight. Removing cryogenic tanks from an Apollo spacecraft was a delicate job. Since it was nearly impossible to separate any one tank from the tangle of pipes and cables that ran from it, the entire shelf, along with all of its associated hardware, would have to be removed. In order to do this, technicians would attach a train to the edge of the shelf, remove the four bolts that are in place, and pull the assembly out. On October 21st 1968, Rockwell Engineers unbolted the tank shelf in the spacecraft and began to lift it carefully from the ship. Unknown to the crane operators, one of the four bolts had been left in place. When the winch motor was activated, the shelf rose only two inches before the bolt caught, the cranes slipped, and the shelf dropped back into place. The jolt caused by the drop was a small one but the procedure for dealing with it was clear. Any accident on the factory floor, no matter how minor, required that the spacecraft components involved be inspected to ensure that they had not suffered any damage. The tanks on the dropped shelf were examined and found to be unharmed. Shortly afterwards, they were removed, upgraded, and reinstalled in the service module, which was to become part of the spacecraft called Apollo. In early 1970, the Saturn Five Booster with Apollo 13 mounted at its tip was taken out to the launch pad and readied for an April lift off. One of the most important milestones in the weeks leading up to an Apollo launch was the exercise known as countdown demonstration test. During the Apollo 13's demonstration test, no significant problems occurred. At the end of the long dress rehearsal, however, the ground crew did report a small anomaly. The cryogenic system, which had to be emptied off its super cold liquids before the spacecraft was shut down, was behaving bulkily. The draining for the cryogenic tanks was not ordinarily complicated. It required engineers simply to pump gaseous oxygen into the tank through one line, forcing the liquids out through another line. Both hydrogen tanks, as well as oxygen tank one emptied easily. But oxygen tank 2 seemed jammed, venting only about 8 percent of its super cold slush and then releasing no more.

Examining the schematics of the tank and its manufacturing history, the engineers at the Cape and at Beech Aircraft believed they knew what the problem was. When the shelf was dropped eighteen months ago, they now suspected, the tank had suffered more damage than the factory technicians at first realised, knocking one of the drain tubes in the neck of the vessel out of alignment. This would cause the gaseous oxygen pumped through the line leading into the tank to leak directly into the line leading out of the tank, disturbing almost none of the liquid oxygen it was supposed to be pumping away. The de-tanking method would be used only during pad tests. During the flight itself, the liquid oxygen contained in the vessel would be channelled out not through the venting tube, but through an entirely different set of tubes leading either to the fuel cells or to the atmospheric system that pressurised the cockpit with breathable air. If the engineers could figure out some way to get the tank emptied today, therefore they could fill it up again on launch day and never have to worry about the fill lines and drain lines again. The technicians came up with very elegant and simple solution. At its present super-cold temperature and relatively low pressure, the liquid in the tank was not going anywhere. However, what would happen if the heaters were used? Why not just flip the warming coils on, cook the slush up, and force the entire load of O_2 out of the vent line?

The alternative would have been to remove the tank altogether and replace it with a new one. However, the latter solution required forty-five hours for replacement plus the time needed for testing and checking it out. This would cause the miss of the launch window, and the whole mission would have been postponed for at least a month.

Unfortunately, none of the launch-pad test crew knew that the wrong thermostat was in the tank, thus, they could not analyse the consequence of leaving the heaters on for too long. The technicians proceeded with their plan on the evening of March 27th, the warming coils in spacecraft's second oxygen tank were switched on. As the large quantity of O_2 was trapped in the tank, the engineers predicted that eight hours is required for the last few wisps of gas to vent away. During that time the temperature in the tank could have climbed above the 80-degree, but the engineers knew they could rely on the thermostat to take care of any problem. When this thermostat reached the critical temperature, however, and tried to open up, the 65 volts surging through it fused it instantly shut. However, the technicians on the Cape launch pad had no way of knowing that the tiny component that was supposed to protect the oxygen tank had welded closed. A single engineer was assigned to oversee the detanking procedure, but all his instrument told him about the cryogenic heater was that the contacts on the thermostat remained shut as they should be, indicating that the tank had not heated up too much. The only possible clue that the system was not functioning properly, was provided by a gauge on the launch pad's instrument panel, that constantly monitored the temperature inside the oxygen tanks. If the readout climbed above 80 degrees, the technicians would know that the thermostat had failed, and would shut the heater off manually.

Unfortunately, the readout on the instrument panel was not able to record temperature above 80 degrees. With so little chance that the temperature inside the tank would ever rise that far, and with 80 degrees representing the bottom of the danger zone, the design team who designed the instrument panel saw no reason to peg the gauge any higher, designating 80 as its upper limit. What engineer on duty that night did not know, could not know, was that with the thermostat fused shut, the temperature inside this particular tank was climbing indeed, up to 1000 degrees mark. All the time the heater was left running, the temperature reading was registering a warm but safe 80 degrees. At the end of eight hours, the last of the troublesome liquid oxygen had cooked away, as was expected, but so too had most of the Teflon insulation that protected the tank's internal wiring. Coursing through the now empty tank was a web of raw, spark-prone copper, soon to be re-immersed in the one liquid likelier than any other to propagate a fire: pure oxygen.

Seventeen days later and nearly 200,000 miles out in space, Jack Swigert, responding to a routine daily request from the ground, switched on the cryogenic fan to stir up the contents of the oxygen tanks. The first two times Swigert had complied with this instrument, the fan had operated normally. This time, however, a spark flew from a naked wire, igniting the remains of the Teflon. The sudden build-up of heat and pressure in the pure-oxygen environment blew off the neck of the tank, the weakest part of the vessel. The 300 pounds of oxygen inside the tank flashed instantly into gas and filled bay four of the service module, blowing out the ship's external panel and causing the bang that so startled the crew. As the curved piece of hull flew past, it collided with the orbiter's high-gain antenna, causing the mysterious channel switching that the communications officer reported at the same moment the astronauts were reporting their bang and jolt.

Though tank one was not directly damaged by the blast, it did share some common plumbing with tank two; as the explosion ripped these delicate pipes away, the undamaged tank found a leak path through the lines and bled its contents away into space. Making matters worse, when the explosion shook the ship, it caused the valves that fed several of the attitude-control thrusters to slam shut, permanently disabling those jets. As the ship rocked from both the tank one venting and the explosion itself, the autopilot began firing the thrusters to try to stabilise the spacecraft's attitude. But with only some of the jets working, the ship control of the half-crippled attitude system, his luck was little better. Within two hours, the spacecraft was drifting and dead.

Chapter 2

Probability Theory

We do not know how to predict what would happen in any given circumstances, and we believe now that it is possible, that the only thing that can be predicted is the probability of different events

Richard Feynman

Probability theory plays a leading role in modern science in spite of the fact that it was initially developed as a tool that could be used for guessing the outcome of some games of chance. Probability theory is applicable to everyday life situations where the outcome of a repeated process, experiment, test, or trial is uncertain and a prediction has to be made.

In order to apply probability to everyday engineering practice it is necessary to learn the terminology, definitions and rules of probability theory. This chapter is not intended to a rigorous treatment of all-relevant theorems and proofs. The intention is to provide an understanding of the main concepts in probability theory that can be applied to problems in reliability, maintenance and logistic support, which are discussed in the following chapters.

2.1.   PROBABILITY TERMS AND DEFINITIONS

In this section those elements essential for understanding the rudiments of elementary probability theory will be discussed and defined in a general manner, together with illustrative examples related to engineering practice.

To facilitate the discussion some relevant terms and their definitions are introduced.

Experiment

An experiment is a well-defined act or process that leads to a single well-defined outcome. Figure 2.1 illustrates the concept of random experiments. Every experiment must:

1. Be capable of being described, so that the observer knows when it occurs.
2. Have one and only one outcome, so that the set of all possible outcomes can be specified.

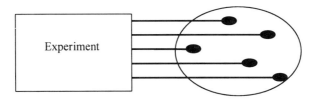

Figure 2.1 Graphical Representation of an Experiment and its outcomes.

Elementary event

An elementary event is every separate outcome of an experiment.

From the definition of an experiment, it is possible to conclude that the total number of elementary events is equal to the total number of possible outcomes, since every experiment must have only one outcome.

Sample space

The set of all possible distinct outcomes for an experiment is called the sample space for that experiment.

Most frequently in the literature the symbol S is used to represent the *sample space*, and small letters, $a,b,c,..$, for elementary events that are possible outcomes of the experiment under consideration. The set S may contain either a finite or an infinite number of elementary events. Figure 2.2 is a graphical presentation of the sample space.

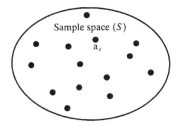

Figure 2.2 Graphical Presentation of the Sample Space

<u>*Event*</u>

Event is a subset of the sample space, that is, a collection of elementary events.

Capital letters A, B, C, …, are usually used for denoting events. For example, if the experiment performed is measuring the speed of passing cars at a specific road junction, then the elementary event is the speed measured, whereas the sample space consists of all the different speeds one might possibly record. All speed events could be classified in, say, four different speed groups: *A* (less than 30 km/h), *B* (between 30 and 50 km/h), *C* (between 50 and 70 km/h) and *D* (above 70 km/h). If the measured speed of the passing car is, say 35 km/h, then the event *B* is said to have occurred.

2.2. ELEMENTARY THEORY OF PROBABILITY

The theory of probability is developed from axioms proposed by the Russian mathematician *Kolmogrov*. In practice this means that its elements have been defined together with several axioms which govern their relations. All other rules and relations are derived from them.

2.2.1 Axioms of Probability

In cases where the outcome of an experiment is uncertain, it is necessary to assign some measure that will indicate the chances of occurrence of a particular event. Such a measure of events is called the *probability of the event* and symbolised by *P(.), (P(A)* denotes the probability of event *A*). The function which associates each event *A* in the sample space *S*, with the probability measure *P(A)*, is called the *probability function* - the probability of that event. A graphical representation of the probability function is given in Figure 2.3.

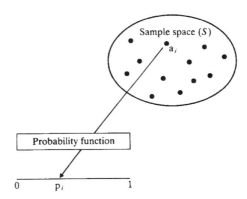

Figure 2.3 Graphical representation of probability function.

Formally, the probability function is defined as:

A function which associates with each event A, a real number, P(A), the probability of event A, such that the following axioms are true:

1. $P(A) > 0$ for every event A,
2. $P(S) = 1$, (probability of the sample space)
3. The probability of the union of mutually exclusive events is the sum of their probabilities, that is

$$P(A_1 \cup A_2 ... \cup A_n) = P(A_1) + P(A_2) + ... + P(A_n)$$

In essence, this definition states that each event A is paired with a non-negative number, probability *P(A)*, and that the probability of the sure event S, or *P(S)*, is always 1. Furthermore, if A_1 and A_2 are any two mutually exclusive events (that is, the occurrence of one event implies the non-occurrence of the other) in the sample space, the probability of their union $P(A_1 \cup A_2)$, is simply the sum of their two probabilities, $P(A_1) + P(A_2)$.

2.2.2   Rules of Probability

The following elementary rules of probability are directly deduced from the original three axioms, using the set theory:

Probability Theory

a) For any event A, the probability of the complementary event, written A', is given by

$$P(A') = 1 - P(A) \tag{2.1}$$

b) The probability of any event must lie between zero and one inclusive:

$$0 \le P(A) \le 1 \tag{2.2}$$

c) The probability of an empty or impossible event, ϕ, is zero.

$$P(\phi) = 0 \tag{2.3}$$

d) If occurrence of an event A implies that an event B occurs, so that the event class A is a subset of event class B, then the probability of A is less than or equal to the probability of B:

$$P(A) \le P(B) \tag{2.4}$$

e) In order to find the probability that A or B or both occur, the probability of A, the probability of B, and also the probability that both occur must be known, thus:

$$P(A \cup B) = P(A) + P(B) - P(A \cap B) \tag{2.5}$$

f) If A and B are mutually exclusive events, so that $P(A \cap B) = 0$, then

$$P(A \cup B) = P(A) + P(B) \tag{2.6}$$

g) If n events form a partition of S, then their probabilities must add up to one:

$$P(A_1) + P(A_2) + ... + P(A_n) = \sum_{i=1}^{n} P(A_i) = 1 \tag{2.7}$$

2.2.3   Joint Events

Any event that is an intersection of two or more events is a joint event.

There is nothing to restrict any given elementary event from the sample space from qualifying for two or more events, provided that those events are not mutually exclusive. Thus, given the event A and the event B, the joint event is $A \cap B$. Since a member of $A \cap B$ must be a member of set A, and also of set B, both A and B events occur when $A \cap B$ occurs. Provided that the elements of set S are all equally likely to occur, the probability of the joint event could be found in the following way:

$$P(A \cap B) = \frac{\text{number of elementary events in A} \cap \text{B}}{\text{total number of elementary events}}$$

2.2.4   Conditional Probability

If A and B are events in a sample space which consists of a finite number of elementary events, the conditional probability of the event B given that the event A has already occurred, denoted by $P(B|A)$, is defined as:

$$P(B|A) = \frac{P(A \cap B)}{P(A)}, \qquad\qquad P(A) > 0 \qquad\qquad (2.8)$$

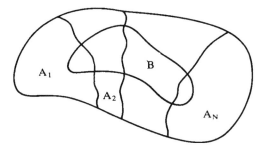

Figure 2.4 Graphical Presentation of the Bayes Theorem

The conditional probability symbol, $P(B|A)$, is read as the probability of *B* given *A*. It is necessary to satisfy the condition that *P(A)>0*, because it does not make sense to consider the probability of *B* given *A* if event *A* is impossible. For any two events *A* and *B,* there are two conditional probabilities that may be calculated:

$$P(B|A) = \frac{P(A \cap B)}{P(A)} \qquad and \qquad P(A|B) = \frac{P(A \cap B)}{P(B)}$$

(The probability of B, given A)   (The probability of A, given B)

One of the important application of conditional probability is due to Bayes theorem, which can be stated as follows:

If $(A_1, A_2, ..., A_N)$ represents the partition of the sample space (N mutually exclusive events), and if B is subset of $(A_1 \cup A_2 \cup ... \cup A_N)$, as illustrated in Figure 2.4, then

$$P(A_i|B) = \frac{P(B|A_i)P(A_i)}{P(B|A_1)P(A_1)+...+P(B|A_i)P(A_i)+...+P(B|A_N)P(A_N)} \quad (2.9)$$

2.3. PROBABILITY AND EXPERIMENTAL DATA

The classical approach to probability estimation is based on the relative frequency of the occurrence of that event. A statement of probability tells us what to expect about the relative frequency of occurrence, given that enough observations are made. In the long run, the relative frequency of occurrence of an event, say A, should approach the probability of this event, if independent trials are made at random over an indefinitely long sequence. This principle was first formulated and proved by James Bernoulli in the early eighteenth century, and is now well-known as *Bernoulli's theorem*:

If the probability of occurrence of an event A is p, and if n trials are made independently and under the same conditions, then the probability that the relative frequency of occurrence of A, (defined as $f(A) = N(A)/n$) differs from p by any amount, however small, approaches zero as the number of trials grows indefinitely large. That is,

$$P(|N(A)/n| - p| > s) \to 0, \qquad\qquad as\ n \to \infty \qquad (2.10)$$

where s is some arbitrarily small positive number. This does not mean that the proportion of $\dfrac{N(A)}{n}$ occurrences among any n trial must be p; the proportion actually observed might be any number between 0 and 1. Nevertheless, given more and more trials, the relative frequency of $f(A)$ occurrences may be expected to become closer and closer to p.

Although it is true that the relative frequency of occurrence of any event is exactly equal to the probability of occurrence of any event only for an infinite number of independent trials, this point must not be over stressed. Even with relatively small number of trials, there is very good reason to expect the observed relative frequency to be quite close to the probability because the rate of convergence of the two is very rapid. *However, the main drawback of the relative frequency approach is that it assumes that all events are equally likely (equally probable).*

2.4.   PROBABILITY DISTRIBUTION

Consider the set of events $A_1, A_2, \ldots, A_n$, and suppose that they form a partition of the sample space S. That is, they are mutually exclusive and exhaustive. The corresponding set of probabilities, $P(A_1), P(A_2), \ldots, P(A_n)$, is a probability distribution. An illustrative presentation of the concept of probability distribution is shown in Figure 2.5.

As a simple example of a probability distribution, imagine a sample space of all Ford cars produced. A car selected at random is classified as a saloon or coupe or estate. The probability distribution might be:

Event	Saloon	Coupe	Estate	Total
P	0.60	0.31	0.09	1.00

All events other than those listed have probabilities of zero

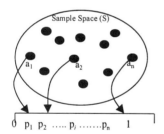

S	a_1	a_2	....	a_n
P	p_1	p_2	....	p_n
Probability Distribution				

Figure 2.5 Graphical representation of Probability Distribution

2.5.   RANDOM VARIABLE

A function that assigns a number (usually a real number) to each sample point in the sample space S is a random variable.

Outcomes of experiments may be expressed in numerical and non-numerical terms. In order to compare and analyse them it is much more convenient to deal with numerical terms. So, for practical applications, it is necessary to assign a numerical value to each possible elementary event in a sample space S. Even if the elementary events themselves are already expressed in terms of numbers, it is possible to reassign a unique real number to each elementary event. The function that achieves this is known as *the random variable*. In other words, a random variable is a real-valued function defined in a sample space. Usually it is denoted with capital letters, such as X, Y and Z, whereas small letters, such as x, y, z, a, b, c, and so on, are used to denote particular values of random variables, see Figure 2.6

If X is a random variable and r is a fixed real number, it is possible to define the event A to be the subset of S consisting of all sample points 'a' to which the random variable X assigns the number r, $A = (a : X(a) = r)$. On the other hand, the event A has a probability $p = P(A)$. The symbol p can be interpreted, generally, as the probability that the random variable X takes on the value r, $p = P(X = r)$. Thus, the symbol $P(X = r)$ represents the probability function of a random variable.

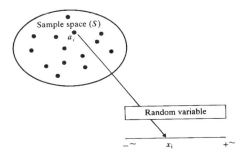

Figure 2.6 Graphical Representation of Random Variable

Therefore, by using the random variable it is possible to assign probabilities to real numbers, although the original probabilities were only defined for events of the set S, as shown in Figure 2.7.

The probability that the random variable X, takes value less than or equal to certain value 'x', is called the *cumulative distribution function, F(t)*. That is,

$$P[X \leq x] = F(x)$$

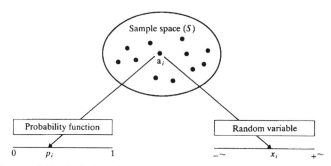

Figure 2.7 Relationship between probability function and a random variable

2.5.1   Types of random variables

Depending on the values, which the random variables can assume, random variables, can be classified as discrete or continuous. The main characteristics, similarities and differences for both types will be briefly described below.

Discrete random variables

If the random variable X can assume only a particular finite or countably infinite set of values, it is said to be a discrete random variable.

There are very many situations where the random variable X can assume only a particular *finite* or *countably infinite* set of values; that is, the possible values of X are finite in number or they are infinite in number but can be put in a one-to-one correspondence with a set of real number.

Continuous random variables

If the random variable X can assume any value from a finite or an infinite set of values, it is said to be a continuous random variable.

Let us consider an experiment, which consists of recording the temperature of a cooling liquid of an engine in the area of the thermostat at a given time. Suppose that we can measure the temperature exactly, which means that our measuring device allows us to record the temperature to any number of decimal points. If X is the temperature reading, it is not possible for us to specify a finite or countably infinite set of values. For example, if one of the finite set of values is 75.965, we can determine values 75.9651, 75.9652, and so on, which are also possible values of X. What is being demonstrated here is that the possible values of X consist of the set of real numbers, a set which contains an infinite (and uncountable) number of values.

Continuous random variables have enormous utility in reliability, maintenance and logistic support as the random variables time to failure, time to repair and the logistic delay time are continuous random variables.

2.6. THE PROBABILITY DISTRIBUTION OF RANDOM VARIABLE

Taking into account the concept of the probability distribution and the concept of the random variable, it could be said that the probability distribution of the random variable is a set of pairs, $\{r_i, P(X = r_i), \ i = 1, n\}$ as shown in Figure 2.8.

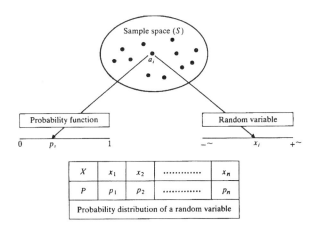

Figure 2.8 Probability Distribution of a Random Variable

The easiest way to present this set is to make a list of all its members. If the number of possible values is small, it is easy to specify a probability distribution. On the other hand, if there are a large number of possible values, a listing may become very difficult. In the extreme case where we have an infinite number of possible values (for example, all real numbers between zero and one), it is clearly impossible to make a listing. Fortunately, there are other methods that could be used for specifying a probability distribution of a random variable:

a) Functional method, where a specific mathematical functions exist from which the probability of any value or interval of values can be calculated.
b) Parametric method, where the entire distribution is represented through one or more parameters known as summary measures.

2.6.1   Functional Method

By definition, a function is a relation where each member of the domain is paired with one member of the range. In this particular case, the relation between numerical values which random variables can have and their probabilities will be considered. The most frequently used functions for the description of probability distribution of a random variable are the probability mass function, the probability density function, and the cumulative distribution function. Each of these will be analysed and defined in the remainder of this chapter.

Probability mass function

This function is related to a discrete random variable and it represents the probability that the discrete random variable, X, will take one specific value x_i, $p_i = P(X = x_i)$. Thus, a probability mass function, which is usually denoted as $PMF(.)$, places a mass of probability p_i at the point of x_i on the X-axis. Given that a discrete random variable takes on only n different values, say $a_1, a_2, ..., a_n$, the corresponding $PMF(.)$ must satisfy the following two conditions:

1. $P(X = a_i) \geq 0$ $for\, i = 1, 2, ..., n$

2. $\sum_{i=1}^{n} P(X = a_i) = 1$ (2.11)

In practice this means that the probability of each value that X can take must be non-negative and the sum of the probabilities must be 1. Thus, a probability distribution can be represented by the set of pairs of values (a_i, p_i), where $i = 1, 2, ..., n$, as shown in Figure 2.9. The advantage of such a graph over a listing is the ease of comprehension and a better provision of a notion for the nature of the probability distribution.

Probability Theory

Figure 2.9 Probability Mass Function

Probability density function

In the previous section, discrete random variables were discussed in terms of probabilities $P(X = x)$, the probability that the random variables take on an *exact* value. However, consider the example of an infinite set for a specific type of car, where the volume of the fuel in the fuel tank is measured with only some degree of accuracy. What is the probability that a car selected at random will have *exactly* 16 litres of fuel? This could be considered as an event that is defined by the interval of values between, say 15.5 and 16.5, or 15.75 and 16.25, or any other interval $\pm 16 \times 0.1i$, where i is not exactly zero. Since the smaller the interval, the smaller the probability, the probability of exactly 16 litres is, in effect, zero. In general, for continuous random variables, the occurrence of any exact value of X may be regarded as having zero probability.

The Probability Density Function, $f(x)$, which represents the probability that the random variable will take values within the interval $x \leq X \leq x + \Delta(x)$, when $\Delta(x)$ approaches zero, is defined as:

$$f(x) = \lim_{\Delta(x) \to 0} \frac{P(x \leq X \leq x + \Delta(x))}{\Delta x} \tag{2.12}$$

As a consequence, the probabilities of a continuous random variable can be discussed only for *intervals* of X values. Thus, instead of the probability that X takes on a specific value, say $'a'$, we deal with the so-called *probability density* of X at $'a'$, symbolised by $f(a)$. In general, the probability distribution of a continuous random variable can be represented by its *Probability Density Function, PDF*, which is defined in the following way:

$$P(a \leq X \leq b) = \int_a^b f(x)dx \tag{2.13}$$

A fully defined probability density function must satisfy the following two requirements:

$$f(x) \geq 0 \qquad\qquad \text{for all } x$$

$$\int_{-\infty}^{+\infty} f(x)dx = 1$$

The *PDF* is always represented as a smooth curve drawn above the horizontal axis, which represents the possible values of the random variable *X*. A curve for a hypothetical distribution is shown in Figure 2.10 where the two points *a* and *b* on the horizontal axis represent limits which define an interval.

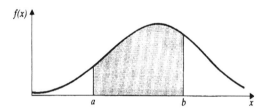

Figure 2.10 Probability Density Function for a Hypothetical Distribution

The shaded portion between *'a'* and *'b'* represents the probability that *X* takes on a value between the limits *'a'* and *'b'*.

Cumulative distribution function

The probability that a random variable *X* takes on a value at or below a given number 'a' is often written as:

$$F(a) = P(X \leq a) \tag{2.14}$$

The symbol $F(a)$ denotes the particular probability for the interval $X \leq a$. The general symbol $F(x)$ is sometimes used to represent the

Probability Theory

function relating the various values of X to the corresponding cumulative probabilities. This function is called the *Cumulative Distribution Function, CDF*, and it must satisfy certain mathematical properties, the most important of which are:

1. $0 \le F(x) \le 1$
2. $if \ a < b, \quad F(a) \le F(b)$
3. $F(\infty) = 1 \qquad and \qquad F(-\infty) = 0$

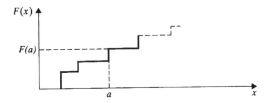

Figure 2.11 Cumulative Distribution Function for Discrete Variable

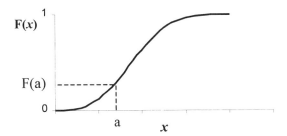

Figure 2.12 Cumulative Distribution Function for Continuous Variable

The symbol $F(x)$ can be used to represent the cumulative probability that X is less than or equal to x. It is defined as:

$$F(a) = \sum_{i=1}^{n} P(X = x_i) \qquad (2.15)$$

For the discrete random variables, whereas in the case of continuous random variables it will take the following form:

$$F(a) = \int_{-\infty}^{a} f(x)dx \qquad\qquad (2.16)$$

Hypothetical cumulative distribution functions for both types of random variable are given in Figures 2.11 and 2.12.

2.6.2   Parametric Method

In some situations it is easier and even more efficient to look only at certain characteristics of distributions rather than to attempt to specify the distribution as a whole. Such characteristics summarise and numerically describe certain features for the entire distribution. Two general groups of such characteristics applicable to any type of distribution are:

a) *Measures of central tendency* (or location) which indicate the typical or the average value of the random variable.
b) *Measures of dispersion* (or variability) which show the spread of the difference among the possible values of the random variable.
In many cases, it is possible to adequately describe a probability distribution with a few measures of this kind. It should be remembered, however, that these measures serve only to summarise some important features of the probability distribution. In general, they do not completely describe the entire distribution.

One of the most common and useful summary measures of a probability distribution is *the expectation* of a random variable, *E(X)*. It is a unique value that indicates a location for the distribution as a whole (In physical science, expected value actually represents the Centre of gravity). The concept of expectation plays an important role not only as a useful measure, but also as a central concept within the theory of probability and statistics.

If a random variable, say *X*, is discrete, then its expectation is defined as:

$$E(X) = \sum_{x} x \times P(X = x) \qquad\qquad (2.17)$$

Where the sum is taken for all the values that the variable *X* can assume. If the random variable is continuous, the expectation is defined as:

$$E(X) = \int_{-\infty}^{+\infty} x \times f(x)dx \qquad\qquad (2.18)$$

Probability Theory

Where the sum is taken over all values that X can assume. For a continuous random variable the expectation is defined as:

$$E(X) = \int_{-\infty}^{+\infty}[1 - F(x)]dx \qquad (2.19)$$

If c is a constant, then

$$E(cX) = c \times E(X) \qquad (2.20)$$

Also, for any two random variables X and Y,

$$E(X + Y) = E(X) + E(Y)$$

Measures of central tendency

The most frequently used measures are:

The mean of a random variable is simply the expectation of the random variable under consideration. Thus, for the random variable, X, the mean value is defined as:

$$Mean = E(X) \qquad (2.21)$$

The median, is defined as the value of X which is midway (in terms of probability) between the smallest possible value and the largest possible value. The median is the point, which divides the total area under the *PDF* into two equal parts. In other words, the probability that X is less than the median is $1/2$, and the probability that X is greater than the median is also $1/2$. Thus, if $P(X \leq a) \geq 0.50$ and $P(X \geq a) \geq 0.50$ then $'a'$ is the *median* of the distribution of X. In the continuous case, this can be expressed as:

$$\int_{-\infty}^{a}f(x)dx = \int_{a}^{+\infty}f(x)dx = 0.50 \qquad (2.22)$$

The mode, is defined as the value of X at which the *PDF* of X reaches its highest point. If a graph of the *PMF (PDF)*, or a listing of possible values of X along with their probabilities is available, determination of the mode is quite simple.

A central tendency parameter, whether it is mode, median, mean, or any other measure, summarises only a certain aspect of a distribution. It is easy to find two distributions which have the same mean but which are not at all similar in any other respect.

Measures of dispersion

The mean is a good indication of the location of a random variable, but *no single value need be exactly like the mean*. A deviation from the mean, D, expresses the measure of error made by using the mean as a particular value:

$$D = x - M$$

Where, x, is a possible value of the random variable, X. The deviation can be taken from other measures of central tendency such as the median or mode. It is quite obvious that the larger such deviations are from a measure of central tendency, the more the individual values differ from each other, and the more apparent the spread within the distribution becomes. Consequently, it is necessary to find a measure that will reflect the spread, or variability, of individual values.

The expectation of the deviation about the mean as a measure of variability, *E(X - M)*, will not work because the expected deviation from the mean must be zero for obvious reasons. The solution is to find the *square* of each deviation from the mean, and then to find the expectation of the squared deviation. This characteristic is known as a *variance of the distribution, V*, thus:

$$V(X) = E(X - Mean)^2 = \sum (X - Mean)^2 \times P(x) \qquad \text{if X is discrete} \quad (2.23)$$

$$V(X) = E(X - Mean)^2 = \int_{-\infty}^{+\infty} (X - Mean)^2 \times f(x)dx \qquad \text{if X is continuous} \quad (2.24)$$

The positive square root of the variance for a distribution is called the *Standard Deviation, SD*.

Probability Theory

$$SD = \sqrt{V(X)} \qquad (2.25)$$

Probability distributions can be analysed in greater depth by introducing other summary measures, known as *moments*. Very simply these are expectations of different powers of the random variable. More information about them can be found in texts on probability.

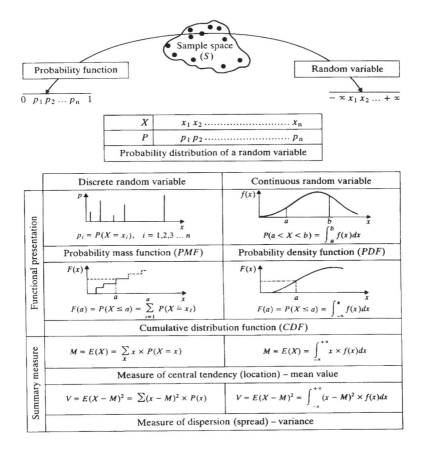

Figure 2.13 Probability System for Continuous Random Variable

Variability

The standard deviation is a measure that shows how closely the values of random variables are concentrated around the mean. Sometimes it is difficult to use only knowledge of the standard deviation, to decide whether the dispersion is considerably large or small, because this will depend on the mean value. In this case the parameter known as coefficient of variation, CV_X, defined as

$$CV_X = \frac{SD}{M} \qquad (2.26)$$

Coefficient of variation is very useful because it gives better information regarding the dispersion. The concept thus discussed so far is summarised in Figure 2.13. In conclusion it could be said that the probability system is wholly abstract and axiomatic. Consequently, every fully defined probability problem has a unique solution.

2.7.   DISCRETE THEORETICAL PROBABILITY DISTRIBUTIONS

In probability theory, there are several rules that define the functional relationship between the possible values of random variable X and their probabilities, $P(X)$. As they are purely theoretical, i.e. they do not exist in reality, they are called *theoretical probability distributions*. Instead of analysing the ways in which these rules have been derived, the analysis in this chapter concentrates on their properties. It is necessary to emphasise that all theoretical distributions represent the family of distributions defined by a common rule through unspecified constants known as *parameters of distribution*. The particular member of the family is defined by fixing numerical values for the parameters, which define the distribution. The probability distributions most frequently used in reliability, maintenance and the logistic support are examined in this chapter.

Among the family of theoretical probability distributions that are related to discrete random variables, the Binomial distribution and the Poisson distribution are relevant to the objectives set by this book. A brief description of each now follows.

2.7.1   Bernuolli Trials

The simple probability distribution is one with only two event classes. For example, a car is tested and one of two events, pass or fail, must occur, each with some probability. The type of experiment consisting of series of independent trials, each of which can eventuate in only one of two outcomes are known as *Bernuolli Trials*, and the two event classes and their associated probabilities a *Bernuolli Process*. In general, one of the two events is called a "success" and the other a "failure" or "nonsuccess". These names serve only to tell the events apart, and are not meant to bear any connotation of "goodness" of the event. The symbol p, stands for the probability of a success, q for the probability of failure $(p + q = 1)$. If 5 independent trials are made $(n = 5)$, then $2^5 = 32$ different sequences of possible outcomes would be observed.

The probability of given sequences depends upon p and q, the probability of the two events. Fortunately, since trials are independent, it is possible to compute the probability of any sequence.

If all possible sequences and their probabilities, are written down the following fact emerges: *The probability of any given sequences of n independent Bernuolli Trials depends only on the number of successes and p*. This is regardless of the order in which successes and failure occur in sequence, the probability is

$$p^r q^{n-r}$$

where r is the number of successes, and $n-r$ is the number of failures. Suppose that in a sequence of 10 trials, exactly 4 success occurs. Then the probability of that particular sequence is $p^4 q^6$. If $p = \frac{2}{3}$, then the probability can worked out from:

$$\left(\frac{2}{3}\right)^4 \left(\frac{1}{3}\right)^6$$

The same procedure would be followed for any r successes out of n trials for any p. Generalising this idea for any r, n, and p, we have the following principle:

In sampling from the Bernuolli Process with the probability of a success equal to p, the probability of observing exactly r successes in n independent trials is:

$$P(r \ successes | n, p) = \binom{n}{r} p^r q^{n-r} = \frac{n!}{r!(n-r)!} p^r q^{n-r} \qquad (2.27)$$

2.7.2   The Binomial Distribution

The theoretical probability distribution, which pairs the number of successes in n trials with its probability, is called the binominal distribution.

This probability distribution is related to experiments, which consist of a series of independent trials, each of which can result in only one of two outcomes: success and or failure. These names are used only to tell the events apart. By convention the symbol p stands for the probability of a success, q for the probability of failure $(p + q = 1)$.

The number of successes, x in n trials is a discrete random variable which can take on only the whole values from 0 through n. The PMF of the Binomial distribution is given by:

$$PMF(x) = P(X = x) = \binom{n}{x} p^x q^{n-x}, \qquad 0 < x < n \qquad (2.28)$$

where:

$$\binom{n}{x} p^x q^{n-x} = \frac{n!}{x!(n-x)!} p^x q^{n-x} \qquad (2.29)$$

The binomial distribution expressed in cumulative form, representing the probability that X falls at or below a certain value $'a'$ is defined by the following equation:

$$P(X \le a) = \sum_{i=0}^{a} P(X = x_i) = \sum_{i=0}^{a} \binom{n}{i} p^i q^{n-i} \qquad (2.30)$$

As an illustration of the binomial distribution, the PMF and CDF are shown in Figure 2.14 with parameters $n = 10$ and $p = 0.3$.

Probability Theory

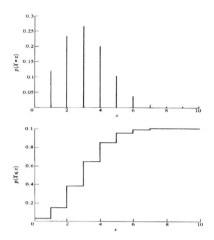

Figure 2.14 PMF and CDF For Binomial Distribution, $n = 10$, $p = 0.3$

$$E(X) = np \qquad (2.31)$$

Similarly, because of the independence of trials, the variance of the binomial distribution is the sum of the variances of the individual trials, or $p(1-p)$ summed n times:

$$V(X) = np(1-p) = npq \qquad (2.32)$$

Consequently, the standard deviation is equal to:

$$Sd(X) = \sqrt{npq} \qquad (2.33)$$

Although the mathematical rule for the binomial distribution is the same regardless of the particular values which parameters n and p take, the shape of the probability mass function and the cumulative distribution function will depend upon them. The *PMF* of the binomial distribution is symmetric if p = 0.5, positively skewed if $p < 0.5$, and negatively skewed if $p > 0.5$.

2.7.3   The Poisson Distribution

The theoretical probability distribution which pairs the number of occurrences of an event in a given time period with its probability is called the Poisson distribution. There are experiments where it is not possible to observe a finite sequence of trials. Instead, observations take place over a continuum, such as time. For example, if the number of cars arriving at a specific junction in a given period of time is observed, say for one minute, it is difficult to think of this situation in terms of finite trials. If the number of binomial trials n, is made larger and larger and p smaller and smaller in such a way that np remains constant, then the probability distribution of the number of occurrences of the random variable approaches the Poisson distribution. The probability mass function in the case of the Poisson distribution for random variable X can be expressed as follows:

$$P(X = x|\lambda) = \frac{e^{-\lambda}\lambda^x}{x!} \qquad \text{where } x = 0, 1, 2, \ldots \qquad (2.34)$$

λ is the *intensity of the process* and represents the expected number of occurrences in a time period of length t. Figure 2.15 shows the *PMF* of the Poisson distribution with $\lambda = 5$

The Cumulative Distribution Function for the Poisson distribution

$$F(x) = P(X \leq x) = \sum_{i=0}^{x} \frac{e^{i}\lambda^x}{i!} \qquad (2.35)$$

The CDF of the Poisson distribution with $\lambda = 5$ is presented in Figure 2.16. Expected value of the distribution is given by

$$E(X) = \sum_{x=0} xP(X = x) = \sum_{x=0} x \frac{e^{-\lambda}\lambda^x}{x!}$$

Applying some simple mathematical transformations it can be proved that:

$$E(X) = \lambda \qquad (2.36)$$

which means that the expected number of occurrences in a period of time t is equal to np, which is equal to λ.

The variance of the Poisson distribution is equal to the mean:

$$V(X) = \lambda \qquad (2.37)$$

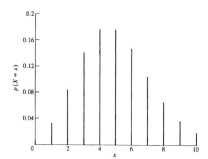

Figure 2.15 PMF of the Poisson Distribution with $\lambda = 5$

Thus, the Poisson distribution is a single parameter distribution because it is completely defined by the parameter λ. In general, the Poisson distribution is positively skewed, although it is nearly symmetrical as λ becomes larger.

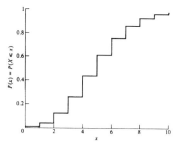

Figure 2.16 CDF of the Poisson Distribution $\lambda = 5$

The Poisson distribution can be derived as a limiting form of the binomial if the following three assumptions were simultaneously satisfied:

1. n becomes large (that is, $n \to \infty$).
2. p becomes small (that is, $p \to 0$).
3. np remains constant.

Under these conditions, the binomial distribution with the parameters n and p, can be approximated to the Poisson distribution with parameter $\lambda = np$. This means that the Poisson distribution provides a good approximation to the binomial distribution if p is very small and n is large.

Since p and q can be interchanged by simply interchanging the definitions of success and failure, the Poisson distribution is also a good approximation when p is close to one and n is large.

As an example of the use of the Poisson distribution as an approximation to the binomial distribution, the case in which $n = 10$ and $p = 0.10$ will be considered.   The Poisson parameter for the approximation is then $\lambda = np = 10 \times 0.10 = 1$.   The binomial distribution and the Poisson approximation are shown in Table 2.2.

The two distributions agree reasonably well. If more precision is desired, a possible rule of thumb is that the Poisson is a good approximation to the binomial if $n / p > 500$ (this should give accuracy to at least two decimal places).

Table 2.2 Poisson Distribution as an Approximation to the Binomial Distribution

	Binomial $P(X = x \vert n = 10, p = 0.1)$	Poisson $P(X = x \vert \lambda = 1)$
0	0.598737	0.606531
1	0.315125	0.303265
2	0.074635	0.075816
3	0.010475	0.012636
4	0.000965	0.001580
5	0.000061	0.000158

2.8.   CONTINUOUS THEORETICAL PROBABILITY DISTRIBUTIONS

It is necessary to emphasise that all theoretical distributions represent the family of distributions defined by a common rule through unspecified constants known as *parameters of distribution*. The particular member of the family is defined by fixing numerical values for the parameters, which define the distribution. The probability distributions most frequently used in reliability, maintainability and supportability engineering are examined in this chapter.   Each of the above mentioned rules define a family of distribution functions.   Each member of the family is defined with a few parameters, which in their own way control the distribution. Parameters of a distribution can be classified in the following three categories (note that not all distributions will have all the three parameters, many distributions may have either one or two parameters):

1. *Scale parameter*, which controls the range of the distribution on the horizontal scale.
2. *Shape parameter*, which controls the shape of the distribution curves.
3. *Source parameter or Location parameter*, which defines the origin or the minimum value which random variable, can have. Location parameter also refers to the point on horizontal axis where the distribution is located.

Thus, individual members of a specific family of the probability distribution are defined by fixing numerical values for the above parameters.

2.8.1 Exponential Distribution

Exponential distribution is fully defined by a single one parameter that governs the scale of the distribution. The probability density function of the exponential distribution is given by:

$$f(x) = \lambda \exp(-\lambda x), x > 0 \qquad (2.38)$$

In Figure 2.17 several graphs are shown of exponential density functions with different values of λ. Notice that the exponential distribution is positively skewed, with the mode occurring at the smallest possible value, zero.

The cumulative distribution of exponential distribution is given by:

$$F(x) = P(X < x) = 1 - \exp(-(\lambda x)) \qquad (2.39)$$

It can be shown that the mean and variance of the exponential distribution are:

$$E(X) = 1/\lambda \qquad (2.40)$$

$$V(X) = (1/\lambda)^2 \qquad (2.41)$$

The standard deviation in the case of the exponential distribution rule has a numerical value identical to the mean and the scale parameter, $SD(X) = E(X) = 1/\lambda$.

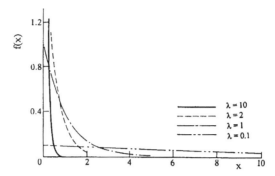

Figure 2.17. Probability density function of exponential distribution for different values of λ

Memory-less Property of Exponential Distribution

One of the unique properties of exponential distribution is that it is the only continuous distribution that has *memory less* property. Suppose that the random variable X measures the duration of time until the occurrence of failure of an item and that it is known that X has an exponential distribution with parameter λ. Suppose the present age of the item is t, that is X > t. Assume that we are interested in finding the probability that this item will not fail for another *s* units of time. This can be expressed using the conditional probability as:

$$P\{X > s+t | x > t\}$$

Using conditional probability of events, the above probability can be written as:

$$P\{X > s+t | X > s\} = \frac{P\{X > s+t \cap X > t\}}{P\{X > t\}} = \frac{P\{X > s+t\}}{P\{X > t\}} \quad (2.42)$$

However we know that for exponential distribution

$$P[X > s+t] = \exp(-\lambda(s+t)) \text{ and } P[X > t] = \exp(-\lambda t)$$

Substituting these expressions in equation (2.42), we get

$$P[X > s + t | X > t] = P[X > s] = \exp(-\lambda s)$$

That is, the conditional probability depends only on the remaining duration and is independent of the current age of the item. *This property is exploited to a great extend in reliability theory.*

2.8.2 Normal Distribution (Gaussian Distribution)

This is the most frequently used and most extensively covered theoretical distribution in the literature. The Normal Distribution is continuous for all values of X between $-\infty$ and $+\infty$. It has a characteristic symmetrical shape, which means that the mean, the median and the mode have the same numerical value. The mathematical expression for its probability density function is as follows:

$$f(x) = \frac{1}{\sigma\sqrt{2\pi}} \exp\left(-\frac{1}{2}\left(\frac{x-\mu}{\sigma}\right)^2\right) \qquad (2.43)$$

Where μ is a location parameter (as it locates the distribution on the horizontal axis) and σ is a scale parameter (as it controls the range of the distribution). μ and σ also represents the mean and the standard deviation of this distribution.

The influence of the parameter μ on the location of the distribution on the horizontal axis is shown in Figure 2.18, where the values for parameter σ are constant. As the deviation of x from the location parameter μ is entered as a squared quantity, *two* different x values, showing the same absolute deviation from μ, will have the same probability density according to this rule. This dictates the symmetry of the normal distribution. Parameter μ can be any finite number, while σ can be any *positive* finite number.

The cumulative distribution function for the normal distribution is:

$$F(a) = P(X \le a) = \int_{-\infty}^{a} f(x) dx$$

where *f(x)* is the normal density function. Taking into account Eq. (2.43) this becomes:

$$F(a) = \int_{-\infty}^{a} \frac{1}{\sigma\sqrt{2\pi}} \exp\left(-\frac{1}{2}\left(\frac{a-\mu}{\sigma}\right)^2\right) dx \qquad (2.44)$$

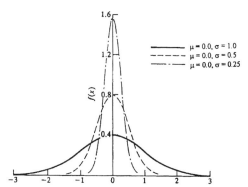

Figure 2.18 Probability density of normal distribution for different σ values

In Figure 2.19 several cumulative distribution functions are given of the Normal Distribution, corresponding to different values of μ *and* σ.

As the integral in Eq. (2.44) cannot be evaluated in a closed form, statisticians have constructed the table of probabilities, which complies with the normal rule for the standardised random variable, Z. This is a theoretical random variable with parameters $\mu = 0$ and $\sigma = 1$. The relationship between standardised random variable Z and random variable X is established by the following expression:

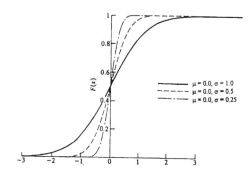

Figure 2.19 Cumulative distribution of normal distribution for different values of μ and σ.

$$z = \frac{x - \mu}{\sigma} \tag{2.45}$$

Making use of the above expression the equation (2.43) becomes simpler:

$$f(z) = \frac{1}{\sigma\sqrt{2\pi}} e^{-\frac{1}{2}z^2} \tag{2.46}$$

The standardised form of the distribution makes it possible to use only one table for the determination of *PDF* for any normal distribution, regardless of its particular parameters (see Table in appendix).

The relationship between *f(x)* and *f(z)* is :

$$f(x) = \frac{f(z)}{\sigma} \tag{2.47}$$

By substituting $\dfrac{x - \mu}{\sigma}$ with z Eq. (2.44) becomes:

$$F(a) = \int_{-\infty}^{z} \frac{1}{\sigma\sqrt{2\pi}} \exp\left(-\frac{1}{2}z^2\right) dz = \Phi\left(\frac{x - \mu}{\sigma}\right) \tag{2.48}$$

where Φ is the standard normal distribution Function defined by

$$\Phi(z) = \int_{-\infty}^{x} \frac{1}{\sqrt{2\pi}} \exp\left(-\frac{1}{2}z^2\right) dx \tag{2.49}$$

The corresponding standard normal probability density function is:

$$f(z) = \frac{1}{\sqrt{2\pi}} \exp\left(-\frac{z^2}{2}\right) \tag{2.50}$$

Most tables of the normal distribution give the cumulative probabilities for various *standardised* values. That is, for a given z value the table provides the cumulative probability up to, and including, that standardised value in a normal distribution. In *Microsoft EXCEL®*, the cumulative distribution function and density function of normal distribution with mean μ and standard deviation σ can be found using the following function.

$F(x) = NORMDIST\ (x,\ \mu,\ \sigma,\ TRUE),\ and\ f(x) = NORMDIST\ (x,\ \mu,\ \sigma,\ FALSE)$

The expectation of a random variable, is equal to the location parameter μ thus:

$$E(X) = \mu \tag{2.51}$$

Whereas the variance is

$$V(X) = \sigma^2 \tag{2.52}$$

Since normal distribution is a symmetrical about its mean, the area between $\mu - k\sigma$, $\mu + k\sigma$ (k is any real number) takes a unique value, which is shown in Figure 2.20.

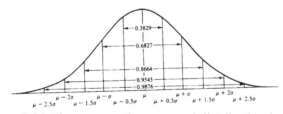

Figure 2.20 The areas under a normal distribution between
$\mu - k\sigma$ and $\mu + k\sigma$

Central Limit Theorem

Suppose X_1, X_2, ... X_n are mutually independent observations on a random variable X having a well-defined mean μ_x and standard deviation σ_x. Let

$$Z_n = \frac{\overline{X} - \mu_x}{\sigma_x / n} \tag{2.53}$$

Where,

$$\overline{X} = \frac{1}{n}\sum_{i=1}^{n} X_i \tag{2.54}$$

and $F_{Z_n}(z)$ be the cumulative distribution function of the random variable Z_n. Then for all z, $-\infty < z < \infty$,

$$\lim_{n\to\infty} F_{Z_n}(z) = F_Z(z) \tag{2.55}$$

where $F_Z(z)$ is the cumulative distribution of standard normal distribution $N(0,1)$. The X values have to be from the same distribution but the remarkable feature is that this distribution does not have to be normal, it can be uniform, exponential, beta, gamma, Weibull or even an unknown one.

2.8.3   Lognormal Distribution

The lognormal probability distribution, can in some respects, be considered as a special case of the normal distribution because of the derivation of its probability function. If a random variable $Y = \ln X$ is normally distributed then, the random variable X follows the lognormal distribution. Thus, the probability density function for a random variable X is defined as:

$$f_X(x) = \frac{1}{x\sigma_l\sqrt{2\pi}} \exp\left(-\frac{1}{2}\left(\frac{\ln x - \mu_l}{\sigma_l}\right)^2\right) \quad \geq 0 \tag{2.56}$$

The parameter μ_l is called the *scale parameter* (see Figure 2.21) and parameter σ_l is called the *shape parameter*. The relationship between parameters μ (location parameter of the normal distribution) and μ_l is defined:

$$\mu = \exp\left(\mu_l + \frac{1}{2}\sigma_l^2\right) \tag{2.57}$$

The cumulative distribution function for the lognormal distribution is defined with the following expression:

$$F_X(x) = P(X \le x) = \int_0^x \frac{1}{x\sigma_l\sqrt{2\pi}} \exp\left(-\frac{1}{2}\left(\frac{\ln x - \mu_l}{\sigma_l}\right)^2\right) dx \quad (2.58)$$

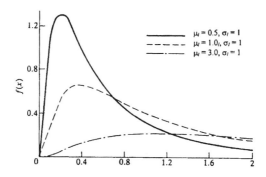

Figure 2.21 Probability density of log-normal distribution

As the integral cannot be evaluated in close form, the same procedure is applied as in the case of normal distribution. Then, making use of the standardised random variable Equation (2.61) transforms into:

$$F_X(x) = P(X \le x) = \Phi\left(\frac{\ln x - \mu_l}{\sigma_l}\right) \quad (2.59)$$

The measures of central tendency in the case of lognormal distributions are defined by the:

(a) Location parameter (Mean)

$$M = E(X) = \exp\left(\mu_l + \frac{1}{2}\sigma_l^2\right) \quad (2.60)$$

(b) Deviation parameter (the variance)

$$V(X) = \exp\left(2\mu_l + \sigma_l\right)^2\left[\exp(\sigma_l^2 - 1)\right] \quad (2.61)$$

2.8.4   Weibull Distribution

This distribution originated from the experimentally observed variations in the yield strength of Bofors steel, the size distribution of fly ash, fibre strength of Indian cotton, and the fatigue life of a *St*-37 steel by the Swedish engineer W.Weibull.   As the Weibull distribution has no characteristic shape, such as the normal distribution, it has a very important role in the statistical analysis of experimental data. The shape of this distribution is governed by its parameter.

The rule for the probability density function of the Weibull distribution is:

$$f(x) = \frac{\beta}{\eta}\left(\frac{x-\gamma}{\eta}\right)^{\beta-1} \exp\left[-\left(\frac{x-\gamma}{\eta}\right)^{\beta}\right] \tag{2.62}$$

where η, β, $\gamma > 0$. As the location parameter v is often set equal to zero, in such cases:

$$f(x) = \frac{\beta}{\eta}\left(\frac{x}{\eta}\right)^{\beta-1} \exp\left[-\left(\frac{x}{\eta}\right)^{\beta}\right] \tag{2.63}$$

By altering the shape parameter β, the Weibull distribution takes different shapes. For example, when $\beta = 3.4$ the Weibull approximates to the normal distribution; when $\beta=1$, it is identical to the exponential distribution. Figure 2.22 shows the Weibull probability density function for selected parameter values.

The cumulative distribution functions for the Weibull distribution is:

$$F(x) = 1 - \exp\left[-\left(\frac{x-\gamma}{\eta}\right)^B\right]$$ (2.64)

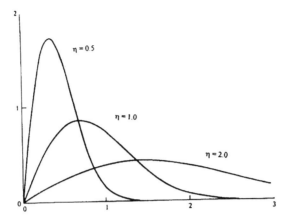

Figure 2.22. Probability density of Weibull distribution with $\beta = 2.0$, $\gamma = 0$, $\eta = 0.5, 1, 2$

For $\gamma = 0$, the cumulative distribution is given by

$$F(x) = 1 - \exp\left[-\left(\frac{x}{\eta}\right)^\beta\right]$$ (2.65)

The expected value of the Weibull distribution is given by:

$$E(X) = \gamma + \eta \times \Gamma\left(\frac{1}{\beta} + 1\right)$$ (2.66)

where Γ is the gamma function, defined as

$$\Gamma(n) = \int_0^\infty e^{-x} \times x^{n-1} dx$$

When n is integer then $\Gamma(n) = (n-1)!$. For other values, one has to solve the above integral to the value. Values for this can be found in Gamma function table given in the appendix. In *Microsoft EXCEL*, Gamma function, $\Gamma(x)$ can be found using the function, *EXP[GAMMALN(x)]*.

The variance of the Weibull distribution is given by:

$$V(X) = (\eta)^2 \left[\Gamma\left(1 + \frac{2}{\beta}\right) - \Gamma^2\left(1 + \frac{1}{\beta}\right) \right] \qquad (2.67)$$

Chapter 3

Reliability Measures

I have seen the future; and it works

Lincoln Steffens

In this chapter we discuss various measures by which hardware and software reliability characteristics can be numerically defined and described. Manufacturers and customers use reliability measure to quantify the effectiveness of the system. Use of any particular reliability measure depends on what is expected of the system and what we are trying measure. Several life cycle decision are made using reliability measure as one of the important design parameter. The reliability characteristics or measures used to specify reliability must reflect the operational requirements of the item. Requirements must be tailored to individual item considering operational environment and mission criticality. In broader sense, the reliability metrics can be classified (Figure 3.1) as: 1. Basic Reliability Measures, 2. Mission Reliability Measures, 3. Operational Reliability Measures, and 4. Contractual Reliability Measures.

Basic Reliability Measures are used to predict the system's ability to operate without maintenance and logistic support. Reliability measures like reliability function and failure function fall under this category.

Mission Reliability Measures are used to predict the system's ability to complete mission. These measures consider only those failures that cause mission failure. Reliability measures such as mission reliability, maintenance free operating period (MFOP), failure free operating period (FFOP), and hazard function fall under this category.

Operational Reliability Measures are used to predict the performance of the system when operated in a planned environment including the combined effect of design, quality, environment, maintenance, support policy, etc. Measures such as Mean Time Between Maintenance (MTBM), Mean Time Between Overhaul (MTBO), Maintenance Free Operating Period (MFOP),

Mean Time Between Critical Failure (MTBCF) and Mean Time Between Unscheduled Removal (MTBUR) fall under this category.

Contractual Reliability Measure is used to define, measure and evaluate the manufacturer's program. Contractual reliability is calculated by considering design and manufacturing characteristics. Basically it is the inherent reliability characteristic. Measures such as Mean Time To Failure (MTTF), Mean Time Between Failure (MTBF) and Failure rate fall under this category.

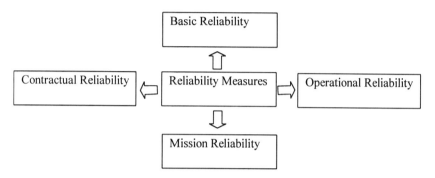

Figure 3.1 Classifications of Reliability Measures

Though we classify the reliability measures into four categories as mentioned above, one may require more than one reliability metric in most of the cases for specifying reliability requirements. Selection of specific measure to quantify the reliability requirements should include mission and logistic reliability along with maintenance and support measures. Currently, many manufacturers specify reliability by using mean time between failure (MTBF) and failure rate. However, MTBF and failure rates have several drawbacks. Recent projects such as Future Offensive Air Systems (FOAS) drive maintenance free operating periods (MFOP) as the preferred reliability requirement.

In the next Section, we define various reliability measures and how to evaluate them in practical problems. All the measures are defined based on the assumption that the time-to-failure (TTF) distribution of the system is known. Procedures for finding the time-to-failure distribution by analysing the failure data that are discussed in Chapter 12.

3.1.   FAILURE FUNCTION

Failure function is a basic (logistic) reliability measure and is defined as the probability that an item will fail before or at the moment of operating time t. *Here time t is used in a generic sense and it can have units such as miles, number of landings, flying hours, number of cycles, etc., depending on the operational profile and the utilisation of the system.* That is, Failure function is equal to the probability that the time-to-failure random variable will be less than or equal a particular value t (in this case operating time, see Figure 3.2a). The failure function is usually represented as *F(t)*.

$F(t)$ = P (failure will occur before or at time t) = P $(TTF \leq t)$

$$= \int_0^t f(u)du \tag{3.1}$$

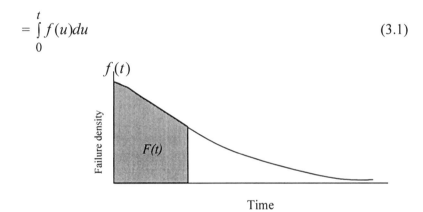

Figure 3.2a. Failure function of a hypothetical distribution

Where $f(t)$ is the probability density function of the time-to-failure random variable *TTF*. Exponential, Weibull, normal, lognormal, Gamma and Gumbel are few popular theoretical distributions that are used to represent failure function. Equation (3.1) is derived by assuming that no maintenance is performed to the system, and gives the probability of failure free operation without maintenance up to time t. However, most of the complex systems will require maintenance at frequent intervals. In such cases, equation (3.1) has to be modified, to incorporate the behaviour of the system under maintenance. Failure functions of few popular theoretical distributions are listed in Table 3.1.

It should be noted that in case of normal distribution the failure function exists between -∞ and +∞, so may have significant value at t ≤ 0. Since negative time is meaningless in reliability, great care should be taken in using normal distribution for the failure function. For μ >> 3σ, probability values for t ≤ 0 can be considered negligible.

Table 3.1 Failure function, $F(t)$, of few theoretical distributions

Distribution	Failure Function, $F(t)$
Exponential	$1 - \exp(-\lambda t) \qquad t > 0, \lambda > 0$
Normal	$\displaystyle\int_0^t \frac{1}{\sigma\sqrt{2\pi}} e^{-\frac{1}{2}\left(\frac{x-\mu}{\sigma}\right)^2} dx \qquad or \qquad \Phi\left(\frac{t-\mu}{\sigma}\right)$ *or NORMDIST(t, μ, σ, TRUE) in EXCEL®*
Lognormal	$\displaystyle\int_0^t \frac{1}{\sigma_l x\sqrt{2\pi}} e^{-\left(\frac{1}{2}\left(\frac{\ln(x)-\mu_l}{\sigma_l}\right)^2\right)} dx \qquad or \qquad \Phi\left(\frac{\ln(t)-\mu_l}{\sigma_l}\right)$ *or NORMDIST(ln(t), μ, σ, TRUE) in EXCEL®*
Weibull	$1 - \exp(-(\frac{t-\gamma}{\eta})^\beta) \qquad \eta, \beta, \gamma > 0, t \geq \gamma$
Gamma	$\displaystyle\frac{1}{\Gamma(\alpha)} \int_0^t \beta^\alpha x^{\alpha-1} e^{-\beta x} dx$

Note that the failure function of normal distribution is defined between 0 and t, since t is greater than 0 for reliability purposes (against the usual limit -∞) Applications of failure function are listed below (Figure 3.2b). Failure functions of various theoretical distributions for different parameter values are shown in Figures 3.3a-3.3c.

Characteristics of failure function

1. Failure function is an increasing function. That is, for $t_1 < t_2$, $F(t_1) \leq F(t_2)$.
2. For modelling purposes it is assumed that the failure function value at time $t = 0$, $F(0) = 0$. However, this assumption may not be valid always. For example, systems can be *dead on arrival*. The value of failure function increases as the time increases and for $t = \infty$, $F(\infty) = 1$.

Applications of failure function

1. *F(t)* is the probability that an individual item will fail by time t.
2. *F(t)* is the fraction of items that fail by time t.
3. 1 - *F(t)* is the probability that an individual item will survive up to time t.

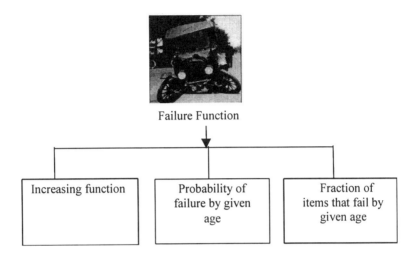

Figure 3.2b. Properties of failure function

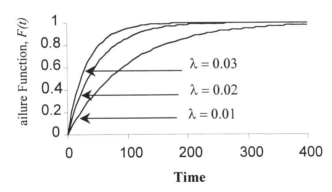

Figure 3.3a: Failure function of exponential distribution for different values of λ

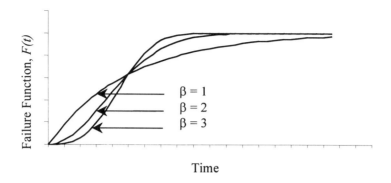

Figure 3.3b Failure function of Weibull distribution for different β values

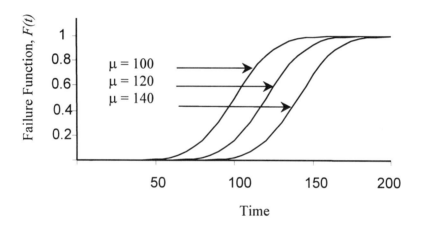

Figure 3.3c Failure function of normal distribution for different μ values

Example 3.1

The time to failure distribution of a sub-system in an aircraft engine follows Weibull distribution with scale parameter $\eta = 1100$ flight hours and the shape parameter $\beta = 3$. Find:

a) Probability of failure during first 100 flight hours.
b) Find the maximum length of flight such that the failure probability is less than 0.05.

SOLUTION:

a) The failure function for Weibull distribution is given by:

$$F(t) = 1 - \exp(-(\frac{t-\gamma}{\eta})^{\beta})$$

It is given that: t = 100 flight hours, η = 1100 flight hours, β = 3 and γ = 0.

Probability of failure within first 100 hours is given by:

$$F(100) = 1 - \exp(-(\frac{100-0}{1100})^3) = 0.00075$$

b) If t is the maximum length of flight such that the failure probability is less than 0.05, we have

$$F(t) = 1 - \exp(-(\frac{t-0}{1100})^3) < 0.05$$

$$= \exp(-(\frac{t}{1100})^3) > 0.95$$

$$= (\frac{t}{1100})^3 > -\ln 0.95 \Rightarrow t = 1100 \times [-\ln(0.95)]^{1/3}$$

Now solving for t, we get t = 408.70 flight hours. The maximum length of flight such that the failure probability is less than 0.05 is 408.70 flight hours.

Example 3.2

The time to failure distribution of a Radar Warning Receiver (RWR) system in a fighter aircraft follows Weibull distribution with scale parameter 1200 flight hours and shape parameter 3. The time to failure distribution of the same RWR in a helicopter follows exponential distribution with scale parameter 0.001. Compare the failure function of the RWR in the fighter aircraft and the helicopter. If the supplier gives a warranty for 750 flight hours, calculate the risk involved with respect to fighter aircraft and the helicopter. (Although we have a same system, the operating conditions have significant impact on the failure function. In this case, RWR in helicopter is subject to more vibrations compared to aircraft).

SOLUTION:

The failure function of RWR on the fighter aircraft is given by:

$$F(t) = 1 - \exp(-(\frac{t}{1200})^3)$$

The failure function of RWR on the helicopter is given by:

$$F(t) = 1 - \exp(-(0.001 \times t))$$

Figure 3.4 depicts the failure function of RWR in fighter aircraft and the helicopter.

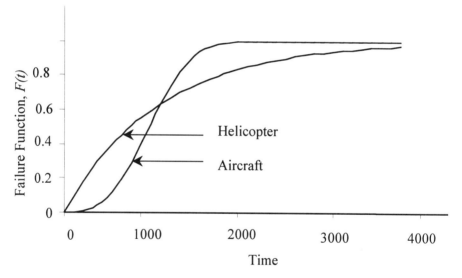

Figure 3.4 Failure function of RWR in fighter aircraft and helicopter

If the supplier provides warranty for 750 flight hours the risk associated with aircraft is given by:

$$F(750) = 1 - \exp(-(\frac{750}{1200})^3) = 0.2166$$

That is, just above 21% percent of RWR are likely to fail if the RWR is installed in the aircraft.

If the RWR is installed in helicopter then the associated risk is given by:

$$F(750) = 1 - \exp(-0.001 \times 750) = 0.5276$$

In the case of helicopter, more than 52% of the RWR's are likely to fail before the warranty period.

3.1.1 Failure function of system under multiple failure mechanisms

It is seldom true that an item's failure is caused by a single failure mechanism. In most of the cases there will be more than one (some times hundreds) mechanism that causes the failure of an item. The expression (3.1) is more appropriate when the failure is caused by a single failure mechanism. However, most of the practical systems fail due to different causes or different failure mechanisms. Assume that the system failure is due to two different failure mechanisms. Let $f_1(t)$ and $f_2(t)$ be the probability density function of the system due to failure mechanism 1 and 2 respectively. Now the probability density function of the time-to-failure of the system caused by either of the failure mechanisms:

$$f(t) = f_1(t)[1 - F_2(t)] + f_2(t)[1 - F_1(t)]$$

where, $F_1(t)$ and $F_2(t)$ the are failure function for failure mechanism 1 and 2 respectively. The failure function of the item under two different failure mechanism is given by:

$$F(t) = \int_0^t \{f_1(x)[1 - F_2(x)] + f_2(x)[1 - F_1(x)]\}dx \qquad (3.2)$$

Example 3.3

Failure of an item is caused by two different failure mechanisms (say failure mechanism A and B). The time-to-failure distribution of the item due to failure mechanism A can be represented by exponential distribution with parameter $\lambda_A = 0.002$ hours. The time-to-failure distribution of the item due to failure mechanism B can be represented by exponential distribution with parameter $\lambda_B = 0.005$ hours. Find the probability that the item will fail before 500 hours of operation.

SOLUTION:
Assume that $f_A(t)$ and $f_B(t)$ represent probability density function of the time-to-failure random variable due to failure mechanism A and B respectively. Thus,

$$f_A(t) = \lambda_A \exp(-\lambda_A t), \quad 1 - F_A(t) = \exp(-\lambda_A t)$$
$$f_B(t) = \lambda_B \exp(-\lambda_B t), \quad 1 - F_B(t) = \exp(-\lambda_B t)$$

Now the failure function of the item is given by:

$$\begin{aligned}
F(t) &= \int_0^t \{\lambda_A \exp(-(\lambda_A + \lambda_B)x) + \lambda_B \exp(-(\lambda_A + \lambda_B)x)dx \\
&= (\lambda_A / \lambda_A + \lambda_B)[1 - \exp(-(\lambda_A + \lambda_B)t] \\
&\quad + (\lambda_B / \lambda_A + \lambda_B)[1 - \exp(-(\lambda_A + \lambda_B)t] \\
&= [1 - \exp(-(\lambda_A + \lambda_B)t]
\end{aligned}$$

Figure 3.5 represents the failure function due to failure mechanism 1, 2 and the system failure function. The probability that the item will fail by 500 hours is given by:

$$F(500) = 1 - \exp(-((0.005 + 0.002) \times 500)) = 0.9698$$

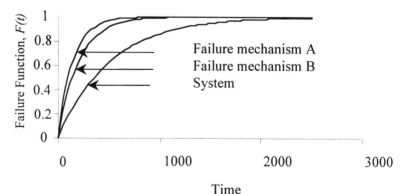

Time
Figure 3.5 Failure function due to different failure mechanisms

3.2.   RELIABILITY FUNCTION

Reliability is the ability of the item to maintain the required function for a specified period of time (or mission time) under given operating conditions. *Reliability function, R(t), is defined as the probability that the system will not fail during the stated period of time, t, under stated operating conditions.*

If *TTF* represents the time-to-failure random variable with failure function (cumulative distribution function) *F(t)*, then the reliability function *R(t)* is given by:

$$R(t) = P\{\text{the system doesn't fail during } [0, t]\} = 1 - F(t) \qquad (3.3)$$

In equation (3.3) we assume that the age of the system before the start of the mission is zero. Thus the equation (3.3) is valid only for new systems or those systems whose failures are not age related (that is, the time-to-failure follows exponential distribution due to *memory less* property of exponential distribution). However, in most of the cases this assumption may not be valid. If the system age is greater than zero at the beginning of the mission, then we have to calculate mission reliability function, which will be discussed later. Figure 3.6 depicts the relation between reliability function and the *TTF* density function. *R(t)* is the area under *TTF* density between t and ∞.

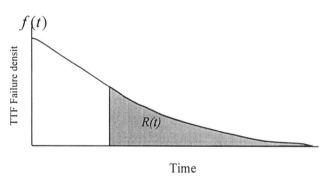

Figure 3.6 Reliability function of a hypothetical probability distribution

Properties of reliability function:

1. Reliability is a decreasing function with time t. That is, for $t_1 < t_2$; $R(t_1) \geq R(t_2)$).
2. It is usually assumed that R (0) = 1. As t becomes larger and larger R(t) approaches zero, that is, R(∞).

Applications of reliability function

1. R(t) is the probability that an individual item survives up to time t.
2. R(t) is the fraction of items in a population that survive up to time t.
3. R(t) is the basic function used for many reliability measures and system reliability prediction.

Reliability function for some important life distributions are given in Table 3.2. Figure 3.7a-c represents reliability function of various theoretical distributions for different parameter values.

Table 3.2. Reliability function, *R(t)*, for popular theoretical distributions

Distribution	Reliability function, R(t)
Exponential	$\exp(-\lambda t)$ $t > 0, \lambda > 0$
Normal	$\Phi(\dfrac{\mu - t}{\sigma}) = 1 - \int\limits_{0}^{t} \dfrac{1}{\sigma\sqrt{2\pi}} e^{-\left(\frac{1}{2}\left(\frac{x-\mu}{\sigma}\right)^2\right)} dx$ *or NORMDIST (μ, t, σ, TRUE) in EXCEL*
Lognormal	$\Phi(\dfrac{\mu_l - \ln t}{\sigma_l}) = 1 - \int\limits_{0}^{t} \dfrac{1}{\sigma_l x\sqrt{2\pi}} e^{-\left(\frac{1}{2}\left(\frac{\ln(x)-\mu_l}{\sigma_l}\right)^2\right)} dx$ or NORMDIST (μ, ln(t), σ, TRUE) in EXCEL
Weibull	$\exp(-(\dfrac{t-\gamma}{\eta})^\beta)$ $\eta, \beta, \gamma > 0, t \geq \gamma$
Gamma	$1 - \dfrac{1}{\Gamma(\alpha)} \int\limits_{0}^{t} \beta^\alpha x^{\alpha-1} e^{-\beta x} dx$

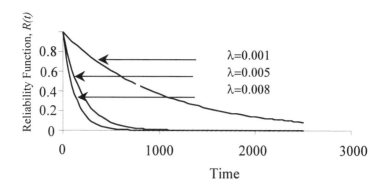

Figure 3.7 a. Reliability function of exponential distribution for different values of λ

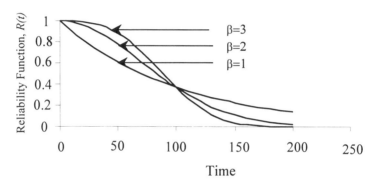

Figure 3.7 b. Reliability function of Weibull distribution for different values of β

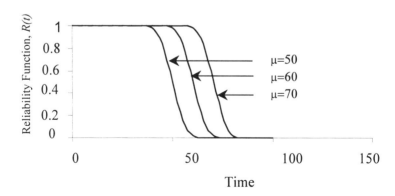

Figure 3.7c. Reliability function of Normal distribution for different values of μ

Example 3.4

Time to failure distribution of a computer memory chip follows normal distribution with mean 9000 hours and standard deviation 2000 hours. Find the reliability of this chip for a mission of 8000 hours.

SOLUTION

Using Table 3.2, the reliability for a mission of 8000 hours is given by:

$$R(t) = \Phi(\frac{\mu - t}{\sigma}) = \Phi(\frac{9000 - 8000}{2000}) = \Phi(0.5) = 0.6915$$

Example 3.5

The time to failure distribution of a steam turbo generator can be represented using Weibull distribution with $\eta = 500$ hours and $\beta = 2.1$. Find the reliability of the generator for 600 hours of operation.

SOLUTION:

Again using Table 3.2, reliability of the generator for 600 hours of operations is given by:

$$R(t) = \exp(-(600 / 500)^{2.1}) = 0.2307$$

3.2.1   Reliability function for items under multiple failure mechanisms

Assume that the failure of the item is caused due to two different failure mechanisms. Let $f_1(t)$ and $f_2(t)$ be the probability density function of the time-to-failure random variable due to failure mechanism 1 and 2 respectively. The probability density function of the time-to-failure of the item is given by caused by either of the failure mechanisms:

$$f(t) = f_1(t)[1 - F_2(t)] + f_2(t) \cdot [1 - F_1(t)]$$

Where $F_1(t)$ and $F_2(t)$ are failure function for failure mechanism 1 and 2 respectively. The Reliability function of the item under two different failure mechanism is given by:

$$R(t) = 1 - F(t) = 1 - \int_0^t \{f_1(x)[1 - F_2(x)] + f_2(x)[1 - F_1(x)]\}dx \quad (3.4)$$

The above result can be extended to obtain expression for reliability function due to more than two failure mechanisms.

Example 3.6

For the example 3.3, find the reliability of the item for 200 hours.

SOLUTION:

Using the expression for failure function obtained in example 3.3, the reliability function can be written as:

$$R(t) = \exp(-(\lambda_A + \lambda_B) \times t)$$
$$R(200) = \exp(-(0.002 + 0.005) \times 200) = 0.2465$$

3.2.2 Mission Reliability Function

In many practical situations, one might be interested in finding the probability of completing a mission successfully. Success probability of hitting an enemy target and returning to the base is an example where mission reliability function can be used. The main difference between reliability function and the mission reliability function is that, in mission reliability we recognise the age of the system before the mission. *Mission reliability is defined, as the probability that the system aged t_b is able to complete mission duration of t_m successfully.* We assume that no maintenance is performed during the mission. The expression for mission reliability MR (t_b, t_m) is given by

$$MR(t_b, t_m) = \frac{R(t_b + t_m)}{R(t_b)} \quad (3.5)$$

where, t_b is the age of the item at the beginning of the mission and t_m is the mission period. If the time to failure distribution is exponential, then the following relation is valid.

$$MR(t_b, t_m) = R(t_m)$$

Application of mission reliability function

1. Mission reliability, $MR(t_a, t_m)$ gives the probability that an individual item aged t_a will complete a mission duration of t_m hours without any need for maintenance.
2. Mission reliability is the appropriate basic reliability measure for ageing items or items whose time-to-failure distribution is other then exponential.

Example 3.7

Time-to-failure distribution of the gearbox within an armoured vehicle can be modelled using Weibull distribution with scale parameter $\eta = 2400$ miles and shape parameter $\beta = 1.25$. Find the probability that that gearbox will not fail during a mission time of 200 miles. Assuming that the age of the gearbox is 1500 miles.

SOLUTION:

Given, $t_b = 1500$ miles and $t_m = 200$ miles

$$MR(t_b, t_m) = \frac{R(t_m + t_b)}{R(t_b)} = \frac{R(1700)}{R(1500)}$$

$$R(1700) = \exp(-(\frac{1700}{2400})^{1.25}) = 0.5221$$

$$R(1500) = \exp(-(\frac{1500}{2400})^{1.25}) = 0.5736$$

$$MR(1500,200) = \frac{R(1700)}{R(1500)} = \frac{0.5221}{0.5736} = 0.9102$$

That is, the gearbox aged 1500 miles has approximately 91% chance of surviving a mission of 200 miles.

3.3.   DESPATCH RELIABILITY

Despatch reliability (DR) is one of popular reliability metrics used by commercial airlines around the world. Despatch reliability is defined as the percentage of revenue departures that do not occur in a delay or cancellation due to technical problems. For most airlines, the delay means that the aircraft is delayed more than 15 minutes. Technical delays occur can be caused due to some unscheduled maintenance. Airlines frequently seek DR guarantees where the aircraft manufactures face penalties if DR levels are not achieved. For commercial airlines despatch reliability is an important economic factor, it is estimated that delay cost per minute for large jets can be as high as 1000 US dollars. The expression for despatch reliability is given by:

$$DR(\%) = \frac{100 - ND_{15} - NC}{100} \times 100\% \tag{3.6}$$

Where,

ND_{15} = Number of delays with more than 15 minutes delay

NC = the number of cancellations

Equation (3.6) is applied only to technical delays. DR is a function of equipment reliability, system and component maintainability, and overall logistic support.

3.4.   HAZARD FUNCTION (HAZARD RATE OR INSTANTANEOUS FAILURE RATE)

Hazard function (or hazard rate) is used as a parameter for comparison of two different designs in reliability theory. Hazard function is the indicator of the effect of ageing on the reliability of the system. It quantifies the risk of failure as the age of the system increases. Mathematically, it represents the conditional probability of failure in an interval t to t + δt given that the system survives up to t, divided by δt, as δt tends to zero, that is,

$$h(t) = \lim_{\delta t \to 0} \frac{1}{\delta t} \cdot \frac{F(t + \delta t) - F(t)}{R(t)} = \lim_{\delta t \to 0} \frac{R(t) - R(t + \delta t)}{\delta t R(t)} \tag{3.7}$$

Note that hazard function, $h(t)$, is not a probability, it is the limiting value of the probability. However, $h(t)\delta t$, represents the probability that the item will fail between ages t and t+δt as δt →0. The above expression can be simplified so that

$$h(t) = \frac{f(t)}{R(t)} \qquad (3.8)$$

Thus, the hazard function is the ratio of the probability density function to the reliability function. Integrating both sides of the above equation, we get:

$$\int_0^t h(x)dx = \int_0^t \frac{f(x)}{R(x)} dx$$

$$= \int_0^t -\frac{R'(x)}{R(x)} dx = -\ln R(t)$$

Thus reliability can be written as:

$$R(t) = \exp\left[-\int_0^t h(x)dx \right] \qquad (3.9)$$

From equation (3.9), it immediately follows that:

$$f(t) = h(t)\exp(-\int_0^t h(x)dx \qquad (3.10)$$

The expression (3.10), which relates reliability and hazard function, is valid for all types of time to failure distribution. Hazard function shows how the risk of the item in use changes over time (hence also called *risk rate)*. The hazard functions of some important theoretical distributions are given in Table 3.3.

Characteristics of hazard function

1. Hazard function can be increasing, decreasing or constant.
2. Hazard function is not a probability and hence can be greater than 1.

Table 3.3. Hazard function, $h(t)$, of few theoretical distributions

Distribution	Hazard function, h(t)
Exponential	λ
Normal	$f(t)/\Phi(\dfrac{\mu-t}{\sigma})$, $f(t)$ is the pdf of normal distribution.
Lognormal	$f_l(t)/\Phi(\dfrac{\mu_l-t}{\sigma_l})$, $f_l(t)$ is the pdf of lognormal distribution.
Weibull	$\dfrac{\beta}{\eta}(\dfrac{t}{\eta})^{\beta-1}$
Gamma	$[\dfrac{\beta^\alpha}{\Gamma(\alpha)}t^{\alpha-1}e^{-\beta t}]/1-\dfrac{1}{\Gamma(\alpha)}\int\limits_0^t \beta^\alpha x^{\alpha-1}e^{-\beta x}dx$

Applications of hazard function

1. h(t) is loosely considered as failure rate at time t (time-dependent)
2. h(t) quantifies the amount of risk a system is under at time t.
3. For h(t) ≤ 1, it is not recommended to carry out preventive maintenance.

Figures 3.8a-c show hazard function of various theoretical distributions for different parameter values.

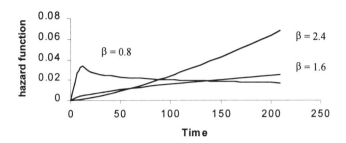

Figure 3.8a Hazard function of Weibull distribution for different values of β

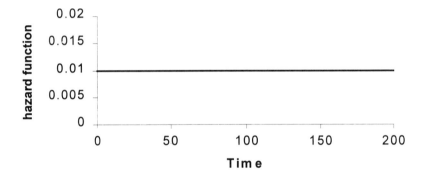

Figure 3.8b Hazard function of exponential distribution

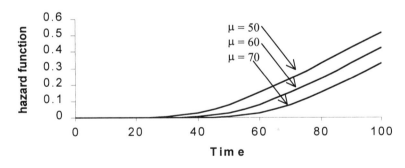

Figure 3.8c Hazard function of normal distribution for different values of μ

Example 3.8

Time to failure distribution of a gas turbine system can be represented using Weibull distribution with scale parameter η = 1000 hours and shape parameter β = 1.7. Find the hazard rate of the gas turbine at time t = 800 hours and t = 1200 hours.

SOLUTION:

The hazard rate for Weibull distribution is given by:

$$h(t) = \frac{\beta}{\eta}(\frac{t}{\eta})^{\beta-1}$$

$$h(800) = \frac{1.7}{1000}(\frac{800}{1000})^{0.7} = 0.00145$$

$$h(1200) = \frac{1.7}{1000}(\frac{1200}{1000})^{0.7} = 0.0019$$

3.4.1   Cumulative hazard function

Cumulative hazard function represents the cumulative hazard or risk of the item during the interval [0,t]. Cumulative hazard function, H(t), is given by:

$$H(t) = \int_0^t h(x)dx \qquad (3.11)$$

Reliability of an item can be conveniently written using cumulative hazard as:

$$R(t) = e^{-H(t)} \qquad (3.12)$$

3.4.2   Cumulative hazard function and the expected number of failures

Consider an item, which upon failure is subject to minimal repair. That is, the hazard rate after repair is same as the hazard rate just before failure. If *N(t)* is the total number of failures by time t, then *M(t)* = *E [N(t)]* is the expected number of failures by time t. It can be shown that under the assumption that the item receives minimal repair[*] ('*as-bad -as-old*') after each failure, then

$$E[N(t)] = M(t) = \int_0^t h(x)dx \qquad (3.13)$$

The   above   expression   can   be   used   to   model   different maintenance/replacement policies. In case of exponential and Weibull time to failure distributions we get the following simple expressions for the expected number of failures of an item subject to minimal repair.

Exponential time to failure distribution

For exponential distribution, the expected number of failures is given by

$$E[N(t)] = \int_0^t h(x)dx = \int_0^t \lambda dx = \lambda t \qquad (3.14)$$

Weibull time to failure distribution

$$E[N(t)] = \int_0^t h(x)dx = \int_0^t \frac{\beta}{\eta}(\frac{x}{\eta})^{\beta-1} dx = (\frac{t}{\eta})^{\beta} \qquad (3.15)$$

Example 3.9

An item is subject to minimal repair whenever it failed. If the time to failure of the item follows Weibull distribution with η = 500 and β = 2. Find: 1. The number of times the item is expected to fail by 1500 hours, and 2. The cost of the item is $ 200. If the cost of minimal repair is $ 100 per each repair, is it advisable to repair or replace the item upon failure.

---

[*] Mathematically minimal repair or '*as bad as old*' means that the hazard rate of the item after repair will be same as the hazard rate just prior to failure.

SOLUTION:

1. The expected number of failures is given by:

$$E[N(t)] = [\frac{t}{\eta}]^\beta = [\frac{1500}{500}]^2 = 3^2 = 9$$

2. Using the above result the cost associated with repair, C_{repair} (t) = 9 × 100 = \$ 900.

If the item is replaced, then the expected number of failures is given by the renewal function, *M(t)* [refer chapter 4], where

$$M(t) = \sum_{i=1}^{\infty} F^i(t)$$

For the above case, the value of M(t) < 4 (The actual calculation of the above function will be discussed in Chapter 4). Thus the cost due to replacement will be less than 4 × 200 = \$ 800. Thus, it is better to replace the item upon failure rather using minimal repair.

3.4.3   Typical Forms of Hazard Function

In practice, hazard function can have different shapes. Figure 3.9 shows most general forms of hazard function. Recent research in the field of reliability centred maintenance (RCM) shows that the hazard rate mostly follows six different patterns. Depending on the equipment and its failure mechanism, one can say that the hazard function may follow any one of these six patterns. *However, one should not blindly assume that hazard rate of any item will follow any one of these six patterns.* These are only possible cases based on some data.

Pattern A is called the *bathtub curve* and consist of three distinct phases. It starts with early failure region (known as burn-in or infant mortality) characterised by decreasing hazard function. Early failure region is followed by constant or gradually increasing region (called useful life). The constant or gradually increasing region is followed by wear out region characterised by increasing hazard function. The reason for such as shape is that the early decreasing hazard rate results from manufacturing defects. Early operation will remove these items from a population of like items. The remaining items have a constant hazard for some extended period of time during which the failure cause is not readily apparent. Finally those items remaining reach a wear-out stage with an increasing hazard rate. One would expect bathtub curve at the system level and not at the part or component level (unless the component has many failure modes which have different *TTF* distribution). It was believed that bathtub curve represents the most general form of the hazard function. However, the recent research shows that in most of the cases hazard function do not follow this pattern.

Pattern B starts with high infant mortality and then follows a constant or very slowly increasing hazard function. Pattern C starts with a constant or slowly increasing failure probability followed by wear out (sharply increasing) hazard function. Pattern D shows constant hazard throughout the file. Pattern E represents a slowly increasing hazard without any sign of wear out. Pattern F starts with a low hazard initially followed by a constant hazard.

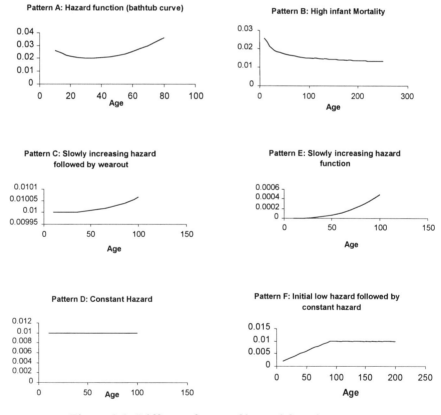

Figure 3.9. Different forms of hazard function

Table 3.4 shows the relationship between failure function, reliability function and hazard function.

Table 3.4. Relationship between F(t), R(t) and h(t)

	F(t)	R(t)	h(t)
F(t)	-----	$1 - R(t)$	$1 - \exp(-\int_0^t h(x)dx)$
R(t)	$1 - F(t)$	------	$\exp(-\int_0^t h(x)dx)$
h(t)	$F'(t)/[1 - R(t)]$	$-R'(t)/R(t)$	-------

3.4.4   Failure rate

Whenever the hazard function is constant, we call it as failure rate. That is, failure rate is a special case of hazard function (which is time dependent failure rate). Failure rate is one of the most widely used contractual reliability measures in the defence and aerospace industry. By definition, it is appropriate to use failure rate only when the time-to-failure distribution is exponential. Also, failure rate can be used only for a non-repairable system. Many defence standards such as MIL-HDBK-217 and British DEF-STAN 00-40 recommend the following equation for estimating the failure rate.

$$Failure\ rate = \frac{Total\ number\ of\ failures\ in\ a\ sample}{Cumulative\ operating\ time\ of\ the\ sample} \quad (3.16)$$

Care should be taken in using the above equation, for good estimation one has to observe the system failure for a sufficiently large operating period.

Applications of failure rate

1. Failure rate represents the number of failures per unit time.
2. If the failure rate is λ, then the expected number of items that fail in [0,t] is λt.
3. Failure rate is one of the popular contractual reliability measures among many industries including aerospace and defence.

3.5.   MEAN TIME TO FAILURE (MTTF)

MTTF represents the expected value of a system's time to first failure. It is used as a measure of reliability for *non-repairable* items such as bulb, microchips and many electronic circuits. Mathematically, MTTF can be defined as:

$$MTTF = \int_0^\infty tf(t)dt = \int_0^\infty R(t)dt \qquad (3.17)$$

Thus, MTTF can be considered as the area under the curve represented by the reliability function, R(t), between zero and infinity. If the item under consideration is repairable, then the expression (3.17) represents mean time to first failure of the item. Figure 3.10 depicts the MTTF value of an item.

For many reliability functions, it is difficult to evaluate the integral (3.17). One may have to use numerical approximation such as trapezium approach to find MTTF value.

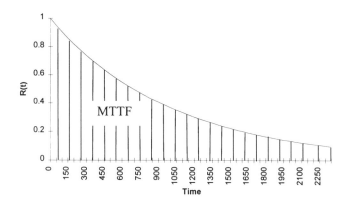

Figure 3.10 MTTF of an item as a function of Reliability

MTTF is one of the most popular measures for specifying reliability of non-repairable items among military and Government organisations throughout the world. Unfortunately there are many misconception about MTTF among reliability analysts. During the *Gulf War,* one of Generals from a defence department said, *'We know exactly how many tanks to send, we measured the distance from the map and divided that by MTTF'.* What many people do not realise is that MTTF is only a measure of central tendency. For example, if the time-to-failure distribution is exponential, then 63% of the items will fail before their age reaches MTTF value.

MTTF is one of the important contractual reliability measures for non-repairable (consumable) items. However, it is important to understand what MTTF value really means. For example let us assume that we have two items A and B with same MTTF (say 500 days). One might think that both the components have equal reliability. However, if the time to failure of the item A is exponential is that of item B is normal then there will be a significant variation in the behaviour of these items. Figure 3.11 shows the cumulative distribution of these two items up to 500 days. The figure clearly shows that items with exponential failure time show higher chance of failure during the initial stages of operation.

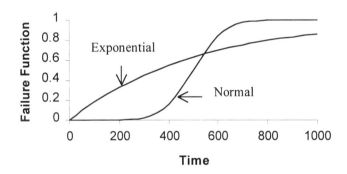

Figure 3.11 Comparison of item with same MTTF

Using the equation (3.17), the MTTF of various failure distributions are listed in Table 3.5.

It is easy to check that if the time to failure of the item is exponential then more than 63% of the items will fail by the time the age of the item reaches MTTF. In the case of normal distribution, it will be 50%.

Applications of MTTF

1. MTTF is the average life of a non-repairable system.
2. For a repairable system, MTTF represents the average time before the first failure.
3. MTTF is one of the popular contractual reliability measures for non-repairable systems.

Table 3.5. MTTF of different time-to-failure distributions

Distribution	MTTF
Exponential	$1/\lambda$
Normal	μ
Lognormal	$\exp(\mu_l + \dfrac{\sigma_l}{2})$
Weibull	$\eta \times \Gamma(1 + \dfrac{1}{\beta})$
Gamma	α / β

3.5.1   Mean Residual Life

In some cases, it may be of interest to know the expected value of the remaining life of the item before it fails from an arbitrary time t_0 (known as, *mean residual life*). We denote this value as MTTF(t_0), which represents the expected time to failure of an item aged t_0. Mathematically, MTTF(t_0) can be expressed as:

$$MTTF(t_0) = \int_{t_0}^{\infty} (t - t_0) f(t|t_0) dt \qquad (3.18)$$

$f(t|t_0)$ is the density of the conditional probability of failure at time t, provided that the item has survived over time t_0. Thus,

$$f(t \mid t_0) = h(t) \times R(t \mid t_0)$$

where, $R(t \mid t_0)$, is the conditional probability that the item survives up to time t, given that it has survived up to time t_0. Now, the above expression can be written as:

$$f(t \mid t_0) = h(t) \times \frac{R(t)}{R(t_0)}$$

The expression for $MTTF(t_0)$ can be written as:

$$MTTF(t_0) = \int_{t_0}^{\infty} (t - t_0) h(t) \frac{R(t)}{R(t_0)} dt \qquad (3.19)$$

substituting for h(t) in the above equation, we have

$$MTTF(t_0) = \int_{t_0}^{\infty} \frac{(t - t_0) f(t)}{R(t_0)} dt = \frac{1}{R(t_0)} \int_{t_0}^{\infty} (t - t_0) f(t) dt$$

The above equation can be written as (using integration by parts):

$$MTTF(t_0) = \frac{\int_{t_0}^{\infty} R(t) dt}{R(t_0)} \qquad (3.20)$$

The concept of mean residual life can be successfully applied for planning maintenance and inspection activities.

Example 3.10

Companies A and B manufacture car tyres. Both the companies claim that the MTTF of their car tyre is 2000 miles. After analysing the field failure data of these two tyres it was found that the time to failure distribution of A is exponential with $\lambda = 0.0005$ and the time to failure distribution of B is normal with $\mu = 2000$ miles and $\sigma = 200$ miles. If the maintenance policy of the Exeter city car rentals is to replace the tyres as soon as it reaches 2000 miles which tyre they should buy:

SOLUTION:

Reliability of the car tyre produced by company A for 2000 miles, $R_A(2000)$, is given by:

$$R_A(2000) = \exp(-0.0005 \times 2000) = 0.3678$$

Reliability of the car tyre produced by company B for 2000 miles, $R_B(2000)$, is given by:

$$R_B(2000) = \Phi(\frac{\mu - 2000}{\sigma}) = \Phi(\frac{2000 - 2000}{200}) = \Phi(0) = 0.5$$

Thus, it is advisable to buy the tyres produced by company B.

Example 3.11

The time to failure of an airborne navigation radar can be represented using Weibull distribution with scale parameter η = 2000 hours and β = 2.1. It was told that the age of the existing radar is 800 hours. Find the expected value of the remaining life for this radar.

SOLUTION:
Using Equation (3.20), The MTTF(800) can be written as:

$$MTTF(800) = \frac{\int\limits_{800}^{\infty} R(t)dt}{R(800)} = \frac{\int\limits_{0}^{\infty} R(t)dt - \int\limits_{0}^{800} R(t)dt}{R(800)}$$

$$MTTF(800) = MTTF(800) = \frac{MTTF - \int\limits_{0}^{800} \exp(-(\frac{t}{2000})^{2.1}) dt}{0.8641}$$

$$MTTF = \eta \times \Gamma(1 + \frac{1}{\beta}) = 2000 \cdot \Gamma(1 + \frac{1}{2.1}) = 1771.2$$

The value of $\Gamma(1+\dfrac{1}{\beta})$ can be found from Gamma function table (see appendix).

Using numerical approximation, $\displaystyle\int_{0}^{800} \exp(-(\dfrac{t}{2000})^{2.1} dt \approx 763.90$

Thus MTTF(800) $\approx$ (1771.2 - 763.90) / 0.8641 = 1165.72 hours

Thus, expected remaining life of the radar aged 800 hours is 1165.72 hours.

3.5.2 MTTF of a maintained system

Assume that an item is subject to preventive maintenance after every T_{pm} units, that is, at T_{pm}, $2T_{pm}$, $3T_{pm}$, etc. The expected time to failure, $MTTF_{pm}$, (MTTF of an subject to preventive) of the item is given by:

$$MTTF_{pm} = \int_{0}^{\infty} R_{pm}(t)dt \qquad (3.21)$$

Using additive property of integration, the above integral can be written as:

$$MTTF_{pm} = \int_{0}^{T_{pm}} R_{pm}(t)dt + \int_{T_{pm}}^{2T_{pm}} R_{pm}(t)dt + \int_{2T_{pm}}^{3T_{pm}} R_{pm}(t)dt+....$$

where R_{pm} (t) is the reliability of the item subject to preventive maintenance. If the item is restored to 'as-good-as-new' state after each maintenance activity, then the reliability function between any two maintenance tasks can be written as:

$$R_{pm}(t) = R[T_{pm}]^{k} R(t), \qquad kT_{pm} \le t \le (k+1)T_{pm}$$

Using the above expression for R_{pm}(t) in the integral (3.21) we have:

$$MTTF_{pm} = \int_0^{T_{pm}} R(t)dt + \int_0^{T_{pm}} R(T_{pm})R(t)dt + \int_0^{T_{pm}} [R(T_{pm})]^2 R(t)dt+...$$

$$= \{1 + R(T_{pm}) + [R(T_{pm})]^2 +....\} \int_0^{T_{pm}} R(t)dt$$

As R(t) ≤ 1, the above expression can be written as:

$$MTTF_{pm} = \frac{\int_0^{T_{pm}} R(t)dt}{1 - R(T_{pm})} = \frac{\int_0^{T_{pm}} R(t)dt}{F(T_{pm})} \tag{3.22}$$

Similar logic can be used to derive the expression for $MTTF_{pm}$ when the repair is not perfect (that is, when the item is not as good as new after maintenance). $MTTF_{pm}$ can be used to quantify the effectiveness of the maintenance action. If $MTTF_{pm}$>MTTF, then one can say that the reliability can be improved by carrying out maintenance. If $MTTF_{pm}$ ≤ MTTF then, the maintenance will not improve the reliability of the item. Figure 3.12 shows $MTTF_{pm}$ values of an item for different T_{pm} whose time-to-failure can be represented using Weibull distribution with η = 200 and β = 2.5. It can be noticed that as the value of T_{pm} increases, the $MTTF_{pm}$ converges to that of corrective maintenance.

Example 3.12

A solid state radar is subject to preventive maintenance after every 400 flight hours. The time to failure of the radar follows exponential distribution with mean life 800 flight hours. Find the $MTTF_{pm}$ of the radar.

SOLUTION:

We have: $T_0 = 500$ flight hours and $(1/\lambda) = 800$

$\lambda = (1/800) = 0.00125$

$$MTTF_{pm} = \frac{\int_0^{400} \exp(-0.00125 \times t)dt}{1 - \exp(-0.00125 \times 400)} = 800$$

There is no improvement in the $MTTF_{pm}$ because the time to failure is exponential. *Thus, preventive maintenance will not improve the reliability of the system, if the time to failure is exponential. This example is used to demonstrate this well known fact mathematically.*

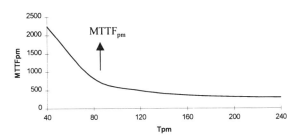

Figure 3.12. $MTTF_{pm}$ of an item for different T_{pm} values

Example 3.13

A manufacturing company buys two machines A and B. The time to failure of machine A can be represented by Weibull distribution with $\eta = 1000$ hours and $\beta = 2$. The time to failure of machine B can be represented by Weibull distribution with $\eta = 1000$ hours and $\beta = 0.5$. The maintenance manager in charge of operation plan to apply preventive maintenance for both the machines for every 200 hours, so that he can improve the expected time to failure of the machines. Check whether the manager's decision is correct.

SOLUTION:

The $MTTF_{pm}$ for machine A is given by:

$$MTTF_{pm} = MTTF_{pm}\frac{\int_0^{200} \exp(-(t/1000)^2)dt}{\exp(-(200/1000)^2)} \approx 5033 \text{ hours}$$

MTTF for machine A is $\eta \times \Gamma(1 + \frac{1}{B}) = 1000 \times \Gamma(1 + \frac{1}{2}) = 886.2$ hours

Thus for machine A, preventive maintenance will improve the mean time to failure of the system.

The MTTF$_{pm}$ for machine B is given by:

$$MTTF_{pm} = \frac{\int\limits_0^{200} \exp(-(t/1000)^{0.5})dt}{\exp(-(200/1000)^{0.5})} \approx 414 \text{ hours}$$

MTTF for machine B is $\eta \times \Gamma(1 + \frac{1}{B}) = 1000 \times \Gamma(1 + \frac{1}{0.5}) = 2000$ hours

Thus for machine B, preventive maintenance will decrease the mean time to failure of the system. Thus, it is better not to apply preventive maintenance for machine B.

3.5.3   Variance of Mean Time To Failure

It is important to know the variance of mean time to failure for better understanding of the item. From definition variance $V(t)$ is given by:

$$V(t) = E(t^2) - [E(t)]^2$$
$$= \int\limits_0^\infty t^2 f(t) - MTTF^2$$

Applying integration by parts:

$$V(t) = 2 \int\limits_0^\infty tR(t)dt - MTTF^2 \qquad (3.23)$$

3.6. MEAN OPERATING TIME BETWEEN FAILURES (MTBF)

MTBF stands for *mean operating time between failures* (wrongly mentioned as *mean time between failures* throughout the literature) and is used as a reliability measure for repairable systems. In British Standard (BS 3527) MTBF is defined as follows:

For a stated period in the life of a functional unit, the mean value of the lengths of time between consecutive failures under stated condition.

MTBF is extremely difficult to predict for fairly reliable items. However, it can be estimated if the appropriate failure data is available. In fact, it is very rarely predicted with an acceptable accuracy. In 1987 the US Army conducted a survey of the purchase of their SINCGARS radios that had been subjected to competitive procurement and delivery from 9 different suppliers. They wanted to establish how the observed Reliability In-service compared to that which had been predicted by each supplier (using MIL-HDBK-217). The output of this exercise is shown in Table 3.6 (Knowles, 1995). It is interesting to note that they are all same radio, same design, same choice of components (but different manufacturers) and the requirement set by the Army was MTBF of 1250 hours with a 80% confidence. Majority of the suppliers' observed MTBF was no where near their prediction.

Table 3.6 SINCGARS radios 217 prediction and the observed MTBF

Vendor	MIL-HDBK-217 (hours)	Observed MTBF (hours)
A	7247	1160
B	5765	74
C	3500	624
D	2500	2174
E	2500	51
F	2000	1056
G	1600	3612
H	1400	98
I	1000	472

Let us assume that the sequence of random variables X_1, X_2, X_3, ...X_n represent the operating time of the item before i-th failure (Figure 3.13). MTBF can be predicted by taking the average of expected values of the random variables X_1, X_2, X_3,..., X_n etc. To determine these expected values it is necessary to determine the distribution type and parameters. As soon as an item fails, appropriate maintenance activities will be carried out. This involves replacing the rejected components with either new ones or ones that have been previously recovered (repaired). Each of these components will have a different wear out characteristic governed by a different distribution. To find the expected value of the random variable X_2 one should take into account the fact that not all components of the item are new and, indeed, those, which are not new, may have quite different ages. This makes it almost impossible to determine the distribution of the random variable X_2 and hence the expected value.

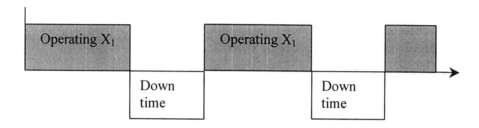

Figure 3.13 operating profile of a generic item

The science of failures has not advanced sufficiently, as yet, to be able to predict failure time distribution in all cases. This is currently done empirically by running a sample of items on test until they fail, or for an extended period, usually under 'ideal' conditions that attempt to simulate the operational environment. Military aircraft-engines, for example, are expected to operate while subjected to forces between -5 and + 9 'g', altitudes from zero to 50000 feet (15000 meters) and speeds from zero to Mach 2+. One has to test the equipment with some new and some old components to find the expected values of the random variables X_2, X_3, etc. In practice most of the testing is done on new items with all new components in pristine condition. The value derived by these type of testing will give the expected value of the random variable X_1 . In practice, the expected value of X_1 is quoted as MTBF. In fact, the expected value of X_1 will give only the Mean Time To First Failure (as the testing is done on new items and the times reflect the time to first failure) and not the MTBF. To calculate MTBF one should consider the expected values of the random variables X_2, X_3, etc.

If the time to failure distribution of the system is exponential then the MTBF can be estimated using the following equation (recommended by MIL-HDBK-217 and DEF-STAN-00-40):

$$MTBF = \frac{T}{n}$$

(3.24)

where, T is the total operating period and `n' is the number of failures during this period. Note that the above relation is valid only for large value of T. If n = 0, then MTBF becomes infinity, thus one should be careful in using the above relation. **The above expression can be used only when sufficient amount of data is available**.

Characteristics of MTBF

1. The value of MTBF is equal to MTTF if after each repair the system is as good as new.
2. MTBF = $1 / \lambda$ for exponential distribution, where λ is the scale parameter (also the hazard function).

Applications of MTBF

1. For a repairable system, MTBF is the average time in service between failures. Note that, this does not include the time spent at repair facility by the system.
2. MTBF is used to predict steady-state availability measures like inherent and operational availability.

3.7.   PERCENTILE LIFE (TTF_P OR $B_{P\%}$)

Percentile life or $B_{p\%}$ is a measure of reliability which is popular among industries. This is the life by which certain proportion of the population (p %) can be expected to have failed. $B_{10\%}$ means the life (time) by which 10% of the items will be expected to have failed. Percentile life is now frequently used among aerospace industries as a design requirement. Mathematically percentile life can be obtained by solving the following equation for t:

$$F(t) = \int_0^t f(x)dx = p\% \qquad (3.25)$$

Assume that $F(t)$ is a exponential distribution with parameter $\lambda = 0.05$, and we are interested in finding B_{10} . Then from above equation we have:

$$1 - \exp(-0.05t) = 0.10 \Rightarrow t = 2.107$$

Thus 2.107 is the B_{10} life for exponential distribution with parameter 0.05. The main application of percentile life lies in prediction of initial spares requirement (*initial spares provisioning, IP*).

3.8.   MAINTENANCE FREE OPERATING PERIOD (MFOP)

Maintenance Free Operating Period is defined as:

The period of operation (for example, for military combat aircraft, a typical MFOP may be 100, 200 or 300 flying hours) during which an item will be able to carry out all its assigned missions, without the operator being restricted in any way due to system faults or limitations, with the minimum of maintenance.

In other words, maintenance free operating period guarantees a certain period of operation without any interruption for unscheduled maintenance. A MFOP (or cycles of MFOP) is usually followed by a maintenance recovery period (MRP*). MRP is defined as the period during which the appropriate scheduled maintenance is carried out.* Since it is almost impossible to give 100 % guaranteed MFOP, we use the concept of *maintenance free operating period survivability* (MFOPS) to measure MFOP. MFOPS is the probability that the part, sub-system or system will survive for the duration of MFOP given that it was in a state of functioning at the start of the period. Note, unlike most warranties, the MFOP will not always apply to new items, indeed, most of the time, the ages of the constituent components will be quite varied and in many cases, unknown. It should be also noted that during MFOP the redundant items are allowed to fail, without causing any unscheduled maintenance.

3.8.1   Maintenance Free Operating Period Survivability Prediction

Let us consider a system with n components connected in series. If the reliability requirement is MFOP of t_{mf} life units, then the corresponding probability that the system will survive the stated MFOP, given that all the components of the system are new is given by:

$$MFOPS(t_{mf}) = \prod_{k=1}^{n} \frac{R_k(t_{mf})}{R_k(0)} \tag{3.26}$$

where $R_k(t_{mf})$ is the reliability of the k-th component for (the first) t_{mf} life units. The equation (3.26) gives the probability for the system to have MFOP of t_{mf} life units during the first cycle. In general, for i-th cycle (here each cycle refers to each t_{mf} life units), the probability the system will have MFOP of t_{mf} life units is given by:

$$MFOPS(t_{mf}) = \prod_{k=1}^{n} \frac{R_k(i \times t_{mf})}{R_k([i-1] \times t_{mf})} \tag{3.27}$$

MFOP of items with Weibull distributed failure times

For a component with failure mode, which can be modelled by the Weibull distribution the probability of surviving t_{mf} units of time given, that the item has survived t units of the time is given by:

$$MFOPS(t_{mf}) = \exp(-\frac{t^\beta - (t + t_{mf})^\beta}{\eta^\beta}) \tag{3.28}$$

where, η is the scale parameter and β is the shape parameter of the Weibull distribution. The MFOP period for a given level of confidence can be calculated by rearranging the above equation as follows:

$$t_{mf} = [t^\beta - \eta^\beta \ln(MFOPS(t_{mf}))]^{1/\beta} - t \tag{3.29}$$

Maximum length of MFOP

The maximum length of MFOP for a stated MFOPS actually represents the design life of that system. Design life denotes the age of the item up to which the reliability of the system is greater than or equal to the designed reliability value. For example assume that the time-to-failure distribution of the item be Weibull with scale parameter η and shape parameter β. Then the MFOP duration for a specified MFOPS requirement is then given by:

$$MFOP = \eta \times \{\ln\left(\frac{1}{MFOPS}\right)\}^{1/\beta} \tag{3.30}$$

Procedure to calculate the number of cycles the system satisfies the required MFOP.

If the required MFOP is say, t_{mf} , life units. It may not be necessary to carry out maintenance recover, after every MFOP. The following steps can be used to find how many such MFOPs can be carried out without any maintenance.

Step 1: Set $i = 1$.

Step 2: Calculate $MFOP(i,\alpha) = \displaystyle\prod_{k=1}^{n} \frac{R_k(i \times t_{mf})}{R_k([i-1] \times t_{mf})}$

Step 3: If $\displaystyle\prod_{k=1}^{i} MFOP(i,\alpha) \le \alpha$, then Go To Step 5.

Step 4: $i = i + 1$, Go To Step 2.

Step 5: Number of cycles is i -1. STOP.

Example 3.14

For a computer to be used in a space station, it was required that the MFOP duration for the memory unit should be at least 10000 hours at 95% confidence. It was also required that the memory unit should be screened for 500 hours at different temperature cycles. Two memory chips were available in the market that can be used in the computer to be installed in the space station. The time to failure of a computer memory chip 1 follows Normal distribution with μ = 12000 hours and σ = 1000. The time to failure of a computer memory chip 2 follows Weibull distribution with η = 12000 hours and $\beta = 2.2$. Find which chip will satisfy the requirement.

SOLUTION:

Since the memory unit is subject to temperature screening, the age of the memory unit when put into mission will be 250 hours.

Case 1. Memory chip 1.

MFOPS for the memory chip 1 for the duration of 10000 hours is given by:

$$\text{MFOPS} = \frac{\Phi(\dfrac{12000-10000}{1000})}{\Phi(\dfrac{12000-500}{1000})} = \frac{\Phi(2.0)}{\Phi(11.5)} = 0.9772$$

Case 2. Memory chip 2

MFOPS for the memory chip 2 for the duration of 10, 000 hours is given by:

$$MFOPS = \frac{\exp(-(10000/12000)^{2.2})}{\exp(-(1000/12000)^{2.2})} = 0.5141$$

Since MFOPS for chip 1 is greater than 0.95, it will satisfy the MFOP requirement.

3.9.   SOFTWARE RELIABILITY CHARACTERISTICS

Software is one of the most complex systems ever built by human. In the recent years, software has become a critical part of many systems. For systems such as air traffic control, space shuttle, fighter aircraft and automated guided missiles it is usually the software that has significant impact on the mission success.   Software is a core element of today's defence systems. Virtually, all major military systems are dependent on the correct operation of defence systems' software. For example, approximately 80% of Euro Fighter 2000 functionality is provided by software (Glen Griffiths, 1997). The B1-B bomber aircraft has 1.2 million lines of code and fighter aircraft F-16 has seven flight computers and 135, 000 lines of code (Edward Koss, 1988). Until recently it was assumed that the software is 100% reliable. However, this confidence has changed in the recent years. The following few examples explain the importance of software reliability engineering.

1. During A310 avionics systems development 45000 hours were spent for the software development against the predicted 20000 hours. 450 modifications were made during the development of this software. The poor reliability was one of the main reason for the excess money and time spent on the software development. One prediction shows that, by year 2015 the entire US Department of Defence Budget will go to software if the current trend of software continues (Hess, 1988).
2. Software developed for the Apollo series of lunar flight is one of the most well planned programs. However, most of the faults of the Apollo program were due to software related failures. A software error in the on board computer of the Apollo 8 erased part of the computer's memory.
3. An error in FORTRAN statement resulted in loss of first American probe to Venus.
4. American Airlines (AA) lost nearly a $1 billion due to software faults. The world's largest system, Sabre, requires 12 mainframes and is buried deep in a nuclear explosion-proof bunker in the US.
5. During Gulf war, failure of a Patriot missile battery to track and intercept an Iraqi launched Scud missile, subsequently struck a warehouse used as a barracks for US forces at Dhahran, in Saudi Arabia. The fault was traced to a 0.36-second error in the timing of software driven clock used for tracking the incoming missile.

Software behaves entirely different from hardware. Software generally becomes more reliable over time (however it might become obsolete as the technology changes). Fault in the software is caused by defect, which appear randomly in time. Errors in software are introduced during various stages, mainly during: 1. Requirements definition, 2. Design, 3. Program development and 4. Operation / maintenance. Thus, any measure of software reliability must start with the core of the issue, operational software error counts and the rate at which they occur, that is the software failure rate (Koss, 1998).

Software failures are not caused by physical environmental wear. Also the main source of software failure comes from requirement and specification error than machine code error or design error. No imperfections or variations are introduced in making additional copies of software. Unlike hardware, software reliability cannot be improved by introducing identical different versions of the program. However, it is possible to improve by producing independent versions of software to improve the reliability. Software redundancy techniques are discussed in Chapter 11. The following are few basic definitions in software reliability:

Failures: A failure occurs when the user perceives that the program ceases to deliver the expected service.

Faults: The cause of the failure or the internal error is said to be a fault. It is also referred as a bug.

Execution time: The execution time for a software system is the CPU time that is actually spent by the computer in executing the software.

3.9.1   Software Reliability

Software reliability is the probability that software will provide failure-free operation in a fixed environment for a fixed interval of the time. Probability of failure is the probability that the software will fail on the next selected input.

Software reliability is typically measured per some units of time, whereas probability of failure is generally time independent. These two measures can be easily related if one knows the frequency of inputs that are executed per unit of time. Here failure is caused by activation of internal fault or *bug*. One can view software as a function, which maps a space of inputs into a space of outputs. The input space can be partitioned in to two mutually exclusive sets U and D. Where inputs from the set U produce desirable output and inputs from the set D produce incorrect or undesirable output.

Assume that p represents the probability that the software fails for an input selection. Then the reliability of the software for n input selection is given by:

$$R_n = (1 - p)^n \tag{3.31}$$

The equation (3.31) is not suitable for complex software, as the prediction of p is almost impossible.

3.9.2   Mean Time To Failure of Software

Mean time to failure is the average time between failures. Great care should be taken while calculating mean time to failure and reliability of software. For example, if a software is executed on two different computers A and B, and if A is faster than B, then one can expect that the mean time to failure of the software when executed on A will be less than that of B. Thus, while measuring the reliability of a software, it is important to mention the operating system on which the software is executed.

3.9.3   Software failure rate

The primary metric for software reliability is the software failure rate, which is given by the expected number of failures per unit time. Here time usually refers to a computer hour. For example, if the software fails 2 times during 2000 hours of operation then the failure rate of the software is 0.001. Mathematically,

$$\lambda(t) = \frac{n}{T} \qquad\qquad\qquad (3.32)$$

where, n is the number of failures during T hours of execution. Usually the software failure rate is measures per 1000 Lines of Source Code (hence called KSLOC) If faults are removed from the software upon failure, and no new faults are introduced during the repair process, then the failure rate of the software will be a decreasing function. Koss (1988) lists the following characteristics as the prime factors that influence the software failure rate.

1. The execution environment.
2. Cyclic dependencies.
3. Variability of data
4. Execution frequency.

3.9.4   Jelinski-Moranda Model

One of the earliest models proposed, which is still being applied today, is the model developed by Jelinski and Moranda, while working on some Navy projects for McDonnell Douglas. In Jelinski-Moranda model it is assumed that the failure intensity of the software is proportional to the current fault content. The following assumptions are used:

1. The number of faults in the initial software is an unknown, but constant.
2. The fault that causes a failure is removed instantaneously after failure, and no new fault is introduced.
3. The failure intensity remains constant throughout the interval between failure occurrences.
4. Times between software failures are independent and exponentially distributed.

If N_0 denotes the number of faults present in the initial software, then the failure intensity between the interval (i-1)st and i-th failure is given by:

$$\lambda_i = \phi[N_0 - (i-1)], \qquad i = 1,2,...,N_0 \tag{3.33}$$

In equation, (3.33), ϕ is a constant of proportionality denoting the failure intensity contributed by each fault. The distribution of time between failure is given by:

$$P[t_i \leq t] = 1 - \exp[-\phi(N_0 - i + 1)t] \tag{3.34}$$

The parameters N_0 and ϕ can be estimated using standard statistical tools (interested readers please refer to M Xie, 1991).

3.9.5   Goel-Okumoto Model

Goel-Okumoto software reliability model assumes that the cumulative number of faults detected at time t follows a non-homogeneous Poisson process. The failure intensity is given by:

$$\lambda(t) = \alpha\beta \exp(-\beta t) \tag{3.35}$$

Where α and β are parameters to be determined from failure data. The mean value function of Goel-Okumoto Model is given by:

$$m(t) = \alpha[1 - \exp(-\beta t)] \tag{3.36}$$

m(t) gives the expected number of failures during t hours of execution.

Chapter 4
Systems Reliability

'A Bird is an instrument working according to a mathematical law. It lies within the power of man to make this instrument with all its motion'

Leonardo da Vinci

In this chapter, we present methodologies that can be used to evaluate systems reliability using simple mathematical tools. The chapter discusses two approaches that can be used to predict the reliability metrics of the system. First, we study the models that are based on simple probability theory, assuming that the time-to-failure distributions of different components within the system are known. These models can be used only for non-repairable items. The second approach is based on Markov models, for predicting different reliability measures. The models for repairable items will be discussed using the Markov models. Throughout the Chapter, the word 'system' is used to represent the complete equipment and the word 'item' is used as a generic term that stands for subsystem, module, component, part or unit. Any reliability prediction methodology using time-to-failure approach will involve the following steps:

1. Construct the reliability block diagram (RBD) of the system. This may involve performing failure modes and effect analysis (FMEA).
2. Determine the operational profile of each block in the reliability block diagram.
3. Derive the time-to-failure distribution of each block.
4. Derive the life exchange rate matrix (LERM) for the different components within the system.
5. Compute reliability function of each block.
6. Compute the reliability function of the system.

4.1.   RELIABILITY BLOCK DIAGRAM

Reliability block diagram, RBD, of an item is a logical diagrammatic illustration of the system in which each item (hardware/software) within the system is represented by a block. RBD forms a basis for calculation of system reliability measures. Each block within a RBD can represent a component, subsystem, module or system. The structure of a RBD is determined by the effect of failure of each block on the functionality of the system as a whole. A block does not have to represent physically connected hardware in the actual system to be connected in the block diagram. In an RBD the items whose failure can cause system failure irrespective of the remaining items of the system are connected in series. Items whose failure alone cannot cause system failure are connected in parallel. Depending on the item, a RBD can be represented by a series, parallel, series-parallel, r-out-of-n or complex configuration. Construction of RBD requires functional analysis of various parts within the system. Each block within a RBD should be described using time-to-failure distribution for the purpose of calculating system reliability measures. The RBD can also have network structures (e.g. communication systems, water network and Internet). In the following sections we address how to evaluate various reliability measures for different reliability block diagrams.

4.2.   RELIABILITY MEASURES FOR SERIES CONFIGURATION

In a series configuration, all the consisting items of the system should be available or functional to maintain the required function of the system. Thus, failure of any one item of the system will cause failure of the system as whole. Series configuration is probably the most commonly encountered RBD in engineering practice. The RBD of a hypothetical system whose items are connected in series is given in Figure 4.1.

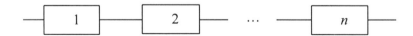

Figure 4.1. Reliability block diagram of a system with series configuration

Reliability function of series configuration

Reliability function of a system with series configuration can be derived from the reliability function of its consisting items. Let $R_S(t)$ represent the reliability function of a series system with n items. Let $R_i(t)$ denote the reliability function of the item i. If TTF_i is the time-to-failure random variable for the item i, then the reliability function of system for 't' hours of operation is given by:

$$R_S(t) = P[\ TTF_1 \geq t,\ TTF_2 \geq t,\ ....,\ TTF_n \geq t\ ] \tag{4.1}$$

The equation (4.1) clearly states that the system under consideration will maintain the required function if and only if all the n items of the system are able to maintain the required function for at least t hours of operation. Assuming that the random variables TTF_i are independent of each other, the expression (4.1) can be written as:

$$R_s(t) = P[TTF_1 \geq t\ ] \times P[TTF_2 \geq t\ ] \times ... \times P[\ TTF_n \geq t\ ]$$
$$= R_1(t) \times R_2(t) \times ... \times R_n(t)$$

Thus, the reliability of a series configuration with n items is given by:

$$R_s(t) = \prod_{i=1}^{n} R_i(t) \tag{4.2}$$

Note that in the above equation (4.2), it is assumed that the connecting media (such as solder joints) between different items is 100% reliable (unless this is specifically included in the RBD). However, this need not be true. In the equation (4.2) time t is used as a generic term. In most case time actually represents age or utilisation of the item under consideration. It can have different units such as hours, miles, landings, cycles etc for different items. One has to normalise the 'time' before calculating the reliability function in such cases. One method of normalising the different life units of the items is using Life Exchange Rate Matrix (LERM), which will be discussed later in this chapter. When the life units of items are different (or different items have different utilisation), we use the following equation to find the reliability of the series system.

$$R_s = P[TTF_1 \geq t_1, TTF_2 \geq t_2, \cdots, TTF_n \geq t_n] = R_1(t_1) \times R_2(t_2) \times \cdots \times R_n(t_n)$$

That is,

$$R_s(t) = \prod_{i=1}^{n} R_i(t_i) \tag{4.3}$$

In equation (4.3), t_i is the age of the item i, which is equivalent to age t of the system. That is, for the system to survive up to age t, the item i should survive up to t_i. Throughout this book we use equation (4.3) unless otherwise specified.

Characteristics of reliability function of a series configuration

1. The value of the reliability function of the system, $R_S(t)$, for a series configuration is less than or equal to the minimum value of the individual reliability function of the constituting items. That is:

$$R_S(t) \leq \underset{i=1,2,..n}{Min} \{R_i(t)\}$$

2. If $h_i(t)$ represent the hazard function of item i, then the system reliability of a series system can be written as:

$$R_s(t) = \prod_{i=1}^{n} \exp(-\int_0^t h_i(x)dx$$

$$= \exp(-\int_0^t [\sum_{i=1}^{n} h_i(x)]dx$$

Example 4.1

A system consists of four items, each of them are necessary to maintain the required function of the system. The time to failure distribution and their corresponding parameter values are given in Table 4.1. Find the reliability of the system for 500 and 750 hours of operation.

Table 4.1 Time to failure distribution and their parameter of the items

Item	Time to failure distribution	Parameter values
Item 1	Exponential	$\lambda = 0.001$
Item 2	Weibull	$\eta = 1200$ hours $\beta = 3.2$
Item 3	Normal	$\mu = 800$ hours $\sigma = 350$
Item 4	Weibull	$\eta = 2000$ hours $\beta = 1.75$

SOLUTION:

From the information given in Table 4.1, the reliability function of various items can be written as:

$$R_1(t) = \exp(-0.001 \times t)$$

$$R_2(t) = \exp[-(\frac{t}{1200})^{3.2}]$$

$$R_3(t) = \Phi(\frac{800-t}{350})$$

$$R_4(t) = \exp[-(\frac{t}{2000})^{1.75}]$$

Since the items are connected in series, the reliability function of the system is given by:

$$R_s(t) = \exp(-0.001 \times t) \times \exp[-(\frac{t}{1200})^{3.2}] \times \Phi(\frac{800-t}{350}) \times \exp[-(\frac{t}{2000})^{1.75}]$$

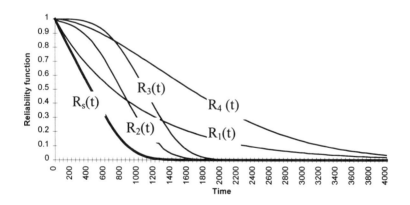

Figure 4.2 Reliability function of the system and its constituent items.

Substituting t = 500 and 750 in the above equation, we get:

R(500) = 0.6065 × 0.9410 × 8043 × 0.9154 = 0.4202

R(750) = 0.4723 × 0.8003 × 0.5568 × 0.8355 = 0.1759

Figure 4.2 shows the reliability function of the system and various items of the system. Note that the system reliability value is always less than or equal to any of the constituting items.

Example 4.2

Avionics system of an aircraft consists of digital auto-pilot, integrated global positioning system, weather and ground mapping radar, digital map display and warning system. Apart from the above items, the avionics system has control software. The time-to-failure distributions of various items are given in Table 4.2. Find the reliability of the avionics system for 100 hours of operation if all the items are necessary to maintain the required function of the avionics system.

Table 4.2 Time-to-failure distribution of various items of the avionics system

Item	Time-to-failure distribution	Parameter values
Digital autopilot	Exponential	$\lambda = 0.003$
Integrated global positioning system	Weibull	$\eta = 1200, \beta = 3.2$
Weather and ground mapping radar	Weibull	$\eta = 1000, \beta = 2.1$
Digital map display	Normal	$\mu = 800, \sigma = 120$
Warning System	Normal	$\mu = 1500, \sigma = 200$
Software	Exponential	$\lambda = 0.001$

SOLUTION:

From the data given in Table 4.2, we can derive the reliability function of various items as follows:

1. Reliability of digital auto-pilot

$$R_1(t) = \exp(-\lambda \times t) \Rightarrow R_1(100) = \exp(-0.003 \times 100) = 0.7408$$

2. Reliability of integrated global positioning system.

$$R_2(100) = \exp(-(t/\eta)^\beta) \Rightarrow R_2(100) = \exp(-(100/1200)^{3.2}) = 0.9996$$

3. Reliability of weather and ground mapping system radar

$$R_3(100) = \exp(-(t/\eta)^\beta) \Rightarrow R_3(100) = \exp(-(100/1000)^{2.1}) = 0.9920$$

4. Reliability of digital map display

$$R_4(100) = \Phi(\frac{\mu-t}{\sigma}) \Rightarrow R_4(100) = \Phi(\frac{800-100}{120}) = \Phi(5.8) = 1$$

5. Reliability of warning system

$$R_5(100) = \Phi(\frac{\mu-t}{\sigma}) \Rightarrow R_4(100) = \Phi(\frac{1500-100}{200}) = \Phi(7) = 1$$

6. Reliability of software

$$R_6(t) = \exp(-\lambda t) \Rightarrow \exp(-0.001 \times 100) = 0.9048$$

Thus, the reliability of the avionics system for 100 hours of operation is given by:

$$R_s(100) = \prod_{i=1}^{6} R_i(100) = 0.7408 \times 0.9996 \times 0.9920 \times 1 \times 1 \times 0.9048 = 0.6646$$

Hazard function of a series configuration

Let $R_S(t)$ denote the reliability function of the system. From definition, the hazard rate of the system, $h_S(t)$, can be written as:

$$h_S(t) = -\frac{dR_S(t)}{dt} \times \frac{1}{R_S(t)} \tag{4.4}$$

Using equation (4.2), the expression for $R_S(t)$ can be written as:

$$R_S(t) = \prod_{i=1}^{n} R_i(t) = \prod_{i=1}^{n} [1 - F_i(t)] \qquad (4.5)$$

where $F_i(t)$ is the failure function of the item i. Differentiating the above expression for reliability function with respect to t, we get:

$$\frac{dR(t)}{dt} = -\sum_{i=1}^{n} f_i(t) \prod_{\substack{j=1 \\ j \neq i}}^{n} [1 - F_i(t)] \qquad (4.6)$$

Substituting equation (4.6) in equation (4.4), we get

$$h_S(t) = \sum_{i=1}^{n} \frac{f_i(t)}{R_i(t)} = \sum_{i=1}^{n} h_i(t) \qquad (4.7)$$

Table 4.3 Hazard rate of series configuration with n items.

Probability density function of i-th item, $f_i(t)$	Hazard function of the system, $h_S(t)$
(Exponential) $\lambda_i \exp(-\lambda_i t)$	$h_S(t) = \sum_{i=1}^{n} \lambda_i$
(Weibull) $\frac{\beta_i}{\eta_i}(\frac{t}{\eta_i})^{\beta_i-1} \exp(-(\frac{t}{\eta_i})^{\beta_i})$	$h_S(t) = \sum_{i=1}^{n} (\frac{\beta_i}{\eta_i})(\frac{t}{\eta_i})^{\beta_i-1}$
(Normal) $\frac{1}{\sigma_i\sqrt{2\pi}} \exp(-\frac{1}{2}(\frac{t-\mu_i}{\sigma_i})^2)$	$h_S(t) = \sum_{i=1}^{n} f_i(t) \Big/ \Phi(\frac{\mu_i - t}{\sigma_i})$

Thus the hazard function of a series system is given by the sum of the hazard function of individual items. Table 4.3 gives hazard function of a series configuration with n item under the assumption that the time-to-failure of the items follows same distribution but have different parameter. Figure 4.3 shows hazard rate of a series system with two items where the time-to-failure of individual items follow Weibull distribution.

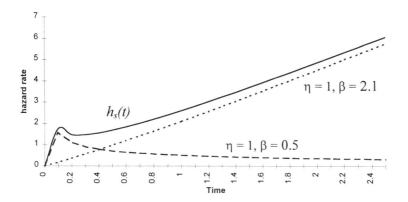

Figure 4.3 Hazard rate of series system with two items with Weibull time-to-failure distribution.

In most cases, the hazard function of a series configuration will be a increasing function. For example, consider a series system with 10 items. Let 9 out of 10 items be identical and have exponential time-to-failure distribution with parameter with rate $\lambda = 0.01$. Now we consider two different cases for the time-to-failure distribution of the remaining one item.

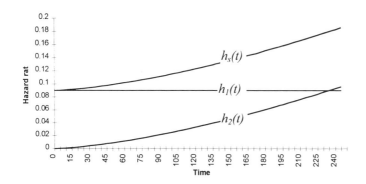

Figure 4.4 Hazard rate the system with 10 items where 9 of them have constant hazard.

Case 1:

Let the time-to-failure of the remaining one item be represented by using Weibull distribution with scale parameter $\eta = 100$ and $\beta = 2.5$. Now the hazard rate of this system is given by:

$$h_s(t) = 9 \times 0.01 + \frac{\beta}{\eta}(\frac{t}{\eta})^{\beta-1}$$

It is obvious from the above expression that the hazard rate of the system is not constant. Figure 4.4 shows the effect of non-constant hazard function on the system hazard function even when most of the items have constant hazard function. In Figure 4.4, $h_1(t)$ represents the hazard rate for the nine items with exponential time-to-failure and $h_2(t)$ represent the hazard rate of the item with Weibull time-to-failure distribution.

Let the time-to-failure of the remaining one item can be represented by using Weibull distribution with scale parameter $\eta = 100$ and $\beta = 0.5$. Now the hazard rate of this system is given by:

$$h_s(t) = 9 \times 0.01 + \frac{\beta}{\eta}(\frac{t}{\eta})^{\beta-1}$$

It is obvious from the above expression that the hazard rate of the system is not constant. Figure 4.5 shows the effect of non-constant hazard function on the system hazard function even when most of the items have constant hazard function. In Figure 4.5, $h_1(t)$ represent the hazard rate for the nine item with exponential time-to-failure and $h_2(t)$ represent the hazard rate of the items with Weibull time-to-failure distribution.

Note: The hazard function of complex repairable system may converge to a constant hazard function under certain conditions (mainly under steady-state conditions). This result proved by Drenick (1961) may not be true for today's highly reliable systems. Thus, one has to be very careful in using constant hazard function and thus exponential time to failure for complex systems. This problem will be further discussed in Chapter 8.

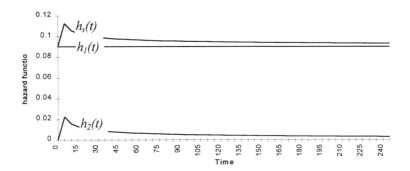

Figure 4.5 Hazard function of the system with 10 items where 9 of them have constant hazard.

Example 4.3

A system has two items A and B connected in series. The time-to-failure of item A follows exponential distribution with parameter $\lambda = 0.002$. The time-to-failure of item B follows Weibull distribution with parameter $\eta = 760$ and $\beta = 1.7$. Find the hazard rate of this system at time $t = 100$ and $t = 500$.

SOLUTION:

Let $h_A(t)$ and $h_B(t)$ represent the hazard rate of item A and B respectively. Since the items are connected in series, the hazard rate of the system, $h_S(t)$ is given by:

$$h_S(t) = h_A(t) + h_B(t) = \lambda + (\frac{\beta}{\eta})(\frac{t}{\eta})^{\beta-1} = 0.002 + (\frac{1.7}{760})(\frac{t}{760})^{0.7}$$

Substituting $t = 100$ and $t = 500$ in the above equation,

$h_S(100) = 0.00254$

$h_S(500) = 0.0036$

Mean time to failure of a series configuration

The mean time to failure, MTTF, of a series configuration, denoted by MTTF$_S$, can be written as:

$$MTTF_S = \int_0^\infty R_S \, dt = \int_0^\infty \prod_{i=1}^n R_i(t) \, dt \tag{4.8}$$

The above integral can be evaluated using numerical integration if the failure distribution is Weibull, normal, lognormal or Gamma. However, in case of exponential distribution the expression for system MTTF$_S$ can be obtained as follows. Assume that the time-to-failure distribution of component i is given by, $1 - \exp(-\lambda_i t)$. Substituting $R_i(t) = \exp(-\lambda_i t)$ in equation (4.8) we have,

$$MTTF_s = \int_0^\infty \prod_{i=1}^n R_i(t) \, dt = \int_0^\infty \prod_{i=1}^n \exp(-\lambda_i t) \, dt = \int_0^\infty \exp(-\sum_{i=1}^n \lambda_i t) \, dt$$

$$MTTF_s = \frac{1}{\sum_{i=1}^n \lambda_i} \tag{4.9}$$

Thus, the *MTTF$_S$* of a series configuration with n items where the time-to-failure of the items are represented by exponential distribution is given by the inverse of the system's hazard function. ***Note that this result is true only when the time-to-failure distribution is exponential.*** The following equation derived using trapezium approximation of equation (4.8) can be used whenever the time-to-failure of at least one item is non-exponential.

$$MTTF_S \approx \frac{h}{2} \times (R[0] + R[M * h]) + \sum_{i=1}^{M-1} h \times R[i \times h] \tag{4.10}$$

Where h is a small value (e.g. 0.01 or 0.1), the value of M is selected such that $R_S(M \times h)$ is almost zero.

Example 4.4

A system consists of three items connected in series. The time-to-failure distribution and their corresponding parameter values are given in Table 4.4. Find the mean time to failure of the system. Compare the value of MTTF$_S$ with mean time to failure of individual items.

Table 4.4 Time-to-failure distribution of different items

Item	Distribution	Parameter values
Item 1	Weibull	$\eta_1 = 10,\ \beta_1 = 2.5$
Item 2	Exponential	$\lambda = 0.2$
Item 3	Weibull	$\eta_2 = 20,\ \beta_2 = 3$

SOLUTION:

Mean time to failure of the system is given by:

$$MTTF_S = \int_0^\infty \prod_{i=1}^3 R_i(t)\,dt$$

$$= \int_0^\infty \exp(-(\frac{t}{\eta_1})^{\beta_1}) \times \exp(-\lambda t) \times \exp(-(\frac{t}{\eta_2})^{\beta_2})\,dt$$

$$MTTF_S = \int_0^\infty \exp(-(\frac{t}{10})^{2.5}) \times \exp(-0.2t) \times \exp(-(\frac{t}{20})^3)\,dt$$

Using numerical integration, the $MTTF_S$ is given by:

$$MTTF_S \approx 3.48$$

Table 4.5 gives the mean time to failure of various items. Note that the mean time to failure of the system is always less than that of the components when the items are connected in series.

Table 4.5 Comparison of *MTTF* of individual items and *MTTF*$_S$

Item 1	Item 2	Item 3	System
$MTTF = 8.87$	$MTTF = 5$	$MTTF = 17.86$	$MTTF_S \approx 3.48$

Characteristics of MTTF$_S$ of series system

1. The $MTTF_S \leq MTTF_i$, where $MTTF_i$ is the mean time to failure of the item i. Thus, the mean time to failure of a system with series RBD will be less than the mean time to time failure of any of its constituting items.

$$MTTF_S \leq \underset{i=1,2,...,n}{Min} \{MTTF_i\}$$

Where $MTTF_i$ denote the mean time to failure of the item i.

2. For complex repairable systems, $MTTF_S$, represents the mean time to first failure.

4.3.   LIFE EXCHANGE RATE MATRIX

Not all the components of the item will have the same utilisation or life unit. In some cases, if the actual mission period is t hours, some items of the system may have to operate more than t hours (in many cases it can be less than *t* hours). An aircraft jet engine will be switched on at least 20 minutes before the actual flight. Thus, for 10 hours flight, the engine may have to operate for more than 10 hours. Operational environment can also change the ageing pattern of different components within a system. For example, the average flight of a domestic flight within Japan is around 30 minutes compared to that of around 3 hours in US. Thus the aircraft used in Japan lands more often than the one in USA. This means that the usage of landing gears, tyres etc of aircraft used in domestic flights in Japan will be much higher than that of USA. It is very common that different items within a system may have different life units such as hour, miles, flying hours, landings, cycles etc. Thus, to find the reliability of a system whose items have different life units it is necessary to normalise the life units. In this section we introduce the concept of life exchange rate matrix, which can be used to describe the exchange rates between various life units.

Life exchange rate matrix (LERM) is a square matrix of size n, where n is the number of items in the system. Let us denote the life exchange rate matrix as R = [r$_{i,j}$], where r$_{i,j}$ is the (i,j) th element in the LERM. Thus, for a system with n items connected in series, the LERM can be represented as:

$$LERM = \begin{bmatrix} r_{1,1} & r_{1,2} & \cdots & r_{1,n} \\ r_{2,1} & r_{2,2} & \cdots & r_{2,n} \\ \cdot & \cdot & \cdot & \cdot \\ \cdot & \cdot & \cdot & \cdot \\ \cdot & \cdot & \cdot & \cdot \\ r_{n,1} & r_{n,2} & \cdots & r_{n,n} \end{bmatrix}$$

The elements of LERM are interpreted as follows:

$r_{i,j}$ denotes that:

1 life unit of $i = r_{i,j} \times 1$ life unit of j.

Any LERM will satisfy the following conditions:

$r_{i,i} = 1$ for all i.

$r_{i,j} = r_{i,k} \times r_{k,j}$ for all i, j, k

$$r_{i,j} = \frac{1}{r_{j,i}}$$

As an example, let us consider a system with three items connected in series (Figure 4.6). Let the life unit of items 1, 2 and 3 be hours, miles and cycles respectively.

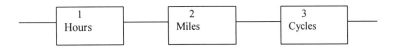

| 1 | 2 | 3 |
| Hours | Miles | Cycles |

Figure 4.6. Series system with three items where each item has different life units

Assume that:

1 hour = 10 miles

1 hour = 5 cycles

Using the above data, it is easy to construct the life exchange rate matrix for the above system. The LERM for the above matrix is:

$$R = \begin{bmatrix} 1 & 10 & 5 \\ 1/10 & 1 & 0.5 \\ 1/5 & 2 & 1 \end{bmatrix}$$

One can easily verify that the above matrix satisfies all three conditions for a life exchange rate matrix. Using the above matrix, one can easily measure reliability characteristics in normalised life unit. For the RBD shown in Figure 4.6, reliability of the system for 5 cycles is given by $R_1(1) \times R_2(1) \times R_3(5)$.

Example 4.5

Reliability block diagram of a system consists of three modules A, B and C connected in series. The time-to-failure of module A follows Weibull distribution with scale parameter $\eta = 100$ hours and $\beta = 3.2$. The time-to-failure of module B follows Normal distribution with parameter $\mu = 400$ cycles and $\sigma = 32$ cycles. The time-to-failure of module C follows exponential distribution with parameter $\lambda = 0.00015$ per mile. It was also noted that, during 1 hour, the module B performs 12 cycles and module C performs 72 miles. Find the probability that the system will survive up to 240 cycles of module B.

SOLUTION:

For the system to survive 240 cycles, module A should survive up to 20 hours and module C should survive up to 1440 miles.

The reliability of individual modules are given by:

$$R_A(t_A) = \exp(-(\frac{t_A}{\eta})^\beta) = \exp(-(\frac{20}{100})^{3.2}) = 0.9942$$

$$R_B(t_B) = \Phi(\frac{\mu - t_B}{\sigma}) = \Phi(\frac{400 - 240}{32}) = 1$$

$$R_C(t_C) = \exp(-\lambda \times t_C) = \exp(-0.00015 \times 1440) = 0.8174$$

The system reliability for 240 cycles is given by:

$$R_S(240) = R_A(20) \times R_B(240) \times R_C(1440) = 0.9942 \times 1 \times 0.8174 = 0.8126$$

4.4.   PARALLEL CONFIGURATION

In a parallel configuration the system fails only when all the items of the system fail. In other words, to maintain the required function only one item of the system is required to function. The reliability block diagram for a system consisting of items connected in parallel is shown in Figure 4.7

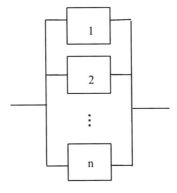

Figure 4.7 Reliability block diagram for a parallel configuration

Parallel components are introduced when the reliability requirements for the system are very high. The use of more than one engine in aircraft is one of the obvious examples of parallel configuration (In practice an aircraft would not be allowed to fly if any of the engine fails. If an engine fails during a flight, the pilot would normally be expected to divert to the nearest airport). However, parallel items will increase cost, complexity and weight of the system. Hence, the number of parallel items required should be carefully determined and if possible optimised.

Reliability function of parallel configuration

Reliability function of a parallel configuration can be obtained using the following arguments. As the system fails only when all the items fail, the failure function, $F_S(t)$, of the system is given by:

$$F_S(t) = P[TTF_1 \le t, TTF_2 \le t, ... TTF_n \le t] \tag{4.11}$$

where TTF_i represents the time-to-failure random variable of item i. Assuming independence among different items, the above expression can be written as:

$$F_S(t) = F_1(t) \times F_2(t) \times ... \times F_n(t) \tag{4.12}$$

where $F_i(t)$ is the time to failure distribution of item i. Substituting $F_i(t) = 1 - R_i(t)$ in equation (4.12), the expression for failure function of a parallel configuration can be written as:

$$F_S(t) = [1 - R_1(t)] \times [1 - R_2(t)] \times ... \times [1 - R_n(t)] \tag{4.13}$$

Now, the reliability function, $R_S(t)$, of a parallel configuration can be written as:

$$R_S(t) = 1 - F_S(t) = 1 - [1 - R_1(t)] \times [1 - R_2(t)] \times ... \times [1 - R_n(t)]$$

or

$$R_S(t) = 1 - \prod_{i=1}^{n} [1 - R_i(t)] \tag{4.14}$$

Characteristics of a parallel configuration

1. The system reliability, $R_S(t)$, is more than reliability of the any of the consisting items. That is,

$$R_S(t) \ge \underset{i=1,...,n}{Max} \{R_i(t)\}$$

2. If $h_i(t)$ represent the hazard rate of item i, then the reliability function of a parallel configuration can be written as:

$$R_S(t) = 1 - \prod_{i=1}^{n} [1 - \exp(-\int_0^t h_i(t)dt]$$

Example 4.6

A fly-by-wire aircraft has four flight control system electronics (FCSE) connected in parallel. The time-to-failure of FCSE can be represented by Weibull distribution with scale parameter η=2800 and β = 2.8. Find the reliability of flight control system for 1000 hours of operation.

SOLUTION:

Reliability function for a parallel system with four identical items is given by:

$$R_S(t) = 1 - \prod_{i=1}^{4}[1 - R_i(t)]$$
$$= 1 - [1 - R(t)]^4$$

where *R(t)* is the reliability function of each item. For t = 1000, *R(t)* is given by:

$$R(t) = \exp(-(t/\eta)^\beta) = \exp(-(1000/2800)^{2.8}) = 0.9455$$

Thus the reliability of flight control system for 1000 hours of operation is given by:

$$R_S(1000) = 1 - [1 - 0.9455]^4 = 0.999991$$

Hazard function of a parallel configuration

Hazard function, $h_S(t)$, of the parallel configuration can be written as:

$$h_S(t) = \frac{-dR_S(t)}{dt} \times \frac{1}{R_S(t)} \tag{4.15}$$

Substituting the expression for $R_S(t)$ from equation (4.14) in the above equation, we get

$$h_S(t) = \{-\frac{d}{dt}[1 - \prod_{i=1}^{n}(1 - R_i(t))]\} \times \frac{1}{[1 - \prod_{i=1}^{n}(1 - R_i(t))]} \tag{4.16}$$

It is easy to verify that the above equation can be written as:

$$h_S(t) = \frac{\sum_{j=1}^{n}\{f_j(t) \times \prod_{i=1,i \neq j}^{n} F_i(t)\}}{1 - \prod_{i=1}^{n}[1 - R_i(t)]} \qquad (4.17)$$

Where, $f_i(t)$ is the probability density function of item i.

Example 4.7

For the flight control system electronics discussed in the example 3.5, find the hazard function of the system at time $t = 100$.

SOLUTION:

Since all the four items are identical, the hazard rate of the system can be written as (using equation (3.15)):

$$h_S(t) = \frac{4 \times f(t) \times [F(t)]^3}{1 - [F(t)]^4}$$

where,

$$f(t) = \frac{\beta}{\eta}(\frac{t}{\eta})^{\beta-1} \exp(-(\frac{t}{\eta})^\beta)$$

$$F(t) = \exp(-(\frac{t}{\eta})^\beta)$$

Substituting $t = 100$, we get

$h_S(t) = 8.0 \times 10^{-8}$

Mean time to failure of parallel configuration

The mean time to failure of a parallel configuration, denoted by $MTTF_S$, can be written as:

$$MTTF_S = \int_0^\infty R_S \, dt = \int_0^\infty \{1 - \prod_{i=1}^{n}[1 - R_i(t)]\} dt \qquad (4.18)$$

For most of the failure distributions one may have to use numerical integration to evaluate the above integral. However, in case of exponential distribution we can get simple expression for system's *MTTF*.

Assume that the time-to-failure distribution of component i is exponential with mean $(1/\lambda_i)$. Then the mean time to failure of the system, $MTTF_S$, is given by:

$$MTTF_s = \int_0^\infty \prod_{i=1}^n R_i(t)dt = \int_0^\infty \{1 - \prod_{i=1}^n [1 - \exp(-\lambda_i t)]\}dt \qquad (4.19)$$

For particular values of n, we can simplify the above integral to derive the expression for the $MTTF_S$.

Case 1: Assume $n = 2$. Equation (4.19) can be written as:

$$MTTF_S = \int_0^\infty \{1 - [(1 - \exp(-\lambda_1 t)) \cdot (1 - \exp(-\lambda_2 t))]\}dt$$

$$= \int_0^\infty [\exp(-\lambda_1 t) + \exp(-\lambda_2 t) - \exp(-(\lambda_1 + \lambda_2)t)]dt$$

$$= \frac{1}{\lambda_1} + \frac{1}{\lambda_2} - \frac{1}{\lambda_1 + \lambda_2}$$

Case 2: Assume $n = 3$, the expression for $MTTF_S$ can be written as:

$$MTTF_S = \int_0^\infty \{1 - [\prod_{i=1}^3 [(1 - \exp(-\lambda_i t))]\}dt$$

$$= \frac{1}{\lambda_1} + \frac{1}{\lambda_2} + \frac{1}{\lambda_3} - \frac{1}{\lambda_1 + \lambda_2} - \frac{1}{\lambda_1 + \lambda_3} - \frac{1}{\lambda_2 + \lambda_3} + \frac{1}{\lambda_1 + \lambda_2 + \lambda_3} \qquad (4.20)$$

4.5.   R-OUT-OF-N SYSTEMS

In an r-out-of-n (or r-out-of-n:G) system, at least r items out of the total n items should maintain their required function for the system to be operational. Following are few examples of r-out-of-n systems:

1. Control software in a space shuttle has four programs. For the successful completion of the mission, at least three of them should maintain the required function and also the output from at least three programs should agree with each other. This is an example of a 3-out-of-4 system.
2. Most of the telecommunication system can be represented as a r-out-of-n systems.

The reliability function of r-out-of-n system can be derived as stated below.

Reliability function of an r-out-of-n system

Consider an r-out-of-n system with identical items. That is, $R_1(t)=R_2(t)=... = R_n(t)$. Then the system reliability, $R_S(t,r,n)$, is given by:

$$R_S(t,r,n) = \sum_{i=r}^{n} \binom{n}{i} [R(t)]^i [1 - R(t)]^{n-i} \qquad (4.21)$$

For the cases when the time-to-failure distribution is exponential or Weibull we have the following expressions for reliability function.

1. Exponential time-to-failure distribution

$$R_S(t,r,n) = \sum_{i=r}^{n} \binom{n}{i} [\exp(-\lambda t)]^i [1 - \exp(-\lambda t]^{n-i}$$

2. Weibull time-to-failure distribution

$$R_S(t,r,n) = \sum_{i=r}^{n} \binom{n}{i} [\exp(-(\frac{t}{\eta})^\beta)]^i [1 - \exp(-(\frac{t}{\eta})^\beta)]^{n-i}$$

However, if the items are not identical then one may have to use other mathematical models such as enumeration to evaluate the reliability. For example consider a 2-out-of-3 system with non-identical items. The reliability function of the system can be derived as follows.

Let E_i denote the event that the item i successfully completes the mission (or survives t hours of operation). Then the reliability function for the system can be written as:

$$R_S(t) = P [\{E_1 \cap E_2\} \cup \{E_1 \cap E_3\} \cup \{E_2 \cap E_3\}]$$

By putting, $A = E_1 \cap E_2$, $B = E_1 \cap E_3$ *and* $C = E_2 \cap E_3$, the above expression can be written as:

$R_S(t,2,3) = P [\{A \cup B \cup C\}]$

$\quad = P(A) + P(B) + P(C) - P(A \cap B) - P(A \cap C) - P(B \cap C)$
$\quad\quad + P(A \cap B \cap C)$

$\quad = P(E_1 \cap E_2) + P(E_1 \cap E_3) + P(E_2 \cap E_3) - 2 P(E_1 \cap E_2 \cap E_3)$

Let $R_i(t)$ represent the reliability function for the item i. Now the above expression can be written as:

$R_S(t, 2, 3) = R_1(t) R_2(t) + R_1(t) R_3(t) + R_2(t) R_3(t) - 2 \times R_1(t) R_2(t) R_3(t)$

The above approach becomes complex when the number of items n increases. However, there are several approaches available to tackle complex r-out-of-n systems with non-identical items. The reliability function of r-1-out-of-n and r-out-of-n system with identical items satisfies the following relation:

$$R_s(t, r-1, n) = \binom{n}{r-1}[R(t)]^{r-1}[1 - R(t)]^{n-r+1} + R_S(t, r, n) \qquad (4.22)$$

Mean Time to Failure of r-out-of-n Systems

The mean time to failure, MTTF, of an r-out-of-n system, $MTTF_S(r,n)$, can be obtained using the following expression:

$$MTTF_S(r,n) = \int_0^\infty R_S(t,r,n)dt$$

One may have to use numerical integration in most of the cases to evaluate the above integral. However, if the time-to-failure distribution is exponential, then the above integral reduces to a simple expression. For example, consider a 2-out-of-3 system with identical items where the time-to-failure distribution of the item is represented by exponential distribution with parameter λ. The reliability function of 2-out-of-3 system with exponential items are given by:

$$R_S(t) = \sum_{i=2}^{3} \binom{3}{i} [\exp(-\lambda t)]^i [1 - \exp(-\lambda t)]^{n-i}$$
$$= 3\exp(-2\lambda t)(1 - \exp(-\lambda t)) + \exp(-3\lambda t)$$

Now the MTTF$_S$ is given by,

$$MTTF_S = \int_0^{\infty} [3\exp(-2\lambda t)(1 - \exp(-\lambda t)) + \exp(-3\lambda t)]dt$$
$$= \frac{5}{6\lambda}$$

Using equation (4.22), we get the following relation between *MTTF$_S$(r-1,n)* and *MTTF$_S$ (r,n)* (Misra, 1992):

$$MTTF_S(r-1,n) = \int_0^{\infty} \binom{n}{r-1} [R(t)]^{r-1}[1 - R(t)]^{n-r+1}dt + MTTF_s(r,n) \quad (4.23)$$

4.6.   SERIES AND PARALLEL CONFIGURATION

In this Section we discuss two types of series and parallel structures, which have wide application in reliability theory.

Model 1. Series-Parallel Configuration

Here the system has a series structure with n items where each item has parallel redundant items. Assume that item *i* has m_i components in parallel. Figure 4.8 shows a series-parallel configuration.

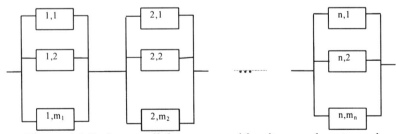

Figure 4.8 Series-parallel structure with *n* items subsystem where subsystem *i* has m_i parallel components

In Figure 4.8, (i,j) represent j-th parallel component of the item i. If $R_{i,j}(t)$ denote the corresponding reliability of the component, then the reliability of item i of the system is given by:

$$R_i(t) = 1 - \prod_{j=1}^{m_i}[1 - R_{i,j}(t)] \tag{4.24}$$

Now the system reliability can be written as:

$$R_S(t) = \prod_{i=1}^{n} R_i(t) = \prod_{i=1}^{n}[1 - \prod_{j=1}^{m_i}(1 - R_{i,j}(t))] \tag{4.25}$$

Model 2. Parallel-Series System

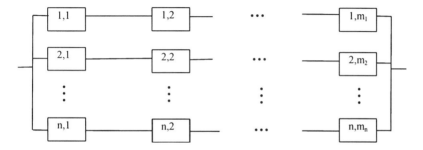

Figure 4.9. Parallel series structure with n sub-systems where subsystem i has m_i components

Assume that the system has n items connected in parallel where each item has components connected in series. An aircraft with more than one engine, is a typical example for these type of configuration. Figure 4.9 shows parallel-series structure.

Since item i has m_i components in series, the reliability of item i is given by:

$$R_i(t) = \prod_{j=1}^{m_i} R_{i,j}(t) \tag{4.26}$$

where $R_{i,j}(t)$ is the reliability function of the component j in item i. Now the reliability of the parallel-series system is given by:

$$R_S(t) = 1 - \prod_{i=1}^{n}[1 - R_i(t)] = 1 - \prod_{i=1}^{n}[1 - \prod_{j=1}^{n} R_{i,j}(t)] \qquad (4.27)$$

4.7.   REDUNDANT SYSTEMS

In systems, redundancy is a means of maintaining system integrity if critical parts of it fail. In some cases this means replicating parts of the system, in others, alternatives are used. A commercial aircraft has to be able to complete a take-off and landing with one of its engines shutdown but, except under very special circumstances, no such aircraft would be allowed to leave the departure gate if any of its engines are not functioning. And yet, ETOPS, extended twin engine operations allows certified twin-engine aircraft (e.g. Boeing 777 and Airbus 330) to fly up to 180 minutes from a suitable landing site. This is based on the probability that even if one of the engines fails that far from land, the other is sufficiently, reliable to make the probability of not reaching a landing site an acceptable risk. It should be noted that in normal flight, i.e. at cruising speed and altitude, the engines are generally doing very little work and usually are throttled back. If an engine fails, it would normally be wind-milled to minimise 'parasitic' drag but, even then, it still offers a considerable resistance and, of course, produces an in-balance which has to be offset by the rudder and other controllable surfaces all of which means the functional engine has to work considerably harder thus increasing its probability of failure.

If the aircraft only had one engine and it failed, the probability of landing safely with no engines is not very high, at least, for fast military jets. In most cases ultimately, if the engine cannot be re-lit, the only option is to eject after directing the aircraft away from inhabited areas, if there is time. With commercial airlines, neither the pilot, the crew nor the passengers have the option of ejecting or baling out if the aircraft suffers a total engine failure (i.e. all engines fail). These aircraft will glide, to a certain extend but, with no power, none of the instruments will function and, there will be no power assistance for the control surfaces or to deploy the landing gear. For this reason, they are fitted with wind turbines that should drop down and start functioning if there is prolonged loss of power. This gives the pilots some control, but even then, large airliners are not going to rise on a thermal, however good the pilot may be.

A Boeing 767, on one of its first flights, had a total engine failure some 1500 miles from its intended destination, Ontario. All attempts to re-light the engines failed simply because it had run out of fuel. There was a total blackout in the cockpit and, even when the co-pilot managed to find a torch (flashlights) all this showed was that none of the instruments were working (being all digital and computer controlled). The pilot, by pure chance, happened to be an extremely accomplished glider pilot and, again by pure chance, the co-pilot happened to be particularly familiar with this part of Canada, some 200 miles outside Winnipeg. For several minutes the pilot manhandled the controls and managed to stop the aircraft from loosing height too quickly. Eventually the wind turbine deployed which gave them enough power for the instruments, radio and power assisted controls to work again. Unfortunately the aircraft had lost too much height to reach Winnipeg but, it had just enough to get to an ex-military runway (used as a strip for drag racing). There was just enough power to lock the main undercarriage down, but not the nose wheel. The *Gimli Glider* as it became known, landed safely with no serious casualties. But, out of eleven other pilots, who later tried to land the aircraft in the same circumstances on a flight simulator all crashed. Had it not been for the 'redundant' wind turbine, it is almost certain even this experienced glider pilot would have crashed killing all on board.

If the Boeing 777, say, was fitted with three or four Rolls-Royce Trent 800s, Pratt & Whitney 4084s or General Electric GE 90's (instead of the two it currently has) then there would be true redundancy since it needs only two to achieve *ETOPS* (Extended Twin-engine Operations). There are, however, a number of problems with this design. Firstly, it would add very significantly to both weight and drag, to the point where it would seriously reduce the payload and range, probably making the aircraft uneconomical to operate and hence undesirable to the airlines. Secondly such an increase in weight and drag would probably mean the normal two engines would provide insufficient thrust therefore either more powerful engines would be needed or, the extra engines would have to be used rendering them no longer truly redundant.

On the Boeing 767, for example, the IFSD (In Flight Shut Down) rate after 10 million hours was less than 0.02 per thousand flying hours (the standard measure in the aerospace industry). And, none of these had led to the loss of a single life, let alone an aircraft with its full complement of passengers and crew. It is quite likely that, in some of the instances, flights would have been diverted from their scheduled destinations to alternatives, for safety reasons. The inconvenience to passengers (and airlines) would have cost the airline but, the amount would, almost certainly, have been significantly less than the loss of revenue resulting from the reduced payload had truly redundant engines been fitted.

In many cases, the redundant items may not be functioning simultaneously as in the case of parallel or r-out-of-n configurations. The redundant items will be turned on only when the main item fails. In some cases, the items may be functioning simultaneously but one of them may be sharing much higher load compared to the other. Such types of systems are called standby redundant systems. Whenever the main item fails, a built-in switch senses the failure and switches on the first standby item. It is important that the switch has to maintain its function. Failure of the switch can cause the system failure. The standby redundant systems are normally classified as *cold standby, warm standby* and *hot standby.*

Cold Standby System

In a cold standby, the redundant part of the system is switched on only when the main part fails. For example, to meet the constantly changing demand for electricity from the 'National Grid' it is necessary to keep a number of steam turbines ready to come on stream whenever there is a surge in demand. The failure of a generator would result in instantaneous reduction in capacity, which would be rectified by bringing one of these 'redundant' turbines up to full power. In the event of a power cut to a hospital, batteries may switch in instantly to provide emergency lighting and keep emergency equipment, e.g. respirators and monitors running. Petrol and diesel generators would then be started up to relieve the batteries and provide additional power.

In a cold standby system, a redundant item is switched on only when the operating item fails. That is, initially one item will be operating and when this item fails, one item from the redundant items will be switched on to maintain the function. In a cold standby, the hazard function of the item in standby mode is zero.

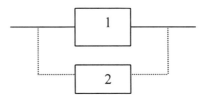

Figure 4.10 Cold standby redundant system

Consider a cold standby system with two identical items (see Figure 4.10). The reliability function of this system can be derived as follows (assuming that the switch is perfect):

$R_S(t)$ = P{The main item survives up to time t}
+ P{The main item fails at time u ($u < t$) and the standby items survives the remaining interval ($t - u$) }

Thus,

$$R_S(t) = R(t) + \int_0^t f(u)R(t-u)du \qquad (4.28)$$

where $f(t)$ is the probability density function of time-to-failure random variable.

As an example consider a cold standby system with two items where the time-to-failure distribution is exponential with parameter λ. Using the equation (4.29) the expression for reliability function is given by:

$$R_S(t) = \exp(-\lambda t) + \int_0^t \lambda \exp(-\lambda u) \times \exp(-\lambda(t-u))du$$

$$= \exp(-\lambda t) + \lambda t \exp(-\lambda t) = \exp(-\lambda t)[1 + \lambda t]$$

For a cold standby system with n identical items with exponential time-to-failure distribution, the expression for reliability function is given by:

$$R_S(t) = \exp(-\lambda t) \sum_{i=0}^{n-1} \frac{(\lambda t)^i}{i!} \qquad (4.29)$$

The equation (4.30) is the cumulative distribution of Poisson distribution with mean λt. One can also derive the expression for non-identical standby units using the arguments presented in equation (4.28). For a cold-standby system with non-identical items, the system reliability function is given by:

$$R_s = R_1(t) + \int_0^t f_1(x)R_2(t-x)dx \qquad (4.30)$$

Where $R_1(t)$ and $f_1(t)$ are the reliability function and failure density function of item 1 and $R_2(t)$ is the reliability function of item 2. Assume that the time-to-failure items 1 and 2 can be modelled using exponential distribution with mean $(1/\lambda_1)$ and $(1/\lambda_2)$ respectively. Using equation (4.31), the reliability function of cold-standby system with non-identical items is with exponential failure time is given by:

$$R_s(t) = \exp(-\lambda_1 t) + \int_0^t \lambda_1 \exp(-\lambda_1 x) \times \exp(-\lambda_2(t-x))dx$$

$$R_s(t) = \exp(-\lambda_1(t) + \frac{\lambda_1}{\lambda_1 - \lambda_2}[\exp(-\lambda_2 t) - \exp(-\lambda_1 t)]$$

The *MTTF* of a cold-standby system can be evaluated by integrating the reliability function between 0 and ∞. The *MTTF* of a cold-standby system with n identical units with exponential failure time is given by:

$$MTTF = \frac{n}{\lambda} \tag{4.31}$$

Equation (4.31) can be easily derived from equation (4.30). For the non-identical MTTF is given by:

$$MTTF_s = \sum_{i=1}^{n} \frac{1}{\lambda_i} \tag{4.31a}$$

Warm Standby System

In a warm standby system, the redundant item will be sharing partial load along with the main item. Thus, in a warm standby, the hazard function of the standby item will be less than that of the main item.

That is, a standby system can deteriorate even when it is not in use. Consider a system with two warm standby items. Assume that $R(t)$ and $R^s(t)$ represent the reliability of the item in operating mode and standby mode respectively. Now the reliability function of the system can be written as:

$$R_S(t) = R(t) + \int_0^t f(x) \times R^s(x) \times R(t - x) du \tag{4.32}$$

For a particular case where $R(t) = \exp(-\lambda t)$ and $R^s(t) = \exp(-\lambda_s t)$ the reliability function of a warm standby system is given by:

$$R_S(t) = \exp(-\lambda t) + \int_0^t \lambda \exp(-\lambda u) \times \exp(-\lambda_s u) \times \exp(-\lambda(t - u)) du$$

$$= \exp(-\lambda t) + \frac{\lambda \exp(-\lambda t)}{\lambda_s}(1 - \exp(-\lambda_s t))$$

Hot Standby System

In a hot standby, the main item and the standby item will be sharing equal load, and hence will have the same hazard rate. Thus, a hot standby can be treated as a parallel system to derive reliability expressions. If $h_o(t)$ and $h_s(t)$ represent the hazard rate of a operating and standby item respectively. The Table 4.6 gives the various redundancies and the properties of hazard rate.

Table 4.6 Types of standby redundancy and the corresponding properties of hazard rate

Type of Redundancy	Properties of hazard rate
Cold Standby	$h_S(t) = 0$
Warm Standby	$h_O(t) > h_S(t)$
Hot Standby	$h_O(t) = h_S(t)$

4.8.   COMPLEX RELIABILITY BLOCK DIAGRAMS

In many cases, the reliability block diagram will have complex combinations of series and parallel blocks. In such cases, one has to reduce the block to either a series structure or a parallel structure before one can predict the reliability characteristics of the system. Reducing a complex reliability structure will involve the following steps:

1. Replace all purely series (parallel) with an equivalent (reliability wise) single block.
2. Repeat step 1 up till the RBD reduces to either a series or parallel structure.
3. Compute the reliability of resulting RBD.

For example, consider the RBD shown in Figure 4.11.

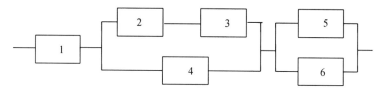

Figure 4.11 Reliability block diagram with combination of series-
parallel structures

The time-to-failure of the six items within the system shown in Figure
4.11 are shown in Table 4.7.

Table 4.7. Time-to-failure of items shown in Figure 4.12

Item	Distribution with parameter values
1	Weibull, $\eta = 450$ hours, $\beta = 2.4$
2	Lognormal $\mu_l = 4.5$, $\sigma_l = 0.75$
3	Weibull, $\eta = 890$ hours, $\beta = 1.75$
4	Exponential, $\lambda = 0.001$
5	Normal $\mu = 800$, $\sigma = 120$
6	Exponential, $\lambda = 0.00125$

The reliability block diagram shown in Figure 4.11 can be evaluated
using the three steps explained above. The RBD in Figure 4.11 can be
replaced by a series structure with three blocks as shown in Figure 4.11a.

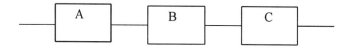

Figure 4.11a Reliability block diagram equivalent to Figure 4.11

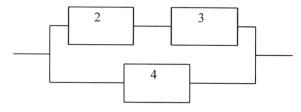

Figure 4.11b RBD equivalent to block B in Figure 4.11

The block A is same as item 1, where block B is equivalent to the RBD shown in Figure 4.11b.

The block B is equivalent to RBD shown in Figure 4.12.c.

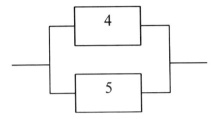

Figure 4.11c. RBD equivalent to block C in Figure 4.11

The expression for reliability function of the system in Figure 4.11 is given by:

$$R_s(t) = R_A(t) \times R_B(t) \times R_C(t)$$

where

$$R_A(t) = R_1(t)$$

$$R_B(t) = 1 - [1 - (1 - R_2(t) \times R_3(t)) \times (1 - R_4(t))]$$

$$R_C(t) = 1 - [1 - (1 - R_5(t)) \times (1 - R_6(t))]$$

For some systems, the reliability block diagram may have more complex configuration than the series/parallel structure as discussed so far. The well-known *'Wheatstone Bridge'* (see Figure 4.12) is an example of such configuration. To find the reliability of such systems one may have to use special tools such as *cut-set, path-set, enumeration* or *the conditional probability approach.* In this Section we illustrate the cut-set approach for evaluating reliability of complex structures.

4.9. CUT SET APPROACH FOR RELIABILITY EVALUATION

Cut-set approach is one of the most popular and widely used methods for predicting reliability of complex structure. The main advantage of cut-set approach is that it is easy to program and most of the commercial software for reliability prediction use cut-set approach to evaluate the reliability of complex structures. *A cut-set is defined as the set of items that, when failed, will cause the system failure. A cut-set with minimum number of items is called minimal cut set.* That is if any item of the minimal cut set has not failed, then the system will not fail. Mathematically, if the set C is a cut set of the system. Then, the set C will be a minimal cut set if for all $c_i \in C$, $C - c_i$ is not a cut set. Here $C - c_i$ represents the set C without the element c_i. The cut set approach to reliability prediction involves identifying all the minimal cut sets of the system.

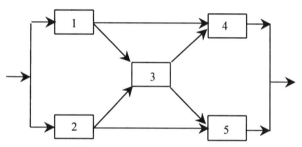

Figure 4.12 Bridge network

In Figure 4.12, the set of items $C = \{1, 2, 3\}$ forms a cut set, since the failure of the items 1, 2 and 3 will cause system failure. However, the set $C = \{1, 2, 3\}$ is not a minimal-cut set since $C - 3 = \{1, 2\}$ still forms a cut set. For the structure shown in Figure 4.12, the minimal cut sets are given by:

$$C_1 = \{1, 2\}, \, C_2 = \{1, 3, 5\}, \, C_3 = \{2, 3, 4\} \text{ and } C_4 = \{4, 5\}$$

Since all the elements of the minimal cut set should fail to cause the system failure, each cut set can be considered as a parallel configuration. Thus, the cut sets C_1, C_2, C_3 and C_4 represent the following structures shown in Figure 4.13.

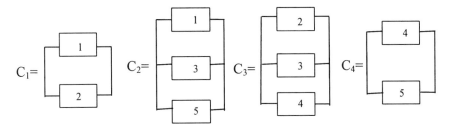

Figure 4.13. Equivalent RBD for minimal cut sets of the system shown in Figure 4.12

Since the system will fail when at least one minimal cut sets fail, the reliability function of the system can be written as:

$$R_S(t) = RC_1(t) \times RC_2(t) \times RC_3(t) \times RC_4(t) \qquad (4.33)$$

where $RC_1(t)$, $RC_2(t)$, $RC_3(t)$ and $RC_4(t)$ are the reliability function of the structures represented by the cut sets C_1, C_2, C_3 and C_4 respectively. If $R_i(t)$ denote the failure function of the items 1, 2, 3, 4 and 5, then we have:

$$RC_1(t) = 1 - F_1(t)F_2(t), \qquad RC_2(t) = 1 - F_1(t)F_3(t)F_5(t)$$
$$RC_3(t) = 1 - F_2(t)F_3(t)F_4(t), \qquad RC_4(t) = 1 - F_4(t)F_5(t)$$

Substituting the above expressions in equation (4.33), we get the failure function for the complex structure shown in Figure 4.12.

In general, cut set approach involves the following steps:

1. Identify all the minimal cut sets of the system.
2. Since all the elements of the minimal cut set should fail to cause the system failure, each cut set can be treated as a parallel configuration.
3. Since failure of any one minimal cut set can cause system failure, different minimal cut sets can be treated as a series configuration.

4.10. CASE STUDY ON AIRCRAFT ENGINES

Aircraft engine is one of the most critical items used in today's aviation industry. In this section, we try to address several reliability measures one may like to know about an engine. There are totally eleven items including the external gearbox, oil tank and filter. The time-to-failure of these items are given in Table 4.8.

Table 4.8. Time-to-failure distribution of various items of the engine

Item no.	Item	Distribution	Parameter Values
01	LP compressor	Weibull	$\eta = 15\,000,\ \beta = 3$
02	LP stage 2 stator	Weibull	$\eta = 5\,000,\ \beta = 2.8$
03	Intermediate casing	Weibull	$\eta = 11\,000,\ \beta = 3$
04	HP compressor	Weibull	$\eta = 12\,000,\ \beta = 3.5$
05	HP NGV	Weibull	$\eta = 8\,000,\ \beta = 3$
06	HP turbine	Weibull	$\eta = 25\,000,\ \beta = 4$
07	LP NGV	Weibull	$\eta = 7\,000,\ \beta = 2.2$
08	LP turbine	Weibull	$\eta = 20\,000,\ \beta = 2.8$
09	Exhaust mixer	Weibull	$\eta = 7\,000,\ \beta = 3$
10	External gear box	Weibull	$\eta = 6\,500,\ \beta = 3$
11	Oil tank and filter	Weibull	$\eta = 5\,000,\ \beta = 3.8$

We are interested in carrying out the following tasks

1. Draw the reliability block diagram of the engine.
2. Find reliability of the engine for 3000 hours of operation.
3. Find the hazard rate of the engine at $t = 3000$ and $t = 7000$ hours.
4. Find the MTTF of different items of the engine and estimate the MTTF of the engine from the MTTF values of the items.
5. Find the MTTF of the engine if all the items are subject to preventive maintenance after every 1000 hours of operation (assume that after maintenance all the items behave as good as new).
6. For an engine of age 5000 hours, find the mission reliability for 1000 hours of operation.
7. Find the MFOPS of the engine for 500 hours of operation for different cycles.

SOLUTION:

1. Since all the item of the engine must maintain their function, the system will have a series configuration as shown below:

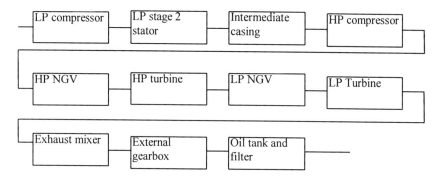

Figure 4.14 Reliability block diagram of the engine

2. Since all the items of the system follow Weibull distribution, the reliability function for each of these items is given by:

$$R(t) = \exp(-(\frac{t}{\eta})^\beta)$$

Substituting the values of η and β for various items in the above equation, the reliability of various items for 3000 hours of operation is given by:

1. Reliability of LP compressor for 3000 hours of operation is given by:

$$R_1(3000) = \exp(-(\frac{3000}{15000})^3) = 0.9920$$

2. Reliability of LP stage 2 stator for 3000 hours of operation is given by:

$$R_2(3000) = \exp(-(\frac{3000}{5000})^{2.8}) = 0.7872$$

3. Reliability of intermediate casing for 3000 hours of operation is given by:

$$R_3(3000) = \exp(-(\frac{3000}{11000})^3) = 0.9799$$

4. Reliability of HP compressor for 3000 hours of operation is given by:

$$R_4(3000) = \exp(-(\frac{3000}{12000})^{3.5}) = 0.9922$$

5. Reliability of HP NGV for 3000 hours of operation is given by:

$$R_5(3000) = \exp(-(\frac{3000}{8000})^3) = 0.9486$$

6. Reliability of HP turbine for 3000 hours of operation is given by:

$$R_6(3000) = \exp(-(\frac{3000}{25000})^4) = 0.9997$$

7. Reliability of LP NGV for 3000 hours of operation is given by:

$$R_7(3000) = \exp(-(\frac{3000}{7000})^{2.2}) = 0.8563$$

8. Reliability of LP turbine for 3000 hours of operation is given by:

$$R_8(3000) = \exp(-(\frac{3000}{20000})^{2.8}) = 0.9950$$

9. Reliability of exhaust mixer for 3000 hours of operation is given by:

$$R_9(3000) = \exp(-(\frac{3000}{7000})^3) = 0.9243$$

10. Reliability of external gearbox for 3000 hours of operation is given by:

$$R_{10}(3000) = \exp(-(\frac{3000}{6500})^3) = 0.9063$$

11. Reliability of oil tank and filter for 3000 hours of operation is given by:

$$R_{11}(3000) = \exp(-(\frac{3000}{5000})^{3.8}) = 0.8662$$

Using the above values of individual reliabilities, the reliability of the system is given by

$$R_S(3000) = \prod_{i=1}^{11} R_i(3000) = 0.4451$$

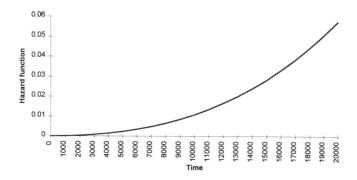

Figure 4.15 hazard function for the engine.
3. Hazard function of the system.

Since all the items of the system follow Weibull time-to-failure, the hazard function is given by:

$$h(t) = (\frac{\beta}{\eta})(\frac{t}{\eta})^{\beta-1}$$

The system hazard function is given by:

$$h_S(t) = \sum_{i=1}^{11} h_i(t)$$

It is easy to verify that the hazard function of the system at t = 3000 and t= 7000 is given by:

$h_S(3000) = 0.000791$ and $h_S(7000) = 0.004796$

Figure 4.15 depicts the hazard function for the engine.

4. The expression for MTTF is given by:

$$MTTF = \eta \times \Gamma(1 + \frac{1}{\beta})$$

By substituting the values of η and β, one can find the *MTTF* of different items. Table 4.9 gives the *MTTF* of different items.

Table 4.9 MTTF of different item of the engine

Item	*MTTF* (in hours)
LP compressor	13 395
LP stage 2 stator	4 450
Intermediate casing	9 823
HP compressor	10 800
HP NGV	7 144
HP turbine	22 650
LP NGV	6 202
LP turbine	17 800
Exhaust mixer	6 251
External gear box	5 804
Oil tank and filter	4 525

Since the lowest MTTF is 4 450 (LP stage 2 stator), the MTTF of engine will be less than 4 450.

5. Mean time to failure of a system subject to preventive maintenance is given by:

$$MTTF_{pm} = \frac{\int_0^{T_P} R_S(t)dt}{1 - R_S(T_P)}$$

It is given that the engine is subject to preventive maintenance every 1000 hours of operation. Thus, $T_P = 1000$ hours. The above expression can be evaluated using numerical integration. The approximate values of $MTTF_{pm}$ is:

$$MTTF_{pm} = \frac{\int_0^{1000} R_S(t)dt}{1 - R_S(1000)} \approx \frac{999.06}{0.0369} \approx 27,075$$

6. The mission reliability of the engine is given by:

$$MR(t_b, t_m) = \frac{R(t_b, t_m)}{R(t_b)}$$

where t_b is the age of the item at the beginning of the mission and t_m is the mission duration. Substituting $t_b = 5000$ and $t_m = 1000$, we have

$$MR(t_b, t_m) = \frac{R(5000 + 1000)}{R(5000)} = \frac{\prod_{i=1}^{11} R_i(6000)}{\prod_{i=1}^{11} R_i(5000)} = \frac{0.0013}{0.02369} = 0.0548$$

7. The maintenance free operating period survivability, MFOPS, for the engine described is given by:

$$MFOPS(t_{mf}) = \frac{R_S(i \times t_{mf})}{R_S([i-1] \times t_{mf})} = \frac{\prod_{i=1}^{11} R_i(i \times t_{mf})}{\prod_{i=1}^{11} R_i([i-1] \times t_{mf})}$$

The above equation can be evaluated for $t_{mf} = 500$ and for i = 1, 2, ... etc. Figure 4.16 shows the MFOPS values for different cycles (note that these values are derived without considering maintenance recovery period MRP).

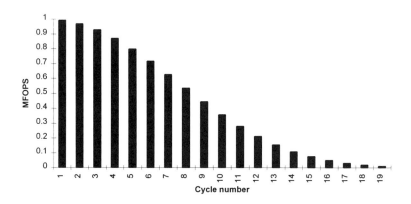

Figure 4.16 MFOPS value for different cycles for the engine

4.11.   SYSTEMS RELIABILITY EVALUATION - STOCHASTIC MODELS

Stochastic modelling is one of the most powerful tools of modern probability that can be used to analyse reliability, maintenance and logistic support problems. It is basically a dynamic model that can be used to analyse random phenomena such as the behaviour of repairable systems, availability, demand for spares etc. The advantage of stochastic modelling is that it allows one to model any system characteristics by incorporating time-to-failure, repair time, repair strategy, maintenance and logistic delay time. In this chapter we discuss various stochastic processes such as Markov process, Poisson Process, renewal process and regenerative process and their applications to reliability engineering.

4.12.   STOCHASTIC PROCESSES

A stochastic process (also known as random process) is a collection of random variables $\{X(t), t \in T\}$, where T is the set of numbers that indexes the random variables X(t). In reliability, it is often appropriate to interpret t as time and T as the range of time being considered.   The set of possible values the stochastic process X(t) can assume is called state. The set of possible states constitutes the state-space, denoted by E. The state-space can be continuous or discrete. For example consider a system with two items connected in parallel as shown in Figure 4.17. Assume that the time-to-failure of the two parallel items are given by two sequence of random variables X_i and Y_i ($i = 1, 2, ...$). Here the subscript i represents the time to i^{th} failure of the items. If the sequence of random variable Z_i represents the ith repair time, then the process $\{ X(t), t \geq 0 \}$ by definition forms a stochastic process. At any time t, it is possible that two, one or none of these two items will be maintaining the required function. Thus, the set $\{0, 1, 2\}$ forms the state-space of the system.

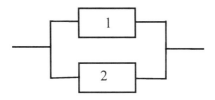

Figure 4.17 System with two identical items connected in parallel

Analysing a system using stochastic processes will involve the following fundamental steps.

1. Identify the time domain T for the system. The time domain T can be discrete or continuous.
2. Identify the state space of the system. The state space can be either discrete or continuous.

Once the process is defined using the family of random variables $\{X(t), t \in T\}$, state space (E) and the parameter set (T), the next step will be to identify the properties of the process that can be used to classify the process and also to analyse the process to extract information. As far as reliability is concerned, processes with a continuous time parameter and discrete state space are important. In this chapter, we discuss the following processes and their applications to reliability engineering.

1. Markov processes
2. Non-homogeneous Poisson Process
3. Renewal processes

Homogeneous Poisson process model is discussed in Chapter 8. Readers who are interested to know more on applications of stochastic process are advised to refer Birolini *(1997)*.

4.13.   MARKOV PROCESS

A stochastic process is said to be a Markov process if the future evolution of the process depends only on the current time and state. That is, the future state of a system is conditionally independent of the past, given the present state and age of the system is known. Thus, to predict the future state one need to know only the present state and age of the system. Mathematically, a stochastic process $\{X(t); t \in T\}$ with state-space E is called a Markov process if it satisfies the condition

$$P[X(t_n + h) = j \mid X(t_n) = i_n, X(t_{n-1}) = i_{n-1}, ..., X(t_0) = i_0]$$
$$= P[X(t_n + h) = j \mid X(t_n) = i_n]$$

(4.34)

for all $(j, i_n, i_{n-1}, ..., i_0) \in$ E. The above property is called *Markov property*. A Markov process with discrete state space is called *Markov Chain*. A Markov process with continuous time and discrete state space is called *continuous time Markov chain (CTMC)*. The conditional probability defined in equation (4.34) is referred as the *transition probability* of Markov process and is defined using the notation $P_{ij}(t_n + h)$

$$P_{ij}(t_n + h) = P[X(t_n + h) = j \mid X(t_n) = i_n]$$

(4.35)

A Markov process is called *time-homogeneous* or *stationary* if the transition probabilities are independent of time t. For a stationary Markov process,

$$P_{ij}(t_n + h) = P_{ij}(t_0 + h) = P_{ij}(h)$$

(4.36)

Thus, the transition from state i to state j in a stationary Markov chain depends only on the duration h. The transition probabilities $P_{ij}(t + h)$ satisfies the following *Chapman-Kolmogrov* equations

$$P_{ij}(t + h) = \sum_{k \in E} P_{ik}(h)P_{kj}(t)$$

(4.37)

In all the models discussed in this Chapter we assume that the Markov process is stationary. It is convenient to use a matrix to represent various state transition probabilities of a Markov process. For example, if a system has n states, we define a matrix **P**, such that

$$\mathbf{P} = [P_{ij}(h)] = \begin{bmatrix} P_{11} & P_{12} & \cdots & P_{1n} \\ P_{21} & P_{22} & \cdots & P_{2n} \\ \cdot & \cdot & \cdot & \cdot \\ \cdot & \cdot & \cdot & \cdot \\ P_{n1} & P_{n2} & \cdots & P_{nn} \end{bmatrix}$$

The matrix **P** is called *Transition Probability Matrix* (TPM) or *Stochastic Matrix*.

Let $\{S_j, j \in E\}$ represent the time spent at state j (sojourn time at state j). The probability that the process will spend more than t hours at state j is, $P[S_j > t]$. Assume that the process has already spend h hours in state j, the probability that it will spend additional t hours in state j is given by:

$$P[S_j > t + h \mid S_j > h] \tag{4.38}$$

Since past is irrelevant in Markov process, the above expression can be written as:

$$P[S_j > t + h \mid S_j > h] = P[S_j > t] \tag{4.39}$$

The only continuous distribution that satisfies the above relation is exponential distribution. The above property of exponential distribution is called *memory-less property.* Thus, in a Markov process, the time spent in any state follows exponential distribution. Thus,

$$P[S_j > t] = \exp(-v_j t) \tag{4.40}$$

where the parameter v_i depends on state i. This is a very important result and limitation of Markov processes. This implies that the Markov process can be applied only when the time-to-failure, repair time and logistic delay time follows exponential distribution.

Transition Rates Between the States of a Markov Process

Since the time spent at any state j of a Markov process follows exponential distribution, the probability that the process remains in state j during a small interval δt is given by:

$$P[S_j > \delta t] = \exp(-v_j \delta t)$$

$$= 1 - \frac{v_j \delta t}{1!} + \frac{(v_j \delta t)^2}{2!} - \dots$$

$$= 1 - v_j \delta t + O(\delta t)$$

where O(δt) represents the terms which are negligible as δ approaches zero. That is,

$$\lim_{\delta t \to 0} \frac{O(\delta t)}{\delta t} = 0$$

Thus, for a small duration of δt, P_{jj} (δt), probability that the process will remain in state *j* for small duration δt is given by:

$$P_{jj}(\delta t) = 1 - v_j \delta t + O(\delta t)$$

Probability that the system will leave state *j* is given by

$$1 - P_{jj}(\delta t) = v_j \delta t + O(\delta t)$$

v_j is the rate at which the process {X(t), t ∈ T} leaves the state *j*. Rearranging the above equation we have,

$$P_{jj}(\delta t) - 1 = -v_j \delta t + O(\delta t)$$

Substituting $\lambda_{jj} = - v_j$ in the above equation, we get

$$P_{jj}(\delta t) - 1 = \lambda_{ii} \delta t + O(\delta t)$$

It is easy to verify that

$$\lim_{\delta t \to 0} \frac{P_{jj} - 1}{\delta t} = \lambda_{jj} \qquad (4.41)$$

The transition probability $P_{ij}(\delta t)$, that is the process will enter state *j* (with probability r_{ij}) after leaving state *i* during a small duration δt is given by:

$$P_{ij}(\delta t) = [1 - P_{ii}(\delta t)] \times r_{ij} = [v_i \delta t + O(\delta t)] \times r_{ij}$$
$$= \lambda_{ij} \delta t + O(\delta t) \tag{4.42}$$

where λ_{ij} is the rate at which the process enters the state j from the state i.

Let $P_j(t) = P[X(t) = j]$, that is $P_j(t)$ denotes that the process is in state j at time t. Now for any δt, we have

$$P_j(t + \delta t) = P[X(t + \delta t) = j]$$
$$= \sum_{i \in E} P[X(t + \delta t) = j \mid X(t) = i] P[X(t) = i]$$

The above expression can be written as

$$P_j(t + \delta t) = \sum_{i \in E} P_{ij}(\delta t) P_{i(t)} \tag{4.43}$$

The above equation (4.43), upon few mathematical manipulation will give a system of differential equation which can be solved to find $P_i(t)$.

From equation (4.43)

$$P_j(t + \delta t) - P_j(t) = \sum_{\substack{i \in E \\ i \neq j}} P_{ij}(\delta t) P_i(t) + P_j(t)[P_{jj}(\delta t) - 1] \tag{4.44}$$

For $\delta t \to 0$, and using equation (4.41) and (4.42), equation (4.44) can be written as:

$$\frac{d}{dt} P_j(t) = \sum_{i \in E} \lambda_{ij} P_i(t) = \sum_{\substack{i \in E \\ i \neq J}} \lambda_{ij} P_i(t) - v_j P_j(t) \tag{4.45}$$

Also

$$\sum_{j \in E} P_j(t) = 1 \tag{4.46}$$

Equation (4.45) is called *Kolmogrov backward equations*, which along with equation (4.46) has a unique solution. Thus, various state probabilities of the process can be obtained by solving the system of differential equations of the form:

$$\frac{d}{dt}P(t) = \Delta P(t) \tag{4.47}$$

where P(t) is a time-dependent N dimensional probability vector and Δ is a square matrix where the element (i,j) represent the rate at which the process enters the state *j* from the state *i*.

Application of Markov Processes to reliability and point availability prediction

The first step in calculating the reliability and availability using Markov modelling is to identify the system up states (states in which the system maintains the required function) and the system down states. The state-space, E, of a Markov chain can be partitioned into two mutually exclusive sets U and D. Where, U is the set system up states (states in which the system maintains the required function) and D is the set of down states (failed state). Now the state probabilities P_j and the time spent on each state can be used to evaluate point availability and the reliability of system using either differential equations or integral equations. The point availability, $A_S(t)$, of the system is given by:

$$A_S(t) = P[X(t) \in U] = \sum_{j \in E} P_j(t) \tag{4.48}$$

$P_j(t)$ can be evaluated by solving the system of differential equations (4.43) and (4.44).

Also, the state probabilities can be evaluated using integral equations. The probabilities $P_j(t)$ can be obtained by solving the following system of integral equations.

For all $i, j \in$ E,

$$P_j(t) = \delta_{ij} \exp(-v_j t) + \sum_{i \in E_0} \int_0^t \lambda_{ji} \exp(-v_j x) P_{ij}(t-x)dx \ , \tag{4.49}$$

where $\delta_{ij} = 0$ for $i \neq j$ and $\delta_{ij} = 1$ for $i = j$.

Example 4.8

Time-to-failure distribution of a repairable item can be represented using exponential distribution with parameter λ. Upon failure the item is repaired and the repair time also follows exponential distribution with parameter μ.

Find

1. the state space of the system.
2. the transition probability matrix.
3. derive the set of differential equations satisfies by the state probabilities.
4. solve the differential equation to find the time-dependent state probabilities.

SOLUTION:

1. The system can be in two states, either operating or failed (thus under repair). Let us denote state 1 as operating state and state 2 as failed state. Thus the state space, E, is given by

$$E = \{1, 2\}$$

Also, the process $\{X(t), t \geq 0\}$ forms a markov process with state space $\{1, 2\}$ since all the times involved in the process follow exponential distribution.

2. Various entries of state transition probability matrix can be derived as follows:

For $\delta t \rightarrow 0$, $P_{11}(\delta t)$, probability that the system will be in state 1 after after a small duration δt is given by:

$$P_{11}(\delta t) = \exp(-\lambda \delta t) \cong 1 - \lambda \delta t$$

On similar argument, one can derive the remaining probabilities, we have

$$P_{12}(\delta t) = \lambda \delta t; \quad P_{21} = \mu \delta t; \quad P_{22} = 1 - \mu \delta t$$

The transition probability matrix **P** is given by:

$$P = \begin{bmatrix} 1-\lambda & \lambda \\ \mu & 1-\mu \end{bmatrix}$$

Figure 4.17 represents the *transition diagram* of the system.

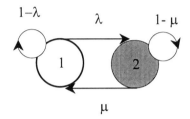

Figure 4.17 Transition diagram for a two-state system

3. The transition rates λ_{ij} is given by:

$\lambda_{11} = - v_1 = - \lambda \quad \lambda_{12} = \lambda$

$\lambda_{21} = \mu \qquad \lambda_{22} = - v_2 = - \mu$

Substituting the above values in equation (4.45) we get

$$\frac{d}{dt} P_1(t) = -\lambda P_1(t) + \mu P_1(t)$$

$$\frac{d}{dt} P_2(t) = \lambda P_1(t) - \mu P_2(t)$$

The above equation represents the Kolmogrov backward equation for the item

4. Since $P_1(t) + P_2(t) = 1$

$$\frac{d}{dt} P_1(t) = -\lambda P_1(t) + \mu(1 - P_1(t))$$

Rearranging the above equation, we get a first order differential equation,

$$\frac{d}{dt}P_1(t) + (\lambda + \mu)P_1(t) = \mu$$

The general solution of the above differential equation is given by

$$P_1(t) = \frac{\mu}{\lambda + \mu} + C \times \exp(-(\lambda + \mu)t)$$

where C is a constant which can be determined using the initial probability $P_j(0)$. Assume that $P_1(0) = 1$. Substituting this in the general solution we get,

$$1 = \frac{\mu}{\lambda + \mu} + C \Rightarrow C = \frac{\lambda}{\lambda + \mu}$$

Thus for the initial condition $P_1(0) = 1$, the probability that the system will be in state 1 (operating state at time $t = 0$) is given by:

$$P_1(t) = \frac{\mu}{\lambda + \mu} + \frac{\lambda}{\lambda + \mu}\exp(-(\lambda + \mu)t)$$

The above equation in fact is the *point availability* of the item. Similarly, $P_2(t)$ is given by

$$P_2(t) = \frac{\lambda}{\lambda + \mu} - \frac{\lambda}{\lambda + \mu}\exp(-(\lambda + \mu)t)$$

For $t \to \infty$, the above equations give,

$$P_1(\infty) = \frac{\mu}{\lambda + \mu} \quad \text{and} \quad P_2(\infty) = \frac{\lambda}{\lambda + \mu}$$

Using the relation, MTTF $= 1 / \lambda$ and MTTR $= 1 / \mu$, one can easily verify that,

$$P_1(\infty) = \frac{MTTF}{MTTF + MTTR}$$

Which is nothing but the *steady-state inherent availability* of the item.

4.14.   NON-HOMOGENEOUS POISSON PROCESS (NHPP)

A counting process $\{N(t), t \geq 0\}$ is said to be a *non-homogeneous Poisson process* with intensity function $\lambda(t), t \geq 0$, if:

1. $N(0) = 0$.
2. $N(t)$ has independent increments.
3. The number of events in any interval t and $t + s$ has a Poisson distribution with mean $[S(t+s) - S(t)]$, that is

$$P[N(t+s) - N(t) = n] = \frac{[S(t+s) - S(t)]^n \exp\{-(S(t+s) - S(t))\}}{n!}$$

(4.50)

Where

$$S(t) = \int_0^t \lambda(x)dx$$

(4.51)

$S(t)$ is the expected number of events in $(0,t)$. Also, $N(t+s) - N(t)$ is Poisson distributed with mean $S(t+s) - S(t)$.

Modelling Repairable items Using Non-Homogeneous Poisson Process Under *'As Bad As Old'* Repair Policy

Consider a repairable item which upon failure is restored in a negligible amount of time and after restoration the condition of the item is identical to that immediately prior to failure (*as bad as old*). That is after repair the hazard function of the system will be same as the value of hazard function, just before repair.

Let $X_1, X_2, \ldots,$ denote the sequence of failure times of a repairable item. That is the sequence $\{X_n, n > 0\}$ are independent and identically distributed random variables with hazard function h(t). That is,

$$h(t) = \frac{f(t)}{R(t)}$$

(4.52)

Where *f(t)* and *R(t)* are probability density function and reliability function of X_n. Assuming that n failures have occurred by time t_n, the probability of $(n+1)^{th}$ failure, $f_{n+1}(t)$, can be written as (Keller, 1984):

$$f_{n+1}(t) = \frac{f(t)}{R_n(t)} = h(t) \text{ the hazard function.}$$

Where, $R_n(t) = \int_{t_n}^{\infty} f(t)dt$

Assume that a failure occurs between *t* and *t+h*. This is possible if and only if some X_n whose value greater than *t* lies between *t* and *t+h*. This probability is basically a hazard function. That is,

$$P[X_n \in (t, t+h) \mid X_n > t] = h(t) \times h + o(h) \tag{4.53}$$

Thus the process $\{N(t), t \geq 0\}$ is a non-homogeneous Poisson process with intensity function h(t). The expected value of N(t) is thus given by the cumulative hazard function

$$H(t) = \int_0^t h(x)dx . \tag{4.54}$$

Thus, if the condition of the item after repair is 'as bad as old', then the expected number of failures for this item during t units of operation is given by its cumulative hazard function *H(t)*.

4.15.   RENEWAL PROCESS

Renewal theory was originally used to analyse the replacement of equipment upon failure, to find the distribution of number of replacement and mean number of replacement. It is the most appropriate tool to predict the demands for consumable items. Let $\{X_n; n = 1, 2, ...\}$ be a sequence of non-negative independent random variables with common distribution F. Let X_n be the time between $(n-1)^{st}$ and n^{th} event. Let:

$$S_0 = 0, \ S_n = \sum_{i=1}^{n} X_i \qquad (4.55)$$

Thus S_n is the time to n^{th} event or *epoch* at which the nth renewal occurs. Let N(t) be the number of renewals by time t.

$$N(t) = Max\{n; \quad S_n \leq t\} \qquad (4.56)$$

Let X_1, X_2, ... are independent and identically distributed random variables with distribution F(t). Then $P\{S_n \leq t\}$ is given by:

$$P\{S_n(t) \leq t\} = F^n(t) \qquad (4.57)$$

where $F^n(t)$ is the n-fold convolution of F(t). That is,

$$F^n(t) = \int_0^t F^{n-1}(x)dF(x) \qquad (4.58)$$

We use the convention that $F^0(t) = 1$ for $t > 0$. $F^n(t)$ represents the probability that the nth renewal occurs by time t. The distribution of N(t) can be derived using the following arguments.

Distribution of N(t)

The counting process, N(t), is called a renewal process. From the definition of N(t) and S_n, we have

$$\{N(t) = n\} \Leftrightarrow \{S_n \leq t, S_{n+1} > t\} \qquad (4.59)$$

$$\begin{aligned}
P[N(t) = n] &= P\{N(t) < n+1\} - P\{N(t) < n\} \\
&= P\{S_{n+1} > t\} - P\{S_n > t\} \qquad (4.60) \\
&= 1 - F^{n+1}(t) - [1 - F^n(t)]
\end{aligned}$$

Thus the probability that the number of renewal by time t is equal to n given by:

$$P\{N(t)=n\} = F^n(t) - F^{n+1}(t) \qquad (4.61)$$

It is difficult to evaluate the above function analytically for many theoretical distributions, however it can be solved using well-known numerical methods.

Renewal Function

The expected number of renewal during specified duration t is given by:

$$E[N(t)] = M(t) = \sum_{i=1}^{\infty} i \times [F^i(t) - F^{i+1}(t)] \qquad (4.62)$$

The above equation can be simplified, and the expected number of renewals (expected number of demands) is given by:

$$M(t) = \sum_{i=1}^{\infty} F^i(t) \qquad (4.63)$$

The above equation is called *renewal function, M(t),* and it gives the number of renewal during (0, t]. Taking the derivative of renewal function we get:

$$m(t) = \frac{d}{dt} M(t) = \sum_{n=1}^{\infty} f^n(t) \qquad (4.64)$$

Where $f^n(t)$ is the derivative of $F^n(t)$. $m(t)\delta t$ is the probability that a renewal occurs during $(t, t+\delta t)$. $m(t)$ is called the *renewal density* or *renewal rate.*

Calculating $F^n(t)$, $P[N(t) = n]$, $M(t)$ and $m(t)$

Exponential Distribution

$$F(t) = 1 - \exp(-\lambda t)$$

When the time to failure distribution is exponential, the renewal process constitutes a Poisson process. Thus, Poisson process is also a special case of renewal process where time to failure is exponential.

$$F^n(t) = 1 - \sum_{i=0}^{n-1} \frac{\exp(-\lambda t) \times (\lambda t)^i}{i!} \tag{4.65}$$

$$P[N(t) = n] = \frac{\exp(-\lambda t) \times (\lambda t)^n}{n!} \tag{4.65}$$

$$M(t) = \lambda t \tag{4.67}$$

$$m(t) = \lambda \tag{4.68}$$

Normal Distribution

By assuming $\sigma \ll \mu$, we have

$$F^n(t) = \Phi(\frac{t - n \times \mu}{\sigma \times \sqrt{n}}), \text{ where } \Phi(t) \text{ is the standard normal distribution.}$$

The distribution of $N(t)$ is given by:

$$P[N(t) = n] = \sum_{n=1}^{\infty} [\Phi(\frac{t - n \times \mu}{\sigma \times \sqrt{n}}) - \Phi(\frac{t - (n+1) \times \mu}{\sigma \times \sqrt{n+1}})]$$

$$M(t) = \sum_{n=1}^{\infty} \Phi(\frac{t - n \times \mu}{\sigma \times \sqrt{n}}) \tag{4.69}$$

For distributions like Weibull, one has to use numerical approximation to find the renewal function. The approximation techniques are discussed in Chapter 8.

Elementary Renewal Theorem

For a distribution function F(t) with F(0) = 0 and finite mean, and if f(x) exists then the following equation is valid

$$\lim_{t \to \infty} \frac{M(t)}{t} = \frac{1}{MTTF} \tag{4.70}$$

The above result is called the *Elementary Renewal Theorem*. This implies that in the steady state, the expected number of failures is given by the ratio of t over the MTTF value.

Chapter 5

Maintainability and Maintenance

*Maintenance is the management of failures and
the assurance of availability*

J Hessburg

Maintainability and maintenance has always been important to the industry as it affects the performance as well as the finance. For commercial airlines, maintenance costs around 10% of the airlines total cost, as much as fuel and travel agents' commission (M Lam, 1995). Operators/users would like their system to be available and safe to operate when required. One should be lucky to find a smiling customer when the system fails and it takes a long time to recover the functionality.

There are several ways that designers can provide maximum utility of their product. One way is to build items/systems that are extremely reliable (and consequently will, almost certainly, have a higher acquisition cost). Another is to design systems that are quick and easy to repair when they fail. Obviously, the main objective of the designer is to provide a reliable and safe item at an affordable price.

Maintenance is the action necessary to sustain and restore the performance, reliability and safety of the item. The main objective of maintenance is to assure the availability of the system for use when required. For aircraft, maintenance forms an essential part of airworthiness. The common objective of aircraft maintenance, civil or military, is to provide a fully serviceable aircraft when it is required by the operator at minimum cost (Knotts, 1996). However, maintenance costs money. The annual maintenance cost of production assets in the United Kingdom is estimated in excess of $13 billion, with $2 billion wasted through inefficient maintenance management practices (Knotts, 1999). Maintenance also accounts for approximately 10% of the organisations' employees and at least 10-15% of its operating costs.

5.1.   CONCEPT OF MAINTAINABILITY

In the previous chapters, we showed that it is important for the operator/user to know the reliability characteristics of the item. We also recognised that it is almost impossible for any item to maintain its function forever, as failure and the degradation of performance is inevitable. Thus, for the user it is equally, or even more important to know:

- When and how often maintenance tasks should be performed

- How they should be performed

- How many people will be needed

- What skills they will need and how much training

- How much the restoration will cost

- How long the system will be down

- What facilities and equipment (special and general) will be required.

All the above information is important as it affects the availability and the life cycle cost of the system. One has to apply a scientific discipline to find answers to these questions.

Maintainability is the scientific discipline that studies complexity, factors and resources related to the maintenance tasks needed to be performed by the user in order to maintain the functionality of a system, and works out methods for their quantification, assessment, prediction and improvement.

Maintainability Engineering is rapidly growing in importance because it provides a very powerful tool to engineers for the quantitative description of the inherent ability of their system/product to be restored by performing specified maintenance tasks. It also contributes towards the reduction of maintenance costs of a system during its utilisation to achieve optimum life cycle cost.

The maintainability engineering function involves the formulation of an acceptable combination of design features, which directly affect maintenance and system support requirements, repair policies, and maintenance resources. Some physical design features such as accessibility, visibility, testability, complexity and interchangeability affect the speed and ease with which maintenance can be performed.

Maintainability studies have the following objectives (R Knotts 1996):
- To guide and direct design decisions
- To predict quantitative maintainability characteristics of a system

– To identify changes to a system's design needed to meet operational requirements

In the technical literature, several definitions for maintainability can be found. For example, the US Department of Defence's MIL-STD-721C (1966) defines maintainability as:

The measure of the ability of an item to be retained in or restored to specified condition when maintenance is performed by personnel having specified skill levels, using prescribed procedures and resources, at each prescribed level of maintenance and repair.

Maintainability can be expressed in terms of maintenance frequency factors, maintenance elapsed times and maintenance cost. Maintainability therefore is an inherent design characteristic dealing with the ease, accuracy, safety, and economy in the performance of maintenance functions. Maintainability requirements are defined in conceptual design as part of system operational requirements and the maintenance concept. Anon (1992) describes maintainability as:

The characteristic of material design and installation that determines the requirements for maintenance expenditures including time, manpower, personnel skill, test equipment, technical data and facilities to accomplish operational objectives in the user's operational environment.

One of the common misperceptions is that maintainability is simply the ability to reach a component to perform the required maintenance task (accessibility). Of course, accessibility is one of the main concerns for many maintenance engineers. Figure 5.1 illustrates an accessibility problem in one of the older twin-engine fighter aircraft, Gloster Javelin. Before an engine could be changed, the jet pipe had to be disconnected and removed. To remove the jet pipe it was necessary for a technician to gain access through a hatch and then be suspended upside down to reach the clamps and pipes which had to be disconnected. The job could only be achieved by touch; the items were outside of the technician's field of view. The technician had to work his way down between the engine and the aircraft's skin, with tools in his hand. For safety reasons, he was held by his ankles, as shown in figure 5.1 (source: R Knotts).

However, there are many other aspects to be considered other than accessibility. Maintainability should also consider factors such as visibility, that is the ability to see a component that requires maintenance action, testability (ability to detect system faults and fault isolation), simplicity and interchangeability. Additionally decision-makers have to be aware of the environment in which maintainers operate. It is much easier to maintain an item on the bench, than at the airport gate, in a war, amongst busy morning traffic, or in any other result-oriented and schedule-driven environment.

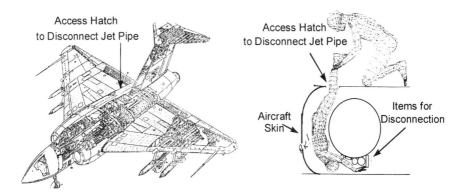

Figure 5.1 Accessibility concern in the Javelin fighter aircraft

Another area to be considered under maintainability is troubleshooting the various modules within the allowed time, i.e. determining whether the system is safe to operate and, if not, what action is needed. For the commercial airlines, there is usually less than an hour at the gate prior to the aircraft's departure to the next destination, whereas for a racing car or weapon system every second could be vital.

To meet these requirements, an easily manageable device is needed which can diagnose with a high degree of accuracy, which modules within the system are at fault. It is now widely accepted that false removals (often referred to as *No Fault Found – NFF)* cost about the same as an actual failure when the component under investigation is removed and replaced. Reducing the number of false removals, therefore, would be a big cost saver.

Devices with these capabilities have been developed in the aerospace, Formula 1 racing car and luxury car industries. For example, the Boeing 777 includes an 'on-board maintenance system' with the objective to assist the airlines to avoid expensive gate delays and flight cancellations. For similar purposes the Flight Control Division of the Wright Laboratory in the USA has developed a fault detection/isolation system for F-16 aircraft, which allows maintainers, novice as well as expert, to find failed components.

In the next section, we discuss the maintainability measures and how these measures can be used for effective maintenance management.

5.2. MEASURES OF MAINTAINABILITY

It is extremely important for the user to have information about the functionality, cost, safety, and other characteristics of the product under consideration at the beginning of its operating life. However, it is equally, or even more important to have information about the characteristics with which to define the maintenance time. Measures of maintainability are related to the ease and economy of maintenance such as; elapsed time that an item spends in the state of failure, man-hours required completing a maintenance task, frequency of maintenance, and the cost of maintenance. As the elapsed time has a significant influence on the availability of the system, operators would like to know the maintenance times; not just the mean time but also the probability that a maintenance task will be completed within a given time. Maintenance elapsed times are even advertised as a marketing strategy.

5.2.1   Maintenance Elapsed-Time

The length of the elapsed time, required for the restoration of functionality, called *time to restore,* is largely determined at an early stage of the design phase. The maintenance elapsed time is influenced by the complexity of the maintenance task, accessibility of the items, safety of the restoration, testability, physical location of the item, as well as the decisions related to the requirements for the maintenance support resources (facilities, spares, tools, trained personnel, etc). It is therefore a function of the maintainability and supportability of the system. It will, of course, also be influenced by other factors during the various stages of the life of the system but any bad decision made (either explicitly or by default) during the design stage will be costly to rectify at a later stage and will significantly affect both the operational costs and system availability.

1. Personnel factors which represent the influence of the skill, motivation, experience, attitude, physical ability, self-discipline, training, responsibility and other similar characteristics related to the personnel involved;
2. Conditional factors which represent the influence of the operating environment and the consequences of failure with the physical condition, geometry, and shape of the item under restoration;
3. Environmental factors which represent the influence of factors such as temperature, humidity, noise, lighting, vibration, time of the day, time of the year, wind, noise, and others such as those similar to the maintenance personnel factors during restoration.

This maintainability measure can be represented using the probability that the maintenance task considered will be completed by a stated time. Since the maintenance elapsed time is a random variable, one can use the cumulative distribution function of the elapsed time to find the percentage of maintenance tasks that will be completed within a specified time.

Mean Time to Repair

One approach for measuring maintainability is through Mean Time to Repair (MTTR). MTTR is the expected value of the item's repair time. With the knowledge of the reliability and maintainability of the sub-systems one can evaluate the maintainability of the system, that is, mean time to repair of the system, $MTTR_s$ (Birolini, 1994).

Assume that the reliability block diagram of the system has a series structure with n items *with no redundancy*. Let $MTTF_i$ and $MTTR_i$ be the mean time to failure and mean time to repair of sub-system i in the system.

Consider an arbitrarily large operating time T. Assuming that the failure rate of the unit is constant, the expected number of failures of unit i in during T is given by:

$$\frac{T}{MTTF_i} \tag{5.1}$$

The mean of total repair time to repair unit i during T is given by:

$$MTTR_i \frac{T}{MTTF_i} \tag{5.2}$$

For the whole system, the mean number of failures is given by:

$$\sum_{i=1}^{n} \frac{T}{MTTF_i} \tag{5.3}$$

For the whole system, the mean of total repair time is given by:

$$\sum_{i=1}^{n} MTTR_i \times \frac{T}{MTTF_i} \tag{5.4}$$

Combining equation (5.3) and (5.4), we get the mean time to repair at the system level, $MTTR_s$, as:

$$MTTR_s = \frac{\sum\limits_{i=1}^{n} \dfrac{MTTR_i}{MTTF_i}}{\sum\limits_{i=1}^{n} \dfrac{1}{MTTF_i}} \tag{5.5}$$

Assuming constant failure rate, that is,

$\lambda_i = \dfrac{1}{MTTF_i}$ and $\lambda_s = \sum\limits_{i=1}^{n} \lambda_i$, equation (5.5) can be written as:

$$MTTR_s = \sum\limits_{i=1}^{n} \frac{\lambda_i}{\lambda_s} MTTR_i \tag{5.6}$$

Example 5.1

The MTTF and MTTR of four sub-systems in a system are given in Table 5.1. Estimate the system level mean time to repair, $MTTR_s$.

Table 5.1 MTTF and MTTR values for a subsystem

Sub-system	MTTF	MTTR
1	200	24
2	500	36
3	340	12
4	420	8

SOLUTION:

Applying equation (5.5), we get:

$$MTTR_s = \frac{\dfrac{24}{200} + \dfrac{36}{500} + \dfrac{12}{340} + \dfrac{8}{420}}{\dfrac{1}{200} + \dfrac{1}{500} + \dfrac{1}{340} + \dfrac{1}{420}} \approx 20 \text{ hours.}$$

Mean Time to Repair – Multi-Indenture Case

Many complex systems are broken down into a number of levels of indenture (LoI). For these systems, recovery of an LoI_i unit is usually achieved by the removal and replacement of LoI_{i+1} items. In many cases, the replacement LoI_{i+1} item will not be the item that was removed. It may be a new (i.e. unused) one or it may be one that was removed from another LoI_i unit and subsequently recovered and put into stock for such an occasion.

Now, for such a system, the time to repair will be the time to remove and refit the units at the next lower level of indenture. The elapsed time will need to take into account logistic delays (i.e. waiting for equipment, personnel, spares and any transport to and from the site at which the maintenance work is to be done). This is discussed in more detail in Chapter 10.

Suppose a system is made up of n levels of indenture and a unit at LoI_i is made up of m_i LoI_{i+1} items. Suppose also that to recovery an LoI_i unit, one of the m_i items is removed and replaced with average times, $MTTRM_{i,j}$ and $MTTRP_{i,j}$ respectively. Let us assume that the probability that item j is rejected given that unit i has been removed is $P_{i,j}$ then over an arbitrarily long operating time T, the expected number of system failures is:

$$\frac{T}{MTTF_1}$$

Where, $MTTF_1$ is the mean time between failures of the system (over time T). Now, the probability that the failure was due to sub-system j is $P_{i,j}$ so the mean time between failures due to sub-system j is

$$MTTF_{1,j} = \frac{1}{\lambda_{1,j}} = \frac{1}{P_{1,j}\lambda_1} = \frac{MTTF_1}{P_{1,j}}$$

Assuming the system reliability block diagram is series and is series and there are no redundancies. The expected number of failures of sub-system j is

$$P_{1,j}\frac{T}{MTTF_1} = \frac{T}{MTTF_{1,j}} = \lambda_{1,j}T$$

The expected time to recover the system given that sub-system j is the cause of its failure is

$$MTTR_{1,j} = MTTRM_{1,j} + MTTRP_{1,j}$$

The expected total time spent recovery the system due to sub-system j failures over time T is then

$$P_{1,j} \frac{MTTR_{1,j}}{MTTF_1} T = \frac{MTTR_{1,j}}{MTTF_{1,j}} T = \lambda_{1,j} MTTR_{1,j} T$$

So, the expected total time spent recovering the system by sub-system exchange is

$$\sum_{j=1}^{m_1} P_{1,j} \frac{MTTR_{1,j}}{MTTF_1} T = \sum_{j=1}^{m_1} \frac{MTTR_{1,j}}{MTTF_{1,j}} T = \sum_{j=1}^{m_1} \lambda_{1,j} MTTR_{1,j} T$$

Where, m_1 is the number of sub-systems. Then the mean time to recover the system (by sub-system exchange per system failure) is

$$MTTR_{1,E} = \sum_{j=1}^{m_1} \frac{\lambda_{1,j}}{\lambda_1} MTTR_{1,j}$$

To determine the total maintenance time, we would have to look at the time spent recovering the sub-systems, by sub-sub-system exchange and so on down to the lowest level components that are recovered in this way and then add on any time spent repairing the lowest level components (parts) if they can be repaired but we will leave this exercise until our next book.

5.2.2 Maintenance Man Hour (MMH)

Although elapsed time is an extremely important maintenance measure, one must also consider the maintenance man-hours, *MMH* (also known as maintenance labour hours). The MMH is an estimate of the expected "spanner-in-hand" time and takes into account all of the maintenance tasks and actions required for each system, sub-system or component recovery. It should be noted that the MMH can be considerably greater than the elapsed time as it is often possible and sometimes even necessary to employ more than one person on any given activity or task.

"Work study" and "time and motion" exercises have generated tables of times for every conceivable maintenance action, from releasing the catches that are used on access panels to inspecting the blades on a turbine using a boroscope to drilling out a stud that has sheared after too much torque has been applied to it, to disconnecting and reconnecting all of the pipes and leads when removing and replacing an engine.

In most cases, these times are based on carrying out these tasks and actions in ideal conditions, i.e. in a properly lit workshop, which is heated and provides shelter from the elements. They are generally done when the components are in pristine condition free from contamination, corrosion or damage. It is also generally assumed that the mechanic carrying out each action will have been properly trained and familiar with the correct procedures. In practice, however, it is very rare for all of these ideal conditions to be met so, the actual times will inevitably be longer than those used in the MMH prediction.

Maintenance man-hours are useful in their own right but very often they are given as a "rate" such as (*MMH*/operating hour), (*MMH*/cycle), (*MMH*/month), and (*MMH*/maintenance task). For example, elapsed times can be reduced (sometimes) by increasing the number of people involved in accomplishing the specific task. However, this may turn out to be an expensive trade-off, particularly when high skill levels are required to perform the tasks. Also, unless it actually requires more than one person to do the job, there is likely to be an "interference factor" which means that the efficiency of each person is reduced. Therefore, a proper balance among elapsed time, labour time, and personnel skills at a minimum maintenance cost is required.

Commercial airlines and air forces use the measure *Maintenance Man-Hour per Flight Hour* (MMH / FH) as an indicator of the maintainability of the aircraft for comparison with other similar aircraft either of an older generation or made by another manufacturer. This measure may be used to decide between alternatives although, in many cases, it will be used to exert pressure on the manufacturer to make improvements. The following expression can be used to evaluate the MMH/FH:

$$MMH / FH = \frac{N_1(t) \times MPMT \times MNC_{pm} + N_2(t) \times MCMT \times MNC_{cm}}{Total \ flying \ hours}$$

(5.7)

Where:

$N_1(t)$ is the total number of preventive maintenance tasks during t hours, and $N_2(t)$ is the total number of corrective maintenance tasks. The value t should be equal to the operational life of the aircraft.

MPMT = Mean preventive maintenance time.

MCMT = Mean corrective maintenance time.

MNC_{pm} = Mean number of crew for preventive maintenance.

MNC_{cm} = Mean number of crew for corrective maintenance.

Note that these estimated mean values should be weighted according to the expected frequency of each maintenance task as we did when calculating MTTR$_s$ above.

A problem with estimating the MMH/FH metric is that it relies on the reliability of the various components of the system, which may be age-related and will, inevitably, depend on the maintenance and support policies. For these reasons, the MMH/FH may not remain constant with aircraft age. The implication of using such a metric is that it is preferential for it to be minimised, however, it may actually be both cheaper and yield a higher level of availability if more time is spent on maintenance, particularly preventative maintenance.

5.2.3 Maintenance Frequency Factors

Maintainability engineering is primarily concerned with designing a system so that it spends a minimum time in maintenance, given that it needs maintaining. Another characteristic of system design pertaining to maintainability is in optimising the mix between preventative and corrective maintenance.

The ideal system design would allow the operators to use the system until just before it fails but, with enough notice of the impending failure so that the operator can choose to perform the necessary maintenance at the most opportune moment. In all but a few cases, prognostics have, as yet, not reached this level of sophistication. An alternative approach is built-in redundancy and fault-tolerant systems. These allow the operators to defer maintenance for a limited period or, in certain circumstances until the backup system fails.

Corrective maintenance can be expensive if the failure causes damage to other parts of the system or if it stops the system from earning its keep. However, redundant components will also add to the cost of the system and may reduce its load-carrying capacity. The spare wheel in cars takes up space that could otherwise be used for carrying luggage, it also increases the gross weight, which will reduce the performance of the car both by reducing its rate of acceleration and increasing the fuel consumption.

It is common practice for motorists to replace tyres before the tread has been completely worn away because it is unsafe to drive on bald tyres. It is also illegal and the penalties can be both expensive and inconvenient. It is also very easy to inspect tyres for wear so it is possible to leave them until the "last minute" or get them replaced when the car is not needed thus minimising the inconvenience or lack of availability.

Brake pads are more difficult to inspect by the owner. As a result, many cars are now fitted with pads that have an in-built electrode, which causes a warning light to be illuminated on the dashboard when it comes into contact with the metallic disc (due to the non-conductive part of the pad being worn away). this generally gives the driver a sufficient warning for him or her to find out what the warning light means and take the necessary corrective action before the brakes become dangerous.

Most motorists have their cam or timing belts replaced within about 1000 miles of the manufacturer's recommended mileage possibly during a routine service (scheduled maintenance) or at the driver/owner's convenience. In this case, the owner has almost certainly no way of knowing how much longer the belt will last and, indeed, it is likely to cost them almost as much to have the belt inspected as it would to have it replaced because of the amount of work involved. In this case, the extent of the damage to the engine if the belt breaks is likely to cost a great deal more than that of replacing the belt early. It would no doubt be possible to devise a monitor that could indicate when the belt was starting to wear but, whether it would be practical in terms of its size, reliability, cost and extra weight is very much open to debate.

Here we have seen four different solutions to the same problem of avoiding failures and hence the need for corrective maintenance. One of the tasks of the maintainability engineer is to determine which, if any of these, or other similar approaches is appropriate taking into consideration the costs and practicalities in each circumstance.

There is clearly a need to strike a balance. Preventative maintenance may cause components to be replaced unnecessarily (or at least prematurely). Allowing a system to run until it fails may maximise the times between maintenance but failures can be expensive to rectify both because of the extent of the damage caused and because of the loss of availability of the system whilst it is being maintained. Prognostics can help but these too have their own problems of reliability and the need for maintenance as well as possibly adding to the weight, complexity and cost of the system.

5.2.4   Maintenance cost factors

For many systems/products, maintenance costs constitute a major segment of the total life-cycle cost. Further, experience has indicated that maintenance costs are significantly affected by design decisions made throughout the early stages of system development. Maintainability is directly concerned with the characteristics of system design that will ultimately result in the accomplishment of maintenance at minimum cost. Thus, one way of measuring maintenance cost is cost per maintenance task, which is the sum of all costs related to elements of logistics support which are required to perform the considered maintenance task.

In addition to the above factors, the frequency with which each maintenance action must be performed is a major factor in both corrective and preventive maintenance. Obviously this is greatly influenced by the reliability of the components but it can also be related to the type and frequency of the maintenance performed. If a component is repaired then it is likely that the time to failure for that component will be less than if it had been replaced by a new one. We will return to the question of repair effectiveness in Chapter 6.

Personnel and human factor considerations are also of prime importance. These considerations include the experience of the technician, training, skill level and number of technicians.

Support considerations cover the logistics system and maintenance organisation required to support the system. They include the availability of spare parts, technical data (manuals), test equipment and required special and general tools.

If a maintenance task requires highly skilled personnel, a clean environment equipped with expensive, special tools then it is unlikely, that it will prove economical to perform this task at first line or, possibly, even at second line. However, if the maintainability engineer had designed the system so that this task could be done by personnel with lower skill levels using standard tooling then it might have allowed the task to be done in the field with a possible reduction in the turnaround (or out-of-service) time. If, the task is only likely to be done once in the system's life during a major overhaul when it would be at a central maintenance unit or returned to the manufacturer then such considerations may be less relevant. For example, there is little to be gained by making it easy to replace a broken cam belt by the side of the road. The damage done to the engine, as a result of a failed cam belt, will mean that the engine will either have to be replaced or overhauled/reconditioned before it is likely to function again.

5.3.   MAINTAINABILITY DEMONSTRATION

The objective of the maintainability demonstration is to show that the various maintenance tasks can be accomplished in the times allotted to them. Generally, the most important issue is whether the system can be recovered by sub-system (or line replaceable unit – LRU) exchange within the specified times. It is a common requirement that each LRU can be removed and replaced without interfering with any other LRU. Some of the early jet fighters were virtually built around the engine so that, in order to replace the engine, it was not so much a question of removing the engine from the aircraft as removing he aircraft from the engine.

A recent innovation on commercial aircraft is to use autonomics, which signal ahead to the destination any detected faults in the mission critical components (i.e. those not on the minimum equipment list). This allows the mechanics to prepare to replace these items as soon as the aircraft has reached the gate. If such replacements can be performed within the 50 min, or so, turnaround time then it will not be necessary to find a replacement aircraft or delay the departure. Anyone who has seen the film *Battle of Britain* or *Reach for the Sky* will recognise the importance of turning fighter aircraft around in minimum time when the airfield may be under attack from enemy bombers and fighters. An aircraft not in the air is bit like a duck out of water, it is particularly vulnerable and do very little to defend itself.

The demonstration is also expected to generate results that can contribute to the whole development process, identifying any remaining deficiencies such as the design of the system and the test equipment, compilation of maintenance manuals, etc. Any maintainability demonstration would involve the following steps:

1. Identify the operation and environmental condition in which the system is likely to be used.
2. Simulate the system failures and perform corrective maintenance action. One should also record the maintenance man-hours required to complete the repair task successfully.

Further, it is an important to take care of the following issues during the demonstration:

1. The test must be on a sample of fixed final build standard.
2. The test conditions must be representative, the equipment/tools, maintenance manuals, lighting and similar factors must be carefully considered.
3. A mix of repairers representative in skills, training, and experience of those who would do the actual repair in service must conduct the repair.

Once we have the recorded repair time data from the above procedure, then it is easy to verify whether the maintainability target has been achieved using the following procedure.

Let t_1, t_2, ..., t_n denote the observed repair times to complete the repair tasks for a sample of n units. For n > 30, the $(1 - \alpha)$ 100 percent confidence limit is given by:

$$MTTR + z_\alpha \frac{s}{\sqrt{n}} \tag{5.8}$$

Where z_α is the z value (standard normal statistic) that locates an area of α to its right and can be found from the normal table. For example, for a 95% confidence limit, the z_α is given by 1.645. MTTR and 's' are given by:

$$MTTR = \frac{1}{n}\sum_{i=1}^{n} t_i, \text{ and } s^2 = \frac{1}{n-1}\sum_{i=1}^{n}(t_i - MTTR)^2$$

If the target maintainability is $MTTR^*$, then to demonstrate that the system has achieved this, we have to show that:

$$MTTR^* \leq MTTR + z_\alpha \frac{s}{\sqrt{n}} \tag{5.9}$$

Whenever the number of repair time data is less than 30, we use t-distribution; in that case, the condition for acceptance is given by:

$$MTTR^* \leq MTTR + t_{\alpha,n-1} \frac{s}{\sqrt{n}} \tag{5.10}$$

The value of $t_{\alpha,n-1}$ can be obtained from the t-distribution table shown given in appendix.

Example 5.2

A maintainability demonstration test is carried out on 20 parts and the accomplished repair times are shown in Table 5.2. If the target MTTR is 20 hours, check whether the system has achieved the target maintainability using 95% confidence level.

Table 5.2. Recorded repair times form a sample of 20 parts in hours

8	6	12	20	24
12	9	17	4	40
32	26	30	19	10
10	14	32	26	18

SOLUTION:

Since the observed number of data, n is less than 30, we use t-statistic. The MTTR and standard deviation, s, are given by:

$$MTTR = \frac{1}{20}\sum_{i=1}^{20} t_i = 18.45 \text{ hours}, \quad s = \sqrt{\frac{1}{19}\sum_{i=1}^{n}(t_i - MTTR)^2} = 10.06 \text{ hours}$$

From the t-distribution table (see appendix) we get, $t_{\alpha, n-1} = 1.729$ ($\alpha = 0.05$, n-1 = 19).

95% upper limit for MTTR is given by:

$$MTTR + t_\alpha \frac{s}{\sqrt{n}} = 18.45 + 1.729 \times \frac{10.06}{4.472} = 22.33$$

Which is greater than 20 hours, which is the target MTTR. Thus the achieved MTTR is significantly greater than the required MTTR and is therefore not acceptable.

5.4.   MAINTENANCE

According to BS 4778, maintenance can be defined as:

The combination of all technical and administrative actions, including supervision actions, intended to retain an item in, or restore it to, a state in which it can perform a required function.

In other words, all actions, which keep the system running and ensure that it is maintained to an acceptable standard in which it is able to operate at the required levels efficiently and effectively. The objectives of maintenance are to:

1. Reduce the consequences of failure.
2. Extend the life of the system, by keeping the system in a proper condition for a longer time. In other words, to increase the "up" time of the system.
3. Ensure that the system is fit and safe to use.
4. Ensure that the condition of the system meets all authorised requirements.
5. Maintain the value of the system.

6. Maintain reliability and achieve a high level of safety.
7. Maintain the system's availability and therefore minimise production and quality losses.
8. Reduce overall maintenance costs and therefore minimise the life cycle cost.

The purpose of maintenance is to keep systems in a state of functioning in accordance with their design and to restore them to a similar state as and when required.

5.5.   MAINTENANCE CONCEPT

The maintenance concept begins with a series of statements defining the input criteria to which the system should be designed. These statements relate to the maintenance tasks that should be performed at each level of maintenance (organisational, intermediate and depot), the test equipment and tools that should be used in maintaining the system, the skill levels of the maintenance personnel that perform the identified tasks, maintenance time constraints, and anticipated maintenance environmental requirements (Knezevic, 1997). A preliminary maintenance concept is developed during the conceptual design stage, is continually updated, and is a prerequisite to system design and development. Maintenance concept at the design phase tends to ensure that all functions of design and support are integrated with each other. The maintenance concept evolved from the definition of system operational requirements delineates [Blanchard et. al., 1995]

• The anticipated level of maintenance
• Overall repair policies
• Elements of maintenance resources
• The organisational responsibilities for maintenance

The maintenance concept serves the following purposes:

1. It provides the basis for the establishment of maintainability and supportability requirements in the system design.
2. It provides the basis for the establishment of requirements for total support which include maintenance tasks, task frequencies and time, personnel quantities and skill levels, spare parts, facilities, and other resources.
3. It provides a basis for detailing the maintenance plan and impacts upon the elements of logistic support.

5.6.   LEVELS OF MAINTENANCE

Complex systems can be considered as made up of several levels of indenture. A combat aircraft that may be considered as the Level 0 (LoI-0), may be thought of as consisting five subsystems: airframe, armament, avionics, propulsion and general. The propulsion system then becomes a LoI-1 item that may consist of the engines, the auxiliary power unit (APU) and various accessories including control units and pumps, each of which may be considered as LoI-2 items. An engine is typically an assembly of a number of modules or LoI-3 items which, in turn, may be made up of sub-assemblies and parts, LoI-4 and 5 respectively.

At the same time, the military typically divides its maintenance and support infrastructure into 3, 4 or 5 echelons, lines or [maintenance] levels. "First Line", or "O-Level" is from where the systems are operated. "Second Line" or "I-Level" is typically the main operational bases from which the squadrons are deployed.   These are usually supported by a depot or maintenance unit at "Third Line" or "D-Level". The contractor, supplier or original equipment manufacturer (OEM) often provides a shadow facility at "Fourth Line" effectively duplicating the Third Line facility's capabilities.

Maintenance levels are concerned with grouping the tasks for each location where maintenance activities are performed. The criteria in which the maintenance tasks selected at each level are; task complexity, personnel skill-level requirements, special maintenance equipment and resources and economic measures. Within the scope of the identified level of maintenance, the manufacturer and the user should define a basic repair policy that may vary from repair/replace a part (LoI-5, say) to replace the entire system. The hierarchies of achieving maintenance tasks are divided into three or four levels.

5.6.1   User level (organisational)

This type of maintenance level is related to all maintenance tasks which are performed on the system whilst it is on deployment or at its operating site.   This would include replenishment tasks, e.g. re-fuelling, re-arming, maintaining oil levels, simple condition and performance monitoring activities, external adjustments and replacement of line replaceable units (LRU). Some minor repairs and routine servicing may also come under this category.

5.6.2 Intermediate level

Intermediate maintenance level is related to all maintenance tasks, which are performed at workshops (mobile, semi-mobile and/or fixed) where the systems would normally be based. Common maintenance tasks accomplished at this level are detailed condition and performance monitoring activities, repair and replacement of major items in a system, major overhaul, system modification, etc. Performing maintenance tasks at this level require higher personnel skills than those at organisational level and additional maintenance resources. Traditionally, a removed LRU would be recovered, generally by module (or shop-replaceable unit – SRU) exchange, at this level.

5.6.3 Depot level

Depot maintenance level is related to all maintenance tasks, which are accomplished beyond the capabilities of intermediate level at remote sites. In the UK system, "Third Line" refers specifically to an operator-owned facility whereas in US nomenclature "D-Level" also includes manufacturer/contractor facilities. Maintenance tasks at this level are carried out by highly skilled specialists at a specialised repair facility or the equipment producer's facility. Maintenance tasks at depot level include complete overhauling and rebuilding of the system, highly complex maintenance actions, etc. They would also include tasks which may only be performed rarely, particularly if they require expensive equipment or are likely to take a long time.

5.6.4 Hole-in-the-Wall

With the move to ever greater efficiency and/or minimal costs, the perceived need to reduce manning levels and the desire of OEMs to increase their revenue by entering the "after-market", the "hole-in-the-wall" concept is gaining in popularity. This is where the only intrusive maintenance task the operator performs is to remove the LRU (at first line). This is then passed through this mythical hole in the wall to the OEM or maintenance contractor in exchange for a replacement (serviceable) LRU. The contractor then takes the LRU away to a convenient location where it is recovered. Such contracts are often funded by fleet hour arrangements such as "power-by-the-hour", see chapter 12.

The advantage to the operators is that they can get on with what they are in business for; putting "bums on seats" or "bombs on target". It is also argued, perhaps more strongly by the OEM than the operator, that having designed and built the LRU, they (the OEM) are the best people to take it apart and repair it. A secondary advantage to the OEM, and again, hopefully to the operator, is that because all of the maintenance is done in one place, the people doing it should become more efficient (as they see the same job more often) and the in-service data (time to failure, cause of failure, items repaired or replaced, etc.) should be consistent and more accurate.

Better data should lead to improved forecasting, reduced logistic delays, more appropriate maintenance policies and, ultimately, to improved designs.

5.7.   MAINTENANCE TASK CLASSIFICATION

All users would like their systems to stay in a state of functioning as long as possible or, at least, as long as they are needed. In order to achieve this, it is necessary to maintain the system's functionality during operation, by performing appropriate maintenance tasks. Thus, maintenance task can be defined as a set of activities that need to be performed, in a specified manner, in order to maintain the functionality of the item/system.

Figure 5.2 shows the process of maintenance task, which is initiated by the need for maintenance due to a reduction, or termination of the item/system functionality. The execution of a maintenance task requires resources such as the right number and skills of personnel, material, equipment, etc. It also requires an appropriate environment in which the maintenance activities can be carried out.

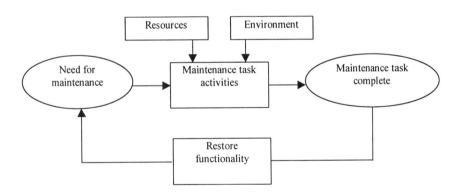

Figure 5.2 Process of maintenance task

Maintenance tasks can be classified into the following three categories:
1. corrective maintenance task

2. preventive (predictive) maintenance task
3. conditional maintenance task

Each maintenance task is briefly discussed in the following sections.

5.7.1   Corrective Maintenance Task

Corrective maintenance task, CRT, is a set of activities, which is performed with the intention of restoring the functionality of the item or system, after the loss of the functionality or performance (i.e. after failure). Figure 5.3 illustrates typical corrective maintenance task activities. The duration of corrective maintenance task, DMT^c, represents the elapsed time needed for the successful completion of the task. Corrective maintenance task is also referred to as an unscheduled or unplanned maintenance task.

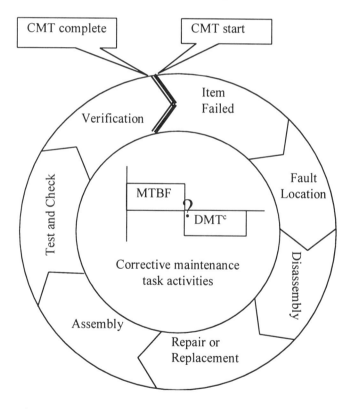

Figure 5.3 Activities of typical corrective maintenance task

5.7.2   Preventive Maintenance Task

Preventive maintenance task, *PMT*, is a maintenance activity that is performed in order to reduce the probability of failure of an item/system or to maximise the operational benefit. Figure 5.4 illustrates the activities of a typical preventive maintenance task. The duration of the preventive maintenance task, DMT^p, represents the elapsed time needed for the successful completion of the task.

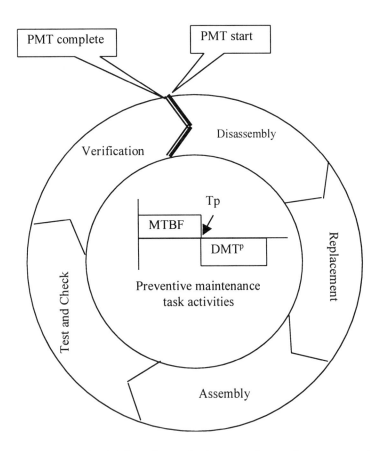

Figure 5.4 Activities of a typical preventive maintenance task

Preventive maintenance task is performed before the transition to the state of failure occurs with the main objective of reducing:

- The probability of the occurrence of a failure
- The consequences of failure

Common preventive maintenance tasks are replacements, renewal and overhaul. These tasks are performed, at fixed intervals based on operating time (e.g. hours), distance (e.g. miles) or number of actions (e.g. landings), regardless of the actual condition of the items/systems.

5.7.3   Conditional (Predictive) Maintenance Task

Conditional maintenance task, *COT*, recognises that a change in condition and/or performance is likely to precede a failure so the maintenance task should be based on the actual condition of the item/system. *COT* does not normally involve an intrusion into the system and actual preventive action is taken only when it is believed that an incipient failure has been detected. Thus, through monitoring of some condition parameter(s) it would be possible to identify the most suitable instant of time at which preventive maintenance tasks should take place.

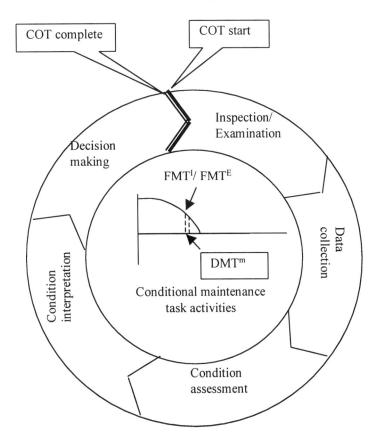

Figure 5.5 Activities of a typical conditional maintenance task.

Figure 5.5 illustrates the activities of a typical conditional maintenance. The duration of conditional maintenance task, DMT^m, represents the elapsed time needed for the successful completion of the task.

In the past, corrective maintenance and preventive maintenance tasks have been popular among maintenance managers. However, in recent years, the disadvantages of these tasks have been recognised by many maintenance management organisations. The need for the provision of safety, and reduction of the maintenance cost have led to an increasing interest in using conditional maintenance task. Waiting until a component fails may maximise the life obtained from that component but, its failure may cause significant damage to other parts of the system and will often occur at inopportune times causing a disruption to the operation and inconvenience to the users. Routine or scheduled preventive maintenance, on the other hand, may be very convenient but is likely to result in an increase in the amount of maintenance needed because parts will be replaced when they have achieved a fraction of their expected life.

5.8.   MAINTENANCE POLICIES

The maintenance policy defines which type of maintenance will (normally) be performed on the various components of the system. It is determined by maintenance engineers, system producers and /or users to achieve high safety, reliability and availability at minimum cost. With respect to the relation of the instant of occurrence of failure and the instant of performing the maintenance task the following maintenance policies exist:

1) *Failure-Based maintenance policy, FBM*, where corrective maintenance tasks are initiated by the occurrence of failure, i.e., loss of function or performance,

2) *Time-Based maintenance policy, LBM*, where preventive maintenance tasks are performed at predetermined times during operation, at fixed length of operational life,

3) *Inspection-Based maintenance policy, IBM*, where conditional maintenance tasks in the form of inspections are performed at fixed intervals of operation, until the performance of a preventive maintenance task is required or until a failure occurs requiring corrective maintenance. Note that the failure could be due to a component of the system that was not being subjected to IBM or it could have happened as a result of some unpredictable external event such as foreign object damage or because the inspection interval was too long or the inspection was ineffective.

4) *Examination-Based maintenance policy, EBM,* where conditional maintenance tasks in the form of examinations are performed in accordance with the monitored condition of the item/system, until the execution of a preventive maintenance task is needed or a failure occurs.

The principal difference between the above maintenance policies occurs at the time when the maintenance task is performed. The advantages and disadvantages of each maintenance policy are briefly described below.

5.8.1   Failure-Based Maintenance Policy

Failure-Based maintenance policy, *FBM*, represents an approach where corrective maintenance tasks are carried out after a failure has occurred, in order to restore the functionality of the item/system considered. Consequently, this approach to maintenance is known as breakdown, post-failure, fire fighting, reactive, or unscheduled maintenance. According to this policy, maintenance tasks often take place in ad hoc manner in response to breakdown of an item following a report from the system user.

A schematic presentation of the maintenance procedure for the failure-based maintenance policy is presented in Figure 5.6. Corrective maintenance task priorities can range from "normal", "urgent" to "emergency". These categories reflect the nature of the response rather than the actual actions done. Failure based maintenance could be the most applicable and effective maintenance policy in situations where:

1. Items for which the loss of functionality does not compromise the safety of the user and/or the environment or the failure has little or no economic consequences (i.e. categories "major" and "minor" see "FMECA" in Chapter 11)
2. systems have built-in redundancy or have been designed to be fault-tolerant

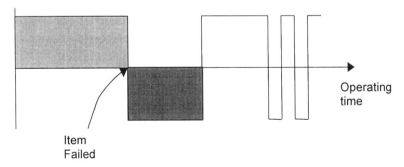

Figure 5.6 Failure-Based Maintenance Policy

Advantages of failure based maintenance

Implementation of FBM to the above situations could lead to full utilisation of the operating life of the item. This means that the non-critical items will have the ability to perform their function(s) for the stated period of time when they operate under stated conditions. This means that coefficient of utilisation, CU, which is the ratio of the Mean Duration of Utilised Life of the item $(MDUL^F)$ to the expected operating life (MTTF), of items considered will have value of 1. The user will get maximum value out the component when the FB maintenance policy is applied.

Disadvantages of failure based maintenance

Despite the advantages of implementing this policy, it has some disadvantages when it is not correctly selected.

1. The failure of an item will generally occur at an inconvenient time.
2. Maintenance activities cannot be planned.
3. It demands a lot of maintenance resources.
4. The failure of an item can cause a large amount of consequential damage to other items in the system.

Analysis of maintenance costs have shown that a repair made after failure will normally be three to four times more expensive than the same maintenance activity when it is well planned [Mobley (1990)].

5.8.2   Time-Based Maintenance Policy

Some failures can lead to economical consequences such as loss of production and therefore a reduction in profit. Some failures may have an impact on the safety of the user, passengers, third parties and environment. Therefore, it is desirable to prevent these failures, if possible, by carrying out maintenance actions before failure occurs.

As the main aim is to reduce the probability of occurrence of failure and avoid the system breakdown, a time-based maintenance policy is performed at fix intervals, which is a function of the time-to-failure distribution of the item considered and in some cases it may be adjusted by the system's user. This policy is very often called age-based, life-based, planned or scheduled maintenance. The reason for that is the fact that the maintenance task is performed at a predetermined frequency, which may be based on, for example, operating times such as, hours, years, miles, number of actions or any other units of use, that make it is possible to plan all tasks and fully support them in advance. A schematic presentation of time-based maintenance procedure is presented in Figure 5.7. The frequency of maintenance task, FMT^L, is determined even before the item has started functioning. Thus, at the predetermined length of operational life specified, preventive maintenance tasks take place. The time-based maintenance policy could be effectively applied to items/systems that meet some of the following requirements:

1. the probability of occurrence of failure is reduced
2. the likely consequences of failure is "catastrophic" (e.g. loss of life or serious injury)
3. the total costs of applying this policy are substantially lower than the alternatives
4. the condition of the system, or its consisting items, cannot be monitored or is impractical or uneconomical.

Advantages of time-based maintenance policy

One of the main advantages of this maintenance policy is the fact that preventive maintenance tasks are performed at a predetermined instant of time when all maintenance support resources could be planned and provided in advance, and potential costly outages avoided. For failures, which could have catastrophic consequences to the user/operator and environment (Chernobyl, Bhopal, Piper Alpha and similar) it may be the only feasible option. Time-based maintenance has many advantages over failure-based maintenance, which are summarised in the following list:

1. Maintenance can be planned ahead and performed when it is convenient from the operational and logistics point of view.
2. The cost of lost production and of consequential damage can be reduced.
3. Downtime, the time that the system is out of service, can be minimised.
4. Safety can be improved.

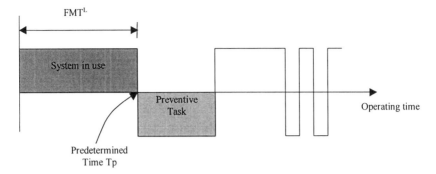

Figure 5.7 Time Based Maintenance Policy

Disadvantages of time -based maintenance policy

In spite of the advantages given above, the time-based maintenance policy has several disadvantages that must be minimised. This policy could be uneconomical because the majority of items are prematurely replaced, irrespective of their condition. In many industries this policy is now only used under special conditions because it is very costly, and also because its efficiency in reducing failures is not always supported by experience. A summary of the disadvantages of time-based maintenance policy is listed below.

1. Time-based maintenance is performed irrespective of the condition of the system. Consequently, a large number of unnecessary tasks will be carried out on a system that could have been operated safely for a much longer time.
2. The tasks may require higher numbers of skilled mechanics.
3. If the time to perform the maintenance is greater than the time the system would normally be idle (eg overnight) then because of the frequency, it could cause higher levels of unavilability.
4. It cannot guarantee the elimination of all failures and will do nothing to reduce non-age-related failures.
5. Increasing the frequency of maintenance tasks may lead to an increase in the probability of human errors in the form of maintenance-induced failures.
6. Reducing the probability of failure by prematurely replacing components means that the coefficient of utilisation of the item/system, CU^L, will have a value much less than one.

5.8.3 Condition Based Maintenance (Predictive Maintenance)

The need for the provision of safety, increased system availability, and reduced maintenance costs have led to an increasing interest in development of alternative maintenance policies. A policy which overcomes many of the disadvantages of the previous maintenance policies (failure-based and time-based), and has proved its ability to extend the operating life of a system without increasing the risk of failure is condition-based maintenance, CBM. CBM is also known as predictive maintenance.

Condition-based maintenance can be defined as: "*Maintenance carried out in response to a significant deterioration in a unit as indicated by a change in the monitored parameters of the unit's condition or performance*" [Kelly & Harris (1978)]. This means that the principle reason for carrying out maintenance activities is the change or deterioration in condition and/or performance, and the time to perform maintenance actions is determined by monitoring the actual state of the system, its performance and/or other condition parameters. This should mean the system is operated in its most efficient state and that maintenance is only performed when it is cost-effective. A schematic presentation of condition-based maintenance procedure is presented in Figure 5.8. This policy is worth applying in situations where:

1- The state of the system is described by one or more condition parameters.
2- The cost of the condition monitoring technique is lower than the expected reduction in overall maintenance costs.
3- There is a high probability of detecting potentially catastrophic failures (before they happen).

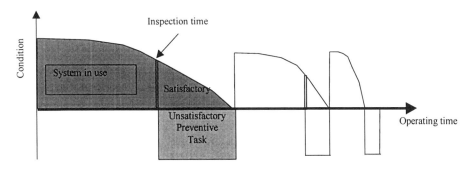

Figure 5.8 Condition based maintenance policy

The condition-based maintenance is a condition or performance-driven preventive maintenance. This means that the timing of the maintenance task is not simply a function of the mean-time-to-failure. The principle of condition-based maintenance therefore is based on the way of monitoring the condition parameters of a system giving three different types of condition monitoring:

1- Inspection

Inspection is generally performed at regular intervals using any of a number of non-destructive test (NDT) procedures which are designed to determine whether the condition of the (inspected) item is satisfactory or unsatisfactory and hence whether further action is required.

2- Examination

This is a condition-monitoring task, which presents a numerical description of the condition of the item at that moment through relevant condition predictors. The results directly affect the scheduling of the next examination. This is possible because of the unique properties and characteristics of the relevant condition predictor.

3 - Performance Trend Monitoring

For propulsion or energy producing systems, in particular, the "performance" may be expressed as a ratio of the output to input, e.g. miles per gallon, kilometres per litre, thrust per kilogram or (mega)watts per tonne. As the system deteriorates, usually through wear but also through damage, these ratios may show signs of decreasing. For systems operating in relatively constant conditions (e.g. constant ambient temperature, pressure and output), consistent changes in the specific fuel consumption (SFC) will almost certainly be indicative of a deterioration in the system which will need some form of maintenance to restore it to an acceptable level. For systems that are operated in an inconsistent manor for which the environmental conditions may be in a constant state of change, the SFC may be subject to considerable noise and hence any deterioration will only be apparent by using sophisticated trending algorithms, such as Kalman Filtering.

8.3.1 Setting up condition-based maintenance policy

In order to implement CBM policy, it is necessary to use the following management steps that are shown in Figure 5.9

Identification and selection of maintenance significant items

The first requirement of implementing CBM is to decide which items of the system should be monitored, since it is likely to be both uneconomical and impractical to monitor them all. Therefore, the first step of the condition-based maintenance decision process is a comprehensive review of all items in a system, in order to identify the maintenance significant items, MSIs. These are items whose failures could be safety-critical, environmentally damaging or revenue sensitive. Thus, each item within the system should be analysed from the point of view of failure, especially the consequences of failure. The most frequently used engineering tools for performing this task is a Failure Mode Effect and Criticality Analysis, FMECA and Reliability Centred Maintenance, RCM (see also Chapters 6 and 11). Care should be taken to ensure that all of the maintenance significant items are identified and listed.

Identification and selection of condition parameters

Once the maintenance significant items are identified it is necessary to determine all monitorable parameters which describe their condition or performance. The condition parameter can be defined as a measurable variable able to display directly or reflect indirectly information about the condition of an item at any instance of operating time. Ideally, maintenance engineers would like to find many condition/parameters which can be monitored and which accurately reflects the condition /performance of the system. In practice there are two distinguishable types of condition parameters which are able to achieve this (Knezevic *et al*, 1995):

Relevant Condition Indicator, RCI

The Relevant Condition Indictor, RCI, is a parameter that describes the condition of an item during its operating time and it indicates the condition of the item at the instant of inspection. The numerical value of RCI represents the local value of the condition of an item/system at the time of inspection. This type of condition parameter is usually related to the performance. However, RCI is not able to predict the future development of the condition of the considered item/system. Typical examples of the RCI are performance, the level of vibration, level of oil, pressure, temperature, etc. It is necessary to stress that the RCI could have an identical value at different instances of operating time.

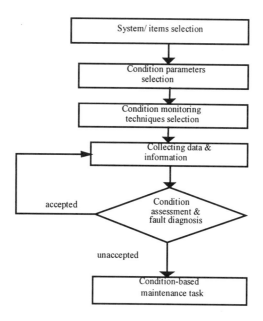

Figure 5.9 Flow of condition based maintenance

Relevant Condition Predictor (RCP)

The Relevant Condition Predictor, RCP, is a parameter, which describes the condition of an item at every instant of operating time. Usually this parameter is directly related to the shape, geometry, weight, and other characteristics, which describe the condition of the item under consideration. The RCP represents the condition of the item/system which is most likely to be affected by a gradual deterioration failure such as wear, corrosion fatigue crack growth. The general principles of the RCP are discussed by Knezevic (1987). Typical examples of RCP are: thickness of an item, crack length, depth of tyre treads, etc. The RCP cannot have identical values at two or more instance of time. The numerical value of the relevant condition predictor at any instant of operating time quantifies the cumulative value of the condition of an item/system at the time of examination.

Selection of condition monitoring technique

Having identified the maintenance significant item and the associated condition parameter(s), the next step is to select the suitable monitoring technique, which will be used to inspect and examine each condition parameter.

The condition monitoring technique is a device used to inspect or examine an item in order to provide data and information about its condition at any instance of operating time. Numerous condition monitoring techniques, for instance, NDT techniques, performance, vibration, etc are available for use by maintenance engineers in order to determine measurable value of condition parameter. It is important to understand the behaviour of the failure that the item exhibits so that the most effective monitoring techniques can be chosen.

The decision as to which condition-monitoring techniques are selected depends greatly on the type of system, the type of condition parameter and, in the end, on cost and safety. Once the decision is made as to which techniques are to be used, it is possible to define the equipment or instrument that will be needed to carry out condition monitoring.

Collecting data and information

The philosophy of condition monitoring is to assess the condition of an item/system by the use of techniques which can range from human sensing to sophisticated instrumentation, in order to determine the need for performing preventive maintenance tasks. With the increased interest in condition monitoring in recent years there have been a number of developments in the techniques that are used to collect data and provide information, which helps maintenance engineers assessing the condition of an item or a system. These developments have made it possible to obtain more reliable information on the condition of the system. In many instances such information is used to insure that the status of the system will continue to be in a functioning state without significant risk of breakdown, and in some instances to make a decision on the timing of when maintenance tasks should be performed. The method of data collection can be classified into the following categories:

On-line data collection and monitoring

On-line data collection and monitoring uses instrumentation fitted to the system which takes continuous measurements of the condition parameters. These may then be analysed by an on-board computer to determine whether there has been a change in the condition of the item/system and whether that change requires any action. The benefit of using on-line monitoring is to reduce the need for human intervention and minimise the probability of a failure occurring between inspections.

Off-line collection and monitoring

Off-line collection and monitoring is periodic measurement of a condition of an item/system or continuous data collection which is analysed remotely. This type of method involves either the collection of data using a portable data collector, or taking a physical sample, for example, lubrication oil samples for analysis of contamination and debris content. Periodic monitoring therefore provides a way of detecting progressive faults in a way that may be cheaper than the on-line system.

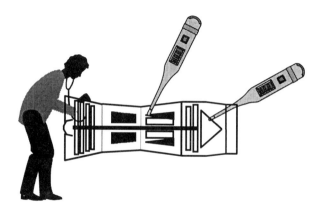

Figure 5.10. Condition monitoring and condition assessment

Condition assessment

The assessment of the condition of an item/system (Figure 5.10) can range from human experience to sophisticated instrumentation. The last few decades have seen a number of developments in the methods which are used to help the maintenance engineers assess and diagnose the condition of an item/system and provide them with information on which to base their decision. Once condition monitoring sensors have been installed and data are being collected, it is necessary to have reliable methods of interpreting the data to identify whether the considered item is undergoing a transition from the normal to abnormal condition and in many cases to identify the causes of the changes.

Effective condition-based maintenance requires a large number of measurements taken continuously or at intervals that assure recognition of change in the condition of the item/system in sufficient time to avoid the need for any corrective action. The volume of data necessary to accurately determine the condition of the item/system can require an excessive amount of time to process and analyse. Consequently, the demand to manipulate and process large amounts of data very quickly has lead to the development of tools such as Artificial Intelligence, AI, to assist engineers to gain maximum value from the data.

In recent years, Artificial Intelligence techniques such as Expert System, Neural Networks and Fuzzy Logic have been applied to the discipline of monitoring and diagnostic systems [Mann et al (1995)]. These techniques extend the power of the computer beyond the usual mathematical and statistical functions by using dialogue and logic to determine various possible courses of action or outcome. By processing information much faster (than humans) the time to assess the condition and diagnose the causes of failures can be reduced. It can analyse situations objectively and will not forget any relevant facts (given that it has been supplied them), therefore the probability of making a wrong assessment or diagnosis may be reduced. Furthermore, it can detect incipient failures through its on-line monitoring of the condition parameters of the system [Lavalle et al (1993)].

Implementation of condition based maintenance

Having identified and listed all the condition parameters of the maintenance significant items, the aim of this step is to implement condition based maintenance. According to the classifications of condition parameter, condition based maintenance could be divided in two policies:

Inspection Based Maintenance Policy

The suitable maintenance policy for items for which their conditions are described by the relevant condition indicator, *RCI* is inspection-based maintenance. The algorithm, which presents the maintenance procedure in this case, is shown in Figure 5.11

Inspection is carried at fixed intervals to determine whether the condition of the item, is satisfactory or unsatisfactory according to the *RCI*. Before the item/system is introduced into service the most suitable frequency of the inspection, FMT^I, and critical value of relevant condition indicator RCI_{cr} has to be determined. Once the critical level is reached, $RCI(FMT^I) > RCI_{cr}$, the prescribed preventive maintenance tasks take place. If the item fails between inspections, corrective maintenance takes place.

Advantages of inspection based maintenance

CBM has the potential to produce large savings simply by allowing items in the system to be run to the end of their useful life. This reduces the equipment down time and minimises both scheduled and unscheduled breakdown situations. By eliminating all unscheduled interruptions to operation and production and only carrying out required maintenance in a carefully controlled manner, it is possible to reduce the maintenance cost, to improve safety, improve the efficiency of the operation and increase the system's availability.

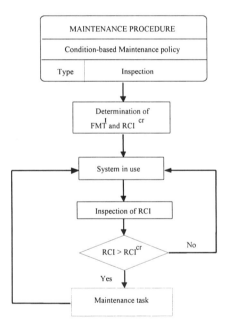

Figure 5.11 Algorithm for inspection based maintenance task

The benefits of inspection based maintenance policy can be summarised as follows:

1. Reduce unplanned downtime, since maintenance engineers can determine optimal maintenance intervals through the condition of constituent items in the system. This allows for better maintenance planning and more efficient use of resources.

2. Improve safety, since monitoring and detection of the deterioration in condition and/or performance of an item/system will enable the user to stop the system (just) before a failure occurs.

3. Extending the operating life of each individual items and therefore the coefficient of life utilisation will be increased compared to time based maintenance

4. Improve availability by being able to keep the system running longer and reducing the repair time.

5. Reduce maintenance resources due to reduction in unnecessary maintenance activities

6. The above benefits will lead to a reduction in maintenance costs

Examination Based Maintenance Policy

The decision for performing the condition-based maintenance tasks is based on the information related to the condition of an item/system established through condition checks during its operational life. This indicates that inspection-based maintenance strategy has achieved the demand for increasing the level of utilisation of an item/system. However, the system availability may not increase, due to an increased number of interruptions of the operation caused by increasing the number of inspections. Therefore, as an alternative, examination based maintenance approach is proposed by Knezevic (1987b) for the determination of maintenance tasks based on relevant condition predictors.

Examination based maintenance provides additional information about the change in condition of the items considered during its operational life. Consequently, examination based maintenance was developed for the control of maintenance procedures [El-Haram 1995]. With more information about the process of change in condition, a higher level of utilisation of the items can be achieved whilst maintaining a low probability of failure during the operation.

It is a dynamic process because the time of the next examination is fully determined by the real condition of the system at the time of examination. Dynamic control of maintenance tasks allows each individual item to perform the requested function with the required probability of failure, as in the case of time-based preventive maintenance but with fuller utilisation of operating life, hence with a reduction of total cost of operation and production.

The critical level of the relevant condition predictor RCP_{cr}, sets the limit above which appropriate maintenance tasks should be performed. The interval between the limit ($RCP_{\lim}$) and critical values depends on the ability of the operator to measure the condition of the item through the

RCP. The item under consideration could be in one of the following three states, according to the numerical value of the RCP,:

1. $RCP_{initial} < RCP(l) < RCP_{cr}$: continue with examinations;
2. $RCP_{cr} < RCP(l) < RCP_{lim}$: preventive maintenance task required;
3. $RCP_{lim} < RCP(l)$: corrective maintenance task, because the failure has already occurred.

In order to minimise interruptions to the operation and maximise the availability of the system, no stoppages occur until the time to the first examination of the condition of the item, FMT_1^E. The result of the examination is given as a numerical value of the relevant condition predictor, $MRCP(FMT_1^E)$, and it presents the real condition of the item at this instant of time. The following two conditions are possible, dependent on the value recorded:

1. $MRCP(FMT_1^E) > RCP_{cr}$, which means that a prescribed maintenance task should take place.
2. $MRCP(FMT_1^E) < RCP_{cr}$, the item can continue to be used.

The question, which immediately arises here, is: when will the next examination have to be done, preserving the required reliability level? The time to the next examination depends on the difference between the RCP_{cr} and $MRCP(FMT_1^E)$. The greater the difference, the longer the (operational) time to the next examination, FMT_2^E. At the predetermined time of the next examination, FMT_n^E, either of the two conditions is possible, and the same procedure should be followed, as shown in Figure 5.12

Advantages of Examination Based Policy

The advantages of the examination-based maintenance policy are:
1. Fuller utilisation of the functional life of each individual system than in case of time -based maintenance;
2. Provision of the required reliability level of each individual system as in case of time-based maintenance;
3. Reduction of the total maintenance cost as a result of extending the realisable operating life of the system and provision of a plan for maintenance tasks from the point of view of logistic support;

4. Increased availability of the item by a reduction of the number of inspections in comparison with inspection-based maintenance.
5. Applicability to all engineering systems. The main difficulties are the selection of a relevant condition predictor and the determination of the mathematical description of the $RCP(l)$.

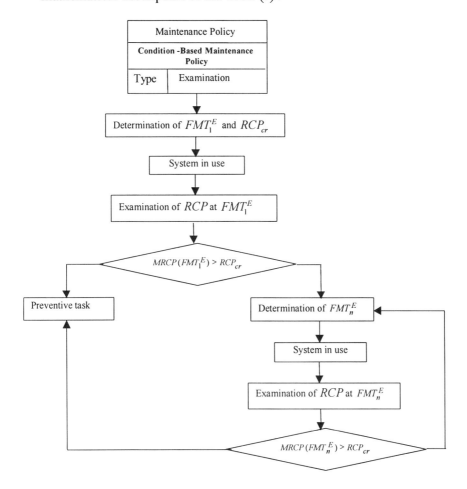

Figure 5.12 Maintenance procedure for examination based maintenance

In practice, it is impossible to eliminate all breakdowns. In some cases, it may not be economical or practical to use examination-based maintenance. Sometimes it is not physically possible to monitor the condition of all maintenance significant items. For these reasons, condition-based maintenance should not be considered to be a stand-alone policy. It should be integrated as a part of the overall maintenance policy. Thus, the optimal selection of maintenance policy for a system should include failure-based, time-based, inspection -based and examination-based maintenance strategies. The reasons for this are summarised below:

1. Not all items in the system are significant; the suitable maintenance policy is therefore, failure-based maintenance.
2. It may not be possible or practical to monitor the condition / performance of the significant items, so the suitable maintenance policy is therefore, time-based maintenance.
3. If the condition parameters of a significant item cannot be described by a relevant condition predictor, then the suitable maintenance policy is inspection-based maintenance
4. For significant items with relevant condition predictors, the most suitable policy is examination-based maintenance.

A maintenance management approach such as reliability centred maintenance could be used to select the most applicable and effective maintenance task for each item in the system

5.9.   MAINTENANCE RESOURCES

It is important to stress that the number of activities, their sequence and the type and quantity of resource required mainly depends on the decisions taken during the design phase of the item/system. The time required to perform a maintenance task will also depend on decisions made during this phase, such as the complexity, testability, accessibility and any special facilities, equipment, tools or resources needed.

Resources required primarily to facilitate the maintenance process will be called *Maintenance Resources, MR..* The resources needed for the successful completion of every maintenance task, could be grouped into the following categories (Knezevic 1997):

1. *Maintenance Supply Support, MSS*: is generic name which includes all spares, repair items, consumables, special supplies, and related inventories needed to support the maintenance process
2. *Maintenance Test and Support Equipment, MTE*: includes all tools, special condition monitoring equipment, diagnostic and check-out equipment, metrology and calibration equipment, maintenance stands and servicing and handling equipment required to support maintenance tasks associated with the item/system. Typically, MTE can be divided into two groups: *special to type equipment (STTE)* and *general (to type) equipment (GTTE)*.

3. *Maintenance Personnel, MP*: required for the installation, check-out, handling, and sustaining maintenance of the item/system and its associated test and support equipment are included in this category. Formal training for maintenance personnel required for each maintenance task should be considered

4. *Maintenance Facilities, MFC*: refers to all special facilities needed for completion of maintenance tasks. Physical plant, real estate, portable buildings, inspection pits, dry dock, housing, maintenance shops, calibration laboratories, and special repair and overhaul facilities must be considered related to each maintenance task

5. *Maintenance Technical Data, MTD*: necessary for check-out procedures, maintenance instructions, inspection and calibration procedures, overhaul procedures, modification instructions, facilities information, drawings and specifications that are necessary in the performance of system maintenance functions. Such data not only cover the system but test and support equipment, transportation and handling equipment, training equipment and facilities

6. *Maintenance Computer Resources, MCR*: refers to all computer equipment and accessories, software, program tapes/disks, data bases and so on, necessary in the performance of maintenance functions. This includes both condition monitoring and diagnostics.

On the other hand, it is important to remember that each task is performed in a specific work environment that could make a significant impact on the safety, accuracy and ease of task completion. The main environmental factors could be grouped as follows:

- space impediment (which reflects the obstructions imposed on maintenance personnel during the task execution which requires them to operate in awkward positions)
- Climatic conditions such as rain/snow, solar radiation, humidity, temperature, and similar situations, which could make significant impact on the safety, accuracy and ease of task completion.
- Platform on which maintenance task is performed (on operational site, on board a ship/submarine, space vehicle, workshops, and similar).

5.10.   MAINTENANCE INDUCED FAILURES

Whenever the cause of failure is related to the maintenance performed on the system, we call it maintenance-induced failure *MIF*. The root cause of MIF is poor workmanship, which might lead to poor spares or material selection, improper use of test equipment, training, working environment etc. A few examples of maintenance-induced failure are discussed in this section.

In 1991, Nigel Mansell lost his chance of becoming the Formula 1 World Champion in Portugal when one of the mechanics during a routine tyre change cross-threaded the retaining nut on the rear offside wheel. The result was that the wheel overtook the car as Nigel was exiting from the pit lane and his chance of victory and of the championship ended at that moment.

An airline pilot had a very lucky escape when he was nearly sucked through a window in the cockpit. The window was removed and replaced during a recently completed maintenance activity. When the cabin was pressurised as the aircraft climbed to cruising altitude, the window blew out. The rapid loss of pressure caused the pilot sat next to the window to be sucked through the hole. A combination of his size and the quick reactions of other members of the crew were all that saved him from a certain death. The cause of the window being blown out was that it had been refitted using under-sized screws.

In 1983, a new Air Canada Boeing 767 flying from Montreal to Edmonton ran out of fuel half way between the two at Gimli near Winnipeg. Although this was not entirely the fault of the refuellers, their miscalculations in converting between imperial and metric units was the final straw in an unfortunate sequence of events. A number of recommendations followed this incident which should mean that it never happens again (provided everyone follows the procedures correctly).

A few years ago, a team of "experienced" mechanics thought they knew how to do a particular maintenance task so did not follow the instructions in the maintenance manuals. The result was a cost of several million pounds sterling and a number of aircraft being out of service for considerably longer than they should have been.

These are extreme examples of what may be considered as "maintenance induced failures". They are also ones where it was relatively easy to determine the cause(s).

One of the major causes for accidental damage to components (from line replaceable units to parts) is the need to remove them in order to access other components. Using CATIA and EPIC (or similar systems) can do a great deal to aid the task of making components accessible and removing interference provided, of course, the design team are aware of these needs and their importance to the operational effectiveness of the aircraft.

Fasteners not properly tightened and locked (where appropriate) can work loose. Similarly, if they are not "captured" then there is a danger of them being "lost" when undone. If they are inside the engine or engine nacelle they may be sucked into the delicate machinery almost certainly causing extensive and expensive damage. Fasteners over-tightened may cause distortion resulting in leaks or damage, which may again have serious consequences. Consistent and sensible use of fasteners can not only reduce such problems but will also reduce the parts list and hence improve the supportability of the aircraft.

Some spare parts may be expensive or difficult to obtain. There may be a temptation to use alternative sources (other than those authorised). In many cases these may be made from inferior materials or to less demanding tolerances and quality standards. The use of such rogue parts may result in premature component failure and, possible, serious damage. Configuration control and full traceability of parts is an essential element of aircraft safety but, until practical electronic tagging of all parts becomes available, it will remain difficult to police effectively.

5.11.   MAINTENANCE COST

The world's airlines spend around $21 billion on maintenance, out of which 21% is spent on line maintenance, 27% on heavy maintenance, 31% on engine overhaul, 16% on component overhaul and the remainder on modifications and conversions (M Lam 1995). Repair and maintenance of building stock in the UK represents over 5% of Gross Domestic Product, or £36 billion at 1996 [Building maintenance information report 254,1996]. Maintenance and repair costs can be two to three times the initial capital costs, over the life of many types of buildings.

If one recognises that maintenance is essentially the management of failure then clearly, this expenditure is primarily the result of poor quality and unreliability. However, since it is impossible to produce a system which will never fail if operated for long enough we must consider ways in which the costs of maintenance can be kept to a minimum whilst ensuring system availability, safety and integrity.

We have already seen that there are many factors which can affect the costs of maintaining a system. Whilst the original design will be a major influencing factor on these costs, the operators and maintainers of the system can, nonetheless, do much to minimise the cost of ownership by adopting the most suitable maintenance policies for the conditions prevailing.

5.11.1   Cost of Maintenance Task

The cost of the maintenance task is the cost associated with each corrective or preventive task, whether time-based or condition-based. The expected corrective maintenance cost is the total cost of maintenance resources needed to repair or replace failed items. Similarly, the expected preventive maintenance cost is the total cost of maintenance resources needed to inspect and/or examine an item before failure takes place and to replace any items rejected. Thus, the total maintenance cost throughout the life of a systems/product is the sum of the corrective and preventive maintenance costs and the overhead costs, which consist of all costs other than direct material, labour and plant equipment. The cost of maintenance task can be divided into two categories:

5.11.2   Direct cost of maintenance task

The direct cost associated with each maintenance task, CMT, is related to the cost of maintenance resources, CMR, which are mentioned in Section 9. This is the cost of the maintenance resources directly used during the execution of the maintenance task, which is defined as:

$$CMT = C_s + C_m + C_p + C_{te} + C_f + C_d \qquad (5.11)$$

Where: C_s = cost of spare parts, C_m = cost of material, C_p = cost of personnel, C_{te} = cost of tools and support equipment, C_f = cost of facilities and C_d = cost of technical data.

5.11.3   Indirect cost of maintenance task

Indirect costs includes as management and administration staff needed for the successful completion of the task and the cost of the consequences of not having the system available which is related to a complete or partial loss of production and/or revenue. It also includes the overhead costs, i.e. salaries of employers, heating, insurance, taxes, facilities, electricity, telephone, IT, training and similar which are incurred while the item is in state of failure (and, of course, not included in the direct costs). These costs should not be neglected, because they could be even higher than the other cost elements.

Cost of lost production and/or revenue, CLR, is directly proportional to the product of the length of the time which the system spends in the state of failure (down time) and the income hourly rate, IHR, which is the money the system would earn whilst in operation. Thus, the cost of lost revenue could be determined according to the following expression:

$$CLR = (DMT + DST) \times IHR = DT \times IHR \qquad (5.12)$$

Where DMT is duration of maintenance task, DST is duration of support task and DT is total down time. Note for systems that are not normally in continuous operation, the downtime should take account of the proportion of the time the system would normally be expected to be operational. In particular, preventative, planned or scheduled maintenance would normally be done when the system would be expected to be idle and would only count as "downtime" for any period that the system would be expected to be operational. Thus, for example, if an airliner is not permitted to fly between the hours of 21:00 and 07:00 then any maintenance tasks undertaken and completed during those 10 hours would not affect the revenue-earning capacity of the aircraft.

5.11.4 Total cost of maintenance task

The total cost of maintenance task is the sum cost of direct and indirect costs, thus:

$$CMT = CMR + CLR \qquad (5.13)$$

Making use of the above equations the expression for the cost of the completion of each maintenance task is defined as:

$$CMT = C_s + C_m + C_p + C_{te} + C_f + C_d + (DMT + DST) \times IHR \quad (5.14)$$

It is necessary to underline that the cost defined by the above expression could differ considerably, due to:

1. Adoption of different maintenance policies
2. The direct cost of each maintenance task
3. Consumption of maintenance resources
4. Duration of maintenance task, DMT^c, DMT^p, DMT^I and DMT^E
5. Frequency of preventive maintenance task, FMT^L, the frequency of inspection, FMT^I and frequency of examination FMT^E
6. Duration of support task, DST^c, DST^p, DST^I and DST^E
7. The expected number of maintenance tasks $NMT(T_{st})$ performed during the stated operational length, L_{st}. For example, in the case of FBM, $NMT(T_{st}) = \dfrac{T_{st}}{MTTF}$
8. Different probability distributions and different values which random variables

$DMT^c, DMT^p, DMT^I, DMT^E, DST^c, DST^p, DST^I$ and DST^E
can take.

9. Indirect costs of maintenance tasks.

Thus, the general expression for the cost of each maintenance task will have different data input for different maintenance policies, as shown below:

$$CMT^c = C_{sp}^c + C_m^c + C_p^c + C_{te}^c + C_f^c + C_d^c + (DMT^c + DST^c) \times IHR^c$$
$$CMT^p = C_{sp}^p + C_m^p + C_p^p + C_{te}^p + C_f^p + C_d^p + (DMT^p + DST^p) \times IHR^p$$

$$CMT^I = C_{sp}^I + C_m^I + C_p^I + C_{te}^I + C_f^I + C_d^I + (DMT^I + DST^I) \times IHR^I$$

$$CMT^E = C_{sp}^E + C_m^E + C_p^E + C_{te}^E + C_f^E + C_d^E + (DMT^E + DST^E) \times IHR^E$$

Where: CMT^c is related to the cost of each maintenance task performed after the failure, CMT^p is cost in the case of time based maintenance CMT^I is cost of inspection based maintenance and CMT^I is cost of examination based maintenance.

The expected total maintenance cost for a stated time, $CMT(T_{st})$, is equal to the product of the maintenance cost for each maintenance task and the expected number of maintenance tasks performed during the stated time, $NMT(T_{st})$, thus:

$$CMT(T_{st}) = CMT^c \times NMT^c(T_{st}) + CMT^p \times NMT^p(T_{st}) + $$
$$CMT^I \times NMT^I(T_{st}) + CMT^E \times NMT^E(T_{st}) \qquad (5.15)$$

5.11.5   Factor Affecting Maintenance Costs

Maintenance cost could be affected by the following factors:

1. Supply responsiveness or the probability of having a spare part available when needed, supply lead times for given items, levels of inventory, and so on.
2. Test and support equipment effectiveness, which is the reliability and availability of test equipment, test equipment utilisation, system test thoroughness, and so on.
3. Maintenance facility availability and utilisation.

4. Transportation times between maintenance facilities.
5. Maintenance organisational effectiveness and personnel efficiency.
6. Durability and reliability of items in the system
7. Life expectancy of system
8. Expected number of maintenance tasks
9. Duration of maintenance and support task
10. Maintenance task resources

In order to reduce maintenance costs, it is necessary that the impact of the above factors should be reduced and/or controlled.

In calculating the various cost elements of maintenance, it is important to recognise that facilities, equipment, and personnel may be used for other tasks. For example, mechanics in the armed forces may be put on guard duty or provide a defence role when not performing maintenance tasks. Thus eliminating all maintenance tasks at first line (or O-Level) may not necessarily lead to a significant reduction in the personnel deployed or, indeed, in the operational costs.

5.12.   AIRCRAFT MAINTENANCE - CASE STUDY

For every commercial airline, maintenance is one of the most important functions to assure safe operation. Federal Aviation Regulation (FAR) require that, no person may operate an aircraft unless the mandatory replacement times, inspection intervals and related procedures or alternative inspection intervals and related procedures set forth in the operations specifications or inspection program has been complied with. All aircraft must follow a maintenance program that is approved by a regulatory authority such as FAA (Federal Aviation Administration, USA) and CAA (Civil Aviation Authority, UK). Each airline develops its own maintenance plan, based on the manufacturer's recommendations and by considering its own operation. Thus, two different airlines may have slightly different maintenance program for same aircraft model used under similar operating conditions. Aircraft maintenance is reliability centred. It is claimed that each aircraft receives approximately 14 hours of maintenance for every hour it flies (R Baker, 1995). Maintenance accounts for approximately 10% of an airline's total costs. On average a typical Boeing 747 will generate a total aircraft maintenance cost of approximately $1,700 per block hour.

Aircraft maintenance can be categorised as:

1. Routine scheduled maintenance.
2. Non-routine maintenance.

3. Refurbishment.
4. Modifications.

Routine Scheduled Maintenance

Scheduled maintenance tasks are required at determinant recurring intervals or due to Airworthiness Directives (AD). The most common routine maintenance is visual inspection of the aircraft prior to a scheduled departure (known as *walk around*) by pilots and mechanics to ensure that there are no obvious problems. Routine maintenance can be classified as:

1. Overnight maintenance.
2. Hard time maintenance.
3. Progressive Inspection.

Overnight maintenance normally includes low level maintenance checks, minor servicing and special inspections done at the end of the working for about one to two hours to ensure that the plane is operating in accordance with Minimum Equipment List. Overnight maintenance provides an opportunity to remedy passenger and crew complaints (M Lam, 1995).

Hard time is the oldest primary maintenance process. Hard time requires periodic overhaul or replacement of affected systems/components and structures and is flight, cycle and calendar limited. That is, as soon as the component age reaches it hard time it is replaced with a new component. Most of the rotating engine units are hard timed. The purpose of hard time maintenance is to assure operating safety of component or system, which have a limited redundancy.

Progressive inspection groups like time related maintenance tasks into convenient 'blocks' so that maintenance workload becomes balanced with time and maintenance can be accomplished in small 'bites' making equipment more available. Grouping maintenance tasks also helps better utilisation of the maintenance facilities. These maintenance task groups are (detailed information can be found in M Lam (1995) and L R Crawford, 1995):

1. *Pre-flight* – Visual inspections carried out by the mechanic and the pilots to ensure that there are no obvious problems.
2. *A Check* – Carried out approximately every 150 flight hours, which includes selected operational checks (general inspection of the interior/exterior of the aircraft), fluid servicing, extended visual inspection of fuselage exterior, power supply and certain operational

tasks. During A check, the aircraft is on ground for approximately 8 to 10 hours and requires approximately 60 labour hours.

3. *B Check* – Occurs about every 750 flight hours and includes some preventive maintenance such as engine oil spectro-analysis, oil-filter are removed and checked, lubrication of parts as required and examination of airframe. Also incorporates A-check. The aircraft could be on ground for 10 hours and will require approximately 200 labour hours.

4. *C Check* – Occurs every 3, 000 flight hour (approximately 15 months) and includes detailed inspection of airframe, engines, and accessories. In addition, components are repaired, flight controls are calibrated, and major internal mechanisms are tested. Functional and operational checks are also performed during C-check. It also includes both A and B checks. The aircraft will be on ground for 72 hours and will require approximately 3,000 labour hours.

5. *D Check* – This is the most intensive form of routine maintenance occurs about 20,000 flight hours (six to eight years). It is an overhaul that returns the aircraft to its original condition, as far as possible. Cabin interiors including seats, galleys, furnishings etc are removed to allow careful structural inspections. The aircraft is on ground for about 30 days and will require approximately 20,000 labour hours.

A and B checks and overnight maintenance are instances of line maintenance (performed upon the aircraft incidental to its scheduled revenue operations), often carried out an airport. C and D checks, however are heavy maintenance that requires special facilities and extensive labour. The task intervals for various checks mentioned above could vary significantly. The recommended time intervals for different aircraft models are given in Table 5.3 (*Aircraft Economics*).

Table 5.3 Different scheduled checks in a commercial aircraft

Aircraft Type	A check Flight hours	B Check Flight hours	C Check Flight hours	D Check Flight hours
Boeing 707	90		450	14,000
Boeing 727	80	400	1,600	16,000
Boeing 737-100	125	750	3,000	20,000
Boeing 747-100	300		3,600	25,000
DC-8	150	540	3,325	23,745
DC-9	130	680	3,380	12,600

Non-routine maintenance refers to the maintenance tasks that has to be performed on regular basis during checks, but which is not specified as routine maintenance task on the job cards of the maintenance schedule. Non-routine maintenance shouldn't be confused with unscheduled maintenance, which is repairs that have to be done as a result of an unexpected failure such as accidental damage (such as bird strike) to critical components or a response to airworthiness directives (AD). As the aircraft age, they require more maintenance due to fatigue and corrosion. The most significant of these aging aircraft airworthiness directives concerns Boeing 747. The fuselage of the Boeing 747 is built in sections as separate entities and then assembled during the aircraft production phase. The fuselage is built in five sections and the points at which these sections are joined are called the production breaks. Section 41 is the section from the nose to just aft of the forward passenger entry (*Maintaining the Boeing 747, Aircraft Economics, 1994*). The modification of Section 41, which is the area ahead of the forward passenger doors, requires approximately 60,000-70,000 man-hours to complete and requires replacement of most of the structural components (L Crawford, 1995).

Chapter 6

Maintenance Optimisation

You can see a lot by observing

Yogi Berra

The objective of maintenance optimisation models is to determine the optimum maintenance tasks that minimise the downtime while providing the most effective use of systems in order to secure the desired results at the lowest possible costs, taking all possible constraints into account. The models can be either quantitative or procedure based such as reliability centred maintenance [Nowlan and Heap, 1978], age related maintenance or total productive maintenance.

The widespread application of preventive maintenance has led to extensive mathematical models in the literature that treat the question of how preventive maintenance should be scheduled. Mathematical models could be either deterministic or stochastic. Preventive maintenance tasks are performed in the belief that they will improve system utilisation. In the literature the majority of the models, which are used to determine the optimal maintenance task are often based on some criterion. The most frequently used criteria for developing maintenance models are:

1. Minimising; maintenance cost, down time and time to repair.
2. Maximising; revenue, profit, time between failure and availability.
3. Achieving required level of reliability and safety.

The development of mathematical models and their algorithm are driven by the needs for adequate maintenance planning which should provide optimal solutions to the following question: when should an item be repaired, replaced, inspected or examined? The mathematical model provides answer to the above question, based on information available and the chosen criteria. In many cases the time when a maintenance task is performed could be based on one or more criteria. In order to analyse the impact of the above criteria and similar measures on the selection of optimal maintenance task, it is necessary to establish a relationship between them. This can be achieved by building a model, which could be either mathematical and/or engineering approach, which defines that relationship and provide a basis for all the analysis necessary.

The traditional method for modelling preventive maintenance is to model the relationship between the preventive maintenance interval and the operating cost per unit time, or the system availability. Example of optimal preventive maintenance models using different optimisation criterion include the following work:

- Barlow and Proschan (1975) presented a model to determine the optimal replacement interval, which was based on minimising the expected cost per unit of time.
- Kelly (1976) suggested a model using revenue rather than availability as an optimisation criterion.
- Handlarski (1980) proposed a model using profits as an optimisation criterion.
- Waston (1970) presented a model to minimise downtime per unit time for a group of components.
- Asher and Kobbacy (1995) proposed the use of a non-homogeneous Poisson process to model preventive maintenance situation with increasing rate of occurrence of failure.
- Knezevic (1987) presented a model using required level of system reliability as an optimisation criterion

However the complexity of modelling preventive maintenance stems from the difficulty of quantifying the effect of performing preventive maintenance at different intervals. In fact, most of the mathematical models assume that the systems behaves either '*as good as new*' or ' *as bad as old*' after preventive maintenance is carried out, which is seldom true, unless the whole system is replaced with a new one.

6.1 PREVENTIVE MAINTENANCE AND RELIABILITY

In Chapter 4, we analytically determined the effect of preventive replacement task on reliability the reliability function by considering the *TTF* distribution. We also proved that for the items, which do not wear out, scheduled replacements do not improve reliability. Indeed they are more likely to induce failures. Items subject to failure mechanisms such as wear, corrosion, fatigue, etc, should be considered for preventive maintenance.

As mentioned Chapter 3, the hazard function could be decreasing, constant or increasing. If an item has a decreasing hazard function, then would have a high infant mortality (or it is improving as time progresses) and therefore, maintenance aimed at restoring it to as new condition is actually disadvantageous and not advisable. That is, when the hazard function is increasing, then any replacement will increase the probability of failure as shown in Figure 6.1a.

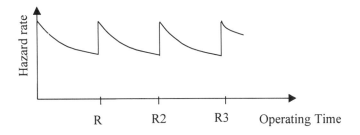

Figure 6.1a Decreasing hazard and Replacement

If an item has a constant hazard function, then its time to failure has an exponential distribution. In other words, the probability of failure during the next time increment remains unchanged throughout the lifetime of the item, indicating that it is "as good as new" no matter how long it has operated. In this case preventive maintenance is irrelevant. That is, replacement will make no difference to the failure probability as shown in Figure 6.1b.

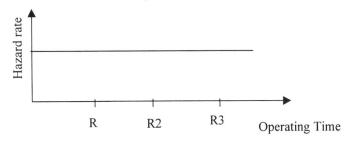

Figure 6.1b Constant hazard and Replacement

If an item has an increasing hazard function, then scheduled replacement at any time will in theory improve reliability of the system. Thus, preventive scheduled maintenance is worthwhile only if the item has an increasing hazard function as shown in Figure 6.1c.

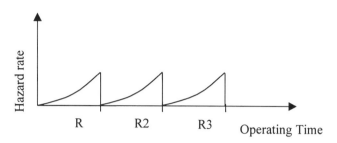

igure 6.1c Increasing hazard and Replacement

The effectiveness and the economy of preventive maintenance can be maximised by taking account of the time-to-failure distributions of the maintained items and of the hazard function trend of the system. In order to optimise preventive replacement, it is therefore necessary to know the following for each item, the time-to-failure distribution, the cost of failure and the cost of scheduled replacement, and the effectiveness of maintenance after scheduled replacement.

6.2 OPTIMAL REPLACEMENT TIMES

The optimal replacement time technique applies to systems with increasing hazard (wear-out condition) only. The costs of wasted item life due to preventive replacement during routine maintenance is balanced against the costs of an unplanned repair. It is necessary to establish the time to failure distribution of the considered item. Then it is possible to balance the additional cost of an unplanned replacement over the convenience of a planned preventive maintenance against the cost of giving up some useful life of the item by replacing it before it fails due to wear-out. This approach depends on good data collection and analysis to identify the distribution of the failure (see Chapter 12). Furthermore, this is only effective when the ration of unplanned maintenance cost to the planned replacement is high. And it usually applies where there is no redundancy. If this is the case then it is necessary to calculate: the failure function (i.e. $1 - R(t)$) in a particular interval times, the cost of the unplanned maintenance and the cost of planned replacements during that interval.

The optimal replacement interval, which minimises the sum of the unplanned maintenance cost and a planned replacements cost, could be based on the following two replacement policies:

- Age Replacement
- Block Replacement

There is a huge literature available on the age replacement policies. Barlow and Proschan (1975) discuss the traditional approach which is to replace the item at failure or at age T, whichever comes first. Extensions of this approach have been treated by Pierskalla et. al. (1976), Valdez-Flores et. al. (1989), Nakagawa et. al. (1983), Berg et. al (1986) Black et. al (1988), Sheu (1994), Vanneste (1992), Dekker (1994) and many more authors deal with an optimal age replacement policy where replacement at failure depends on random cost of minimal repair. These papers provide a fairly comprehensive chronological review of the research performed concerning preventive replacement maintenance.

For the age replacement case, an interval starts at time t = 0 and ends either with a failure or with a replacement at time T, which ever occurs first. The probability of surviving until time t =T is $R(T)$ thus the probability of failing is $(1 - R(T))$. The average duration of all intervals is given by:

$$MTBF = \int_0^T R(t)dt \tag{6.1}$$

Thus the cost per unit time is

$$\frac{[C_u \times (i - R(t)) + C_p \times R(t)]}{\int_0^T R(t)dt} \tag{6.2}$$

where C_u is the cost of unplanned maintenance and C_p is the cost of a planned replacement.

For block replacement case, replacement always occurs at time $t = T$ despite the possibility of failures occurring before $t = T$. For this case the cost per unit time is:

$$\frac{(C_u \times T)}{MTBF \times T} + \frac{C_p}{T} = \frac{C_u}{MTBF} + \frac{C_p}{T} \tag{6.3}$$

Note that, since the failure rate is not constant, the MTBF used in the above equation varies as a function of T.

6.2.1 Optimal Replacement Under Minimal Repair

Some times it may be required to find the optimal replacement time. That is, whenever an item fails it is repaired to 'as-bad-as-old' and is replaced at time t. Let

C = cost of the item

C_o = operating cost per unit of time

C_{tr} = total repair cost

The total replacement cost, $TC_r(t)$, is given by

$$TC_r(t) = C + C_o t + C_{tr} H(t) \qquad (6.4)$$

Assume that the time to failure of the item is Weibull, then the total replacement cost is given by

$$TC_r(t) = C + C_o t + C_{tr} (t/\eta)^\beta \qquad (6.5)$$

The cost of replacement per unit time, TC_r, is given by:

$$TC_r = \frac{C}{t} + C_o + C_{tr} \frac{t^{\beta-1}}{\eta^\beta} \qquad (6.6)$$

To find the optimal replacement time t, we minimise the unit cost of replacement by setting $dTC_r/dt = 0$ and solving for t.

$$\frac{dTC_r}{dt} = -\frac{C}{t^2} + C_{tr} \times (\beta - 1) \times \frac{t^{\beta-2}}{\eta^\beta} = 0$$

The optimal replacement time t^*, is given by:

$$t^* = \left[\frac{C \times \eta^\beta}{C_{tr} \times (\beta - 1)} \right]^{1/\beta} \qquad (6.7)$$

Example 6.1

Time to failure of a global positioning system (GPS) follows Weibull distribution with characteristic life $\eta = 1750$ hours and $\beta = 3.5$. The total cost of repair is $800 and the cost of GPS is $20000. Upon failure, the GPS is minimally repaired. Find the optimal replacement time of the GPS.

$$t^* = \left[\frac{20{,}000 \times 1750^{3.5}}{800 \times 2.5} \right]^{1/3.5} = 3378.72 \text{ hours}$$

6.3 REPAIR VS REPLACEMENT

Before a system enters service, indeed, when it is still at the concept stage, it is necessary to consider which parts of the system are going to be repaired or discarded and where this will be done. Most complex systems are in some way repairable but there are a few exceptions. Putting satellites, manned or unmanned, into space requires the use of rockets. In almost all cases, these rockets are used only once and no attempt is ever made to repair them. The space shuttle is a major exception. Its two solid fuel boosters are jettisoned two minutes into the flight and are subsequently recovered from the ocean and reused up to twenty times, according to Dyson (1992). Their main rockets form part of the winged orbiter and are also reused. The main liquid oxygen and liquid hydrogen tank is jettisoned after 390 second and breaks up on re-entry. The orbiter is expected to return to earth intact, many of the ceramic tiles which act as a heat shield are lost during re-entry but, these are expendable.

The part of the system, which has entered a state of failure, can usually be removed and replaced with a similar part, which is in the state of functioning. The removed part may then be either repaired or discarded. A repair may restore the system to a 'same-as-new' but is much more likely to restore it to a condition, which is only marginally better than 'same-as-old'. An exception to this may occur when the replacement part is 'newer' then the removed one or is an improved design (in terms of reliability). Replacing a worn tyre on a car will make that part of the vehicle more reliable, but not the reliability of the car as a whole. Simply changing a worn tyre, even for a new one, is unlikely to restore the system to a 'same-as-new' condition.

Although most complex systems are usually repairable, they are likely to reach a stage in their lives when it becomes uneconomical to try to repair them. An aircraft which has suffered a 'controlled flight into terrain (crashes), a car which has hit a wall at high speed, a 20 year old television set which has blown a (cathode ray) tube may all be repairable (technically). But, the cost of doing a repair is likely to be higher than the cost of replacement. In each of these cases, one of the most important factors was age, the second of course was cost. Age manifests itself in a number of ways: wear, corrosion, and obsolescence.

The effectiveness of repair plays a major role in deciding whether a component is repairable or discardable. A repair is classified as '*as-good-as-new*' (or '*same as new*') when the value of the hazard function after repair completion is the same as that of a new item. If the value of the hazard function after the repair completion is same as the value just before failure, the repair is called '*as-bad-as-old*' or *minimal repair* (or '*same as old*'). If the value of hazard function after repair completion lies between the 'as-good-as-new' and 'as-bad-as-old', then it is called '*imperfect repair*'. Figure 6.2 illustrates the three levels of repair effectiveness using hazard function.

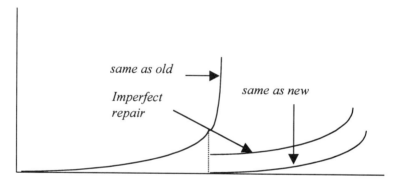

Figure 6.2. Repair effectiveness and hazard function

The decision on whether to repair or to replace individual item or complete systems could be based on the following:

- The age of the item and how does it compare to the expected life or MTBF. If the age is well within the life expectancy than the tendency will be considered repair, if the life expectancy is running out or has been exceeded the tendency will be replace.
- Spare parts availability. In areas of rapid technological change, repair is clearly not feasible if spare parts can not be obtained. This may dictate replacement even when condition assessment indicates repair.
- The trend in maintenance cost. Life cycle cost analysis is useful to demonstrate the cost benefits within an acceptable period in which replacement can offer. Apart from maintenance cost, replacement can

be justified on the ground of other cost savings for example, the saving in energy cost from the use of a modern high efficiency system.

- Financial constrains which could have a great affect on maintenance decisions that are related to repair or replace. While system designer and maintenance engineers may suggest that replacement of an item or a system is preferable, financial constraint may dictate that a repair is carried out. It is for this reason that replacement decisions should be prioritised as being essential, required or advisable.

As mentioned above that, in most cases, the primary reason for deciding to replace an item instead of repair is based on the cost. The cost of repair will depend on the following elements.

1. The expected number of failures over the life of the system. The expected number of failures is computed using renewal function, $M(t)$, or cumulative hazard function, $H(t)$. Whenever the repair is 'as-good-as-new', renewal function is used. If the repair is 'as-bad-as-old', then cumulative hazard function is used.
2. Fixed cost of repair, F_r, involve the cost of maintenance facilities, test and support equipment, training maintenance personnel, technical manuals etc.
3. Variable cost to repair a failure, C_r, involve the labour cost, transportation and handling etc.
4. The percentage of failures, p ($0 \le p < 1$) that cannot be repaired. Under such circumstances one has to replace the item.

Assume that 'C' represents the cost of the item. The total repair cost under 'as-good-as-new' repair policy is given by:

$$F_r + M(t) \times C_r \times (1-p) + p \times [C + C_d] \times M(t) \qquad (6.8)$$

Replacement cost will depend on the following elements:

5. Fixed cost of replacement, F_d, involving facilities, test and equipment, training, technical manuals and inventory costs. Note that the equipment and skill level required to replace an item will be much less that the required for repairing an item.
6. Cost to replace an item, C_d, (personnel cost, transportation and handling).

Again assuming that 'C' represents cost of the item. The replacement cost is given by:

$$F_d + [C + C_d] \times M(t) \qquad\qquad (6.9)$$

It will be cheaper to repair an item rather than replacing it if the following inequality is true.

$$F_r + M(t) \times C_r \times (1 - p) + p \times [C + C_d] \times M(t) \le F_d + [C + C_d] \times M(t) \qquad (6.10)$$

However, many military and commercial organisations decide to repair rather than discard when the repair cost is less than certain percentage (say 60%) of the replacement cost.

Example 6.2

Automatic flight control (AFC) system costs about $150 000 to buy. Also it is known that the time to failure of the AFC follows a normal distribution with mean 1200 flying hours and standard deviation 200 hours. About 90% of the failure modes of AFC are repairable. However, it will cost $400 000 towards fixed cost of repair (that is setting up facilities, equipment, tools etc). Also, on average each repair costs $12 000. Fixed cost for replacement is $220 000 and each replacement costs $2000. For a period of 15 000 hours, find whether it is beneficial to repair or replace assuming that the repair restores the system to 'as-good-as-new' state.

SOLUTION:

We have the following information:

Fixed cost of repair, F_r = $400 000

Cost of repair, C_r = $12 000

Cost of unit part, C = $150 000

Fixed cost of replacement, F_d = $220 000

Cost of replacing the part, C_d = $ 2000

Also it is known that the percentage of failures that cannot be repaired is 10%. From the above information, the expected cost of repair for 15 000 hours of operation is given by:

Total repair cost is given by

$$F_r + M(15000) \times C_r \times 0.9 + M(15000) \times 0.1 \times [C_d + C]$$

where, $M(15000) = \sum_{n=1}^{\infty} \Phi(\frac{15000 - n \times 2000}{\sqrt{n} \times 200}) \approx 12$

Substituting the value of $M(15000)$, the total cost of repair is \$712 000. The total cost of replacement is given by

$$F_d + [C + C_d] \times M(t) = \$2\,044\,000$$

The total repair cost is less than that of replacement cost, thus it is beneficial to repair the item rather than to replace.

6.4 RELIABILITY CENTRED MAINTENANCE

Reliability Centred Maintenance, *RCM*, has its roots in the airline industry in the late 1960s in conjunction with the introduction of the Boeing 747. Federal Aviation Administration (*FAA*) requirements for maintenance, resulted in a maintenance plan with a very extensive set of maintenance tasks. These maintenance tasks were extensive that the airlines probably would not have been able to operate the 747 profitably. It was becoming apparent that it was simply not possible to reduce the failure rates of much of their items using time-based preventive maintenance such as replacements or overhauls. As a result, the Federal Aviation Authority formed a Maintenance Steering Group, *MSG*, consisting operator, manufacturer and regulator. Their task was to research a maintenance area of particular importance to various systems, so that a logical and generally applicable approach could be used for developing maintenance strategies that could ensure the maximum safety and reliability and provide the minimum maintenance cost. The decision diagram was developed and presented in 1967 at the AIAA commercial Aircraft Design and Operations meeting. Subsequent improvements were then embodied in a document on maintenance evaluation. This document became known as MSG-1. Using of the technique led to the further improvements, which were, formed a document known as MSG-2 and finally MSG-3 which were issued in 1970 and 1980 respectively. The commercial aircraft models, Boeing 767, 757, 747-400, 737NG and 777 use *MSG-3* methods.

A major milestone in the history of RCM was a report commissioned by the United States Department of Defence and United Airlines and prepared by Stanley Nowlan and Howard Heap (1978). Nowlan and Heap developed the principles of RCM by looking at the impact of preventive maintenance on the prevention of the failure and its consequences.

The RCM approach later attracted the attention of the U.S military and naval aircraft and then became the approved and required method for defining maintenance programmes for most of Air Force and Army. In the 1980s, observing the cost-benefits form the airline industry, the military and other industries such as nuclear power, chemical, automotive, manufacturing, oil and gas, construction etc. who were also faced with requirements for intensive maintenance programmes, started applying RCM approach.

Why RCM

During the development of the Boeing 747, batches of the engine aircraft were tested to determined the failure patterns, the results are display in Figure 6.2a. The study has established that there are actually six failure relate patterns as illustrated in Figure 6.2. The most two common failure patterns show a failure rate decreasing with age before going into a period of random failure and a totally random failure pattern which represent 82% of the items. The study also shows that 68% of items start with burn-in, which drops eventually to a constant probability of failure (note that the percentages are based on a sample data, and should not be generalised). This means that after the burn-in period, there is no relationship between reliability and operating age. In these cases, unless there is an age-related failure, time-based preventive maintenance do nothing to reduce the probability of occurrence failure. In fact, it can increase the incidence of failure by introducing burn-in into otherwise stable systems. Thus, since the majority of failure patterns do not exhibit pronounced wear-out period, maintenance responses must be aimed primarily at detecting potential failures or hidden failures leading to functional failures. For the system, which exhibits definite wear-out patterns, maintenance responses must also include removal and replacement of items within a specified age limit but only after the exact condition is confirmed with inspection or examination. RCM process was therefore developed on the basis of the following concept:

- There are systems, which do not generally experience a wear-out phase (indicated by increasing of failure rates). This means, that the hazard rate of those systems does not change with age. Time-based preventive task will not prevent such type of failures.

- Not all failures affect the system function(s) equally. Therefore, the emphasis in preventive maintenance should be on maintaining important system functions and must done to mitigate the failure consequences.

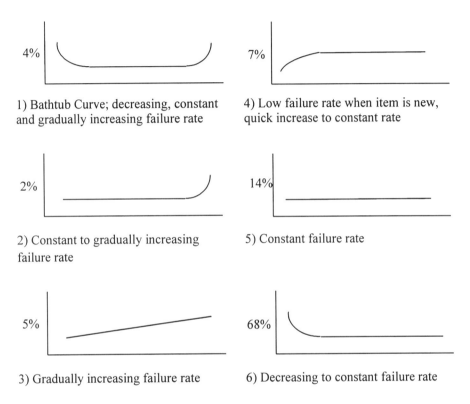

1) Bathtub Curve; decreasing, constant and gradually increasing failure rate

4) Low failure rate when item is new, quick increase to constant rate

2) Constant to gradually increasing failure rate

5) Constant failure rate

3) Gradually increasing failure rate

6) Decreasing to constant failure rate

Figure 6.2a. Patterns of failure

RCM is a systematic approach for selecting applicable and effective preventive maintenance tasks for each item in a system taking into consideration failure consequences. Applicable means that if the task is performed, it will accomplish one of the reasons such as prevent or mitigate failure, or detect a hidden failure. Effective means that the selected maintenance task will be the least expensive task. RCM is therefore a logical process used to determine what must be done to ensure that any system continues fulfil its intended functions [Moubray, 1997).

RCM ensures that preventive maintenance requirements are based on the failure consequences of the system and allow it to realise its inherent reliability. Only applicable and effective maintenance tasks are used to prevent failures. If an appropriate task does not exist, no preventive maintenance will be performed. The item will be redesigned to eliminate the failure if the failure mode has health or safety consequences. RCM focuses on the functionality of system in the desired operating environment. By focusing on the function, maintenance tasks are selected to improve reliability and availability of the system. The implementation of appropriate maintenance strategy allows the system to be operated reliably for the full life cycle of the system.

6.4.1 RCM Process

The RCM process is used to identify system functions, the way these functions fail, and the consequences of the failures and apply this information to develop appropriate maintenance tasks to prevent system failures. The primary objective of RCM is to preserve system functions taking into account the objectives of maintenance such as minimising costs, meeting safety and environmental goals and meeting operational goals. However, the additional objectives of RCM are:

- To eliminate ineffective preventive maintenance tasks
- To focus maintenance effort on failures that may affect health, safety, environment, economic and operation and any other business related consequences.
- To increase system availability
- To ensure system achieves inherent level of reliability
- To achieve the above mentioned goals at minimum operation, maintenance and support costs

The *RCM* analysis is best initiated during the system design phase. However, it can be effectively utilised for an existing system as maintenance planning evaluation and continuous improvement methodology. The RCM process begins with a failure mode and effects analysis (FMEA), which identifies the significant system failure modes in a systematic and structured manner. The process then requires the examination of each significant failure mode to determine the optimum maintenance task to reduce or avoid the severity of each failure. The chosen maintenance task must take into account cost, safety, environmental and operational consequences. The effects of parameter such as redundancy, spares costs, maintenance personnel costs, system ageing and condition and repair times must be taken into account. Figure 6.3 illustrates a typical RCM process. However, the process to perform the RCM analysis varies somewhat among the practitioners and the system users around the world. The basic RCM steps, however, are quite common to all applications. The RCM process comprises the following steps:

1. System selection.
2. Perform Failure Modes and Effects Analysis (*FMEA*).
3. RCM decision logic process. Identification of failure consequences.
4. Selection of maintenance task.

5. The above steps are briefly described below.

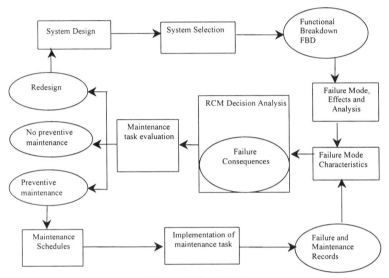

Figure 6.3. *RCM* process

System Selection

All systems in principle may benefits from RCM analysis. However, the RCM team should start with the systems that they assume will benefit most from the analysis. The team should also identify the level of assembly (plant, system, subsystem) at which the analysis should be conducted. They should always try to keep the analysis at the highest practical indenture level. The lower the level, the more difficult it is to define performance standards. Therefore, before a decision to perform an RCM analysis, the following questions should be asked when selecting the system to analyse [Brauer et al. 1987]:

• Will an improvement in preventive maintenance reduce costs and improve reliability and safety?

- Does the current maintenance strategy include a large proportion of time-based maintenance tasks that could easily be replaced by condition-based maintenance tasks?
- Is there a known design problem that is causing failures and incur high maintenance and support costs?

Once a decision to perform the RCM analysis is made and the system is selected, it is necessary to collect as much relevant data and information as possible. The data and information that is required to support the RCM analysis may include:

- Design information including drawings and technical specifications of the system
- Operating performance of the system such as performance requirements and operating profile.
- Historical maintenance data form the maintenance management system. Examples of such data are downtime, cost of maintenance, and all preventive and corrective maintenance tasks performed on the system.
- Reliability data, such as MTBF, failure rate, preventive maintenance task frequency.
- Support data, such as support costs, level of support, etc.

Failure Mode and Effects Analysis

FMEA is a systematic approach to identify all possible ways in which failure of a system can occur together with its causes and thus the failure's potential effect on the system. It is performed to determine how each item in the system is likely to fail and what will happen if it does. FMEA is described in details in Mil-Std 1629A. If, as is usually the case, the FMEA is extended to include an evaluation of the failure criticality- an assessment of the severity of the failure effect and its probability of occurrence. Thus, this procedure is the result of two steps which, when combined, provide the FMECA.

- Failure Modes and Effects Analysis (FMEA)
- Criticality Analysis (CA)

The FMECA activity is an integral part of reliability, maintainability, ILS and RCM. In this chapter we briefly introduce FMEA to provide information which is needed to implement RCM. The FMECA will be discussed in Chapter 11. In general, the objectives of FMEA are to:

- Assist in selecting design alternatives with high reliability and high safety potential during early design phase.
- Ensure that potential failure modes and their effects on operational success of the system have been identified.
- Provide a basis for quantitative reliability, maintainability and availability analyses.
- Provide historical documentation for future reference to aid in analysis of field failure and consideration of design changes.
- Assist in the objective evaluation of design requirements related to redundancy, failure detection systems, fail-safe characteristics, and automatic and manual override.
- Provide data information for the implementation of RCM

Before implementing FMEA and RCM, it is essential that functional failure is fully defined and understood. Unless the failed state of an item is defined, it is difficult to determine failure effects, failure consequences and appropriate maintenance tasks. The definition of failures must be established at the system, subsystem and possibly even lower levels. There are two ways in which a failure or the required function can be terminated. First, the required function is immediately terminated. Such termination usually occurs suddenly without previous indication of damage and independent of the age and condition of the item. Secondly, the required function is gradually terminated due to a change in condition, performance or any other measurable condition parameters of the item, with possibility to observe some deterioration before functional failure takes place. The distinction between the two categories will help in selecting the appropriate maintenance task.

The FMEA procedure might be termed a "what if" approach in that it starts at the item level and asks what if this item fails, the effects are then traced to the system level. The FMEA procedure is carried out and documented using an appropriate worksheet. Varity of different FMEA worksheets are in use. A commonly used FMEA worksheet is shown in Figure 6.4. The following steps are usually completed when performing FMEA:

1. Building functional block diagram
2. Identification of all possible failure modes of each item
3. Identification of all possible causes of each failure mode
4. Identification of the effects of each failure mode on the item, on the subsystem and finally on the system as a whole.

Functional Block Diagram

Prior to the FMEA, it is necessary to gather all the necessary information on design, operating characteristics, system requirements, description and definition of the function(s) of a system and its associated items. A function can be defined as the requirements that a system or an item must accomplish. Function can be divided into *primary function(s)* "the main reason why an item is required" and *secondary function(s)* "the function(s) which an item may be expected to accomplish its primary function(s)" [Moubrey, 1997]. The diagram, which represents the major functions that a system performs, is called a Functional Block Diagram, *FBD*. The functional block diagram is constructed by dividing the system into functional blocks. Each block then further broken down into progressively lower levels of indenture.

System/subsystem/item			Page No. _____ of_____ Date _____ Revision _____						
Indenture Level			Prepared By_____						
Intended Use									
Ident No	Item functional ident.	Function	Failure modes	Failure modes causes	Failure Effects			Failure detection method	Remarks
					LE	NHL	EE		

Figure 6.4 Typical FMEA worksheet

FBD aims to explain the functions of the system and their interrelationships and to help visualise the functional relationship of the various items to each other, to the higher levels of indenture (the system) and to the end plant. It is also used to develop the relationship and the functional flow sequence between the primary and the secondary functions and the inputs and the outputs of each function. The result of this step is a list of function(s) of each item in the system, which forms the basic information, required to start the FMEA process.

Identification of Failure Mode

The second step of FMEA analysis is to identify all possible failure modes of each item in the system. A failure mode can be defined as the manner in which an item fails. Many factors such as material of an item, the way it constructed/manufactured, environmental conditions, and the way it is operated could have an effect on the way in which the item will lose its function(s). For example, two items with the same function but made from different materials, constructed/ manufactured and operated differently will have different failure modes. The number of failure modes a system can have depending on the complexity, the operating circumstances and the level at which it is being analysed. Therefore, the identification of failure modes is an important step in the development of FMEA and RCM analysis.

Identification of Failure Causes

The cause of failure mode refers to the reasons for the failure mode to occur. The objective of this step is to identify all the likely reasons why the failure mode occurred. Since preventing the failure mode means eliminating or at least controlling its causes, it is therefore necessary that all possible causes of each failure mode need to be identified. Clearly, the more precise the description of the causes of failure, the more understanding we have for deciding how it may be eliminated or accommodated. A failure mode of an item could be the results of one of many different causes.

Identification of Failure Effects

The failure effects are the impact of each failure mode on the element function(s). The objective of this step is to identify what happens when each failure mode occurs. Failure effect answers the question "what is the impact a failure mode has on an item function(s) and ultimately on the whole system?" It is necessary to note that failure effects are not the same as failure consequences, which answer the question "why does a failure mode matter?" The impact of failure mode on an item function(s) and on the next higher indenture level (subsystem) and ultimately on the end level (system) should be identified, evaluated and documented. In this step, the analyst considers each failure mode and determines the effects that it will have on the overall system function(s). The general way of identifying the effects of a failure mode is to assume that no preventive maintenance will be carried out, which is as if nothing was being done to prevent the failure mode. The effect of an item failure depends upon the function(s) of the item in the system. For example, two valves may have the same design specifications but the effect of a failure will depend upon what the valve controlling. The failure mode under consideration may impact several indenture levels in addition to the indenture level under analysis. The failure effects can be divided into three levels:

Local effect (Item level): Local effects concentrate specifically on the impact of a failure mode on the operation and function(s) of the item in the indenture level under consideration. The purpose of defining local effects is to provide a basis for evaluating compensating provisions and for recommending corrective actions.

Next Higher Level (Subsystem level): Next higher level effects concentrate on the impact of a failure mode on the operation and function(s) of the items in the next higher indenture level (subsystem) above the indenture level under consideration.

End effect (System level): End effects evaluate and define the total effect a failure mode has on the operation, and function(s), of the whole system. The end effects evaluate the total impact of a failure mode on the function(s) of the system.

6.4.2 RCM Decision Logic Process

The RCM methodology analyses the consequences of each failure mode, which are taken from FMEA, and identifies an applicable and effective maintenance task by using the principle that a maintenance task is worth doing if its deals successfully with the consequences of the failure mode which it meant to prevent. RCM is based on decision logic process, which involves the evaluation of each failure mode for determination of its consequences and evaluation of each consequence for selection of applicable and effective maintenance tasks that can prevent the failure mode and avoid its consequence. The RCM decision logic process is designed to lead, through the use of standard assessment questions, to the most effective maintenance task combinations.

Identification of Failure Consequences

Failure consequence answers the question "why does a failure mode matter?" The identification of failure consequences is the heart of RCM decision process, because RCM addresses the consequences of failure rather than failure itself. It is not whether a failure occurs but what happens when it occurs, which is important to the user and to the business. After the significant items failure modes have been properly identified through the FMEA, a series of questions, which are part of the RCM decision process can be answered. The answers to the following questions determine the consequence for each failure and identify which branch of the decision process to follow during maintenance task evaluation.

1. Can the user detect the failure?
2. Does the failure mode have an affect on health of the user?
3. Does the failure mode have an affect on safety and the environment?
4. Is the cost of failure and its consequential damage greater that the cost of preventing the failure?
5. Dose the failure mode have an affect on the operational performance?
6. Dose the failure mode have an affect on the appearance? In some assets such as building the appearance consequence could have a big affect on the business.

The above questions are asked for each failure mode, and the answers which are in a simple 'yes' or 'no' format, are recorded on a RCM decision logic worksheet. An Example of RCM decision worksheet is shown in Figure 6.5. The answer to the first question will help determine if the failure is evident or hidden. A 'yes' means that the failure is evident, whereas, a 'no' means that the failure is hidden. Hidden failures are those failures in which the user will not be aware of the loss of their function under normal circumstance without special monitoring. The functional failure of an item is considered hidden to the operator or the user if either of the following situations exist [Smith, 1993]:

- The item has a function, which is normally active whenever the system is used, but there is no indication to the operator/user when that function ceases to perform.
- The item has a function, which is normally inactive and there is no prior indication to the operator/user that the function will not perform when called upon.

The consequence of a hidden failure is an increased risk of a multiple failure. In high risk plants, protective devices can be installed on systems or items where hidden failure might occur. Once the hidden failure is identified, it is necessary to analyse it by providing an answer to the following question "does the hidden failure have an effect on user health and safety and the environment? A "yes" answer indicates that the hidden failure have an effect on user health, safety and the environment. The effect will result either when the failure occurs or when the function is called upon. A "no" answer indicates the failure has non-health and safety and the environment hidden failure consequences, which only involve other failure consequences such as economic.

Evident failures are those failures in which the user under normal circumstances will find out about when they occur   The failure consequences of evident failure could be divided into health, safety, environment, economics, operations and appearance. The RCM decision logic for identifying failure mode consequences is shown in Figure 6.6.

This is generally where RCM decision logic structure tend to differ, because it is often necessary to tailor the structure to suit the particular sector of industry. For example, the commercial aircraft industry has to give safety a very high priority. The nuclear power and oil and gas industries have to give safety and environment a very high priority. The Military may give equal emphasis to safety as to performance and availability. Other industries such as manufacturing and construction will primarily be concerned with the cost within the health and safety legislation. Consequently, RCM decision logic structure which was developed for commercial aircraft, nuclear power, oil and gas industries, differ from those found in military standards and also from the production and construction industries.

The answer to the questions two through six in the decision logic diagram will identify the consequences of evident failure, which could be one of the following consequences, which are briefly described below.

Health consequence

This category is for failure modes whose occurrence has a direct effect on the health of the user and/or those whom their health could be affected by the failure or if it causes damage which could lead to the breach of any health regulations. If the second question yields a 'yes' the failure mode is placed in the health list. A 'no' takes RCM team to the third question.

Safety consequence

This category is for failure modes whose occurrence results in possible death or harm of a person, either operate or make use of the system and/or damage or distortion of the system. If the third question yields a 'yes' the failure mode is placed in the safety list. A 'no' takes RCM team to the fourth question.

Economic consequence

This category is for failure modes whose occurrence could have an economical significant effect due to the cost of maintenance, which is the cost of repairing the actual failure and cost of losing the revenue or production or unavailability of the system. If the cost of failure and the cost of its consequential damage is greater than the cost of preventing the failure then the considered failure mode has an economical consequence. If the fourth question yields a 'yes' the failure mode is placed in the economical consequences list. A 'no' takes RCM team to the fifth question.

Operational consequence

This category is for failure modes whose occurrence could have an effect on the operational performance of the system or serviceability of a part of the system. If the fifth question yields a 'yes' the failure mode is placed in the operational consequences list. A 'no' takes us to the last question of failure consequences decision diagram.

Appearance consequence

A failure mode has appearance consequence if its occurrence results in changing the quality of the original cosmetic of an item or a system [El-haram et al. 1997]. This consequence is for failure modes whose occurrence affects non-operational performance, but involves user satisfaction.

This type of consequence might be tolerable to some user until the next target of opportunity arises to restore the item to original appearance specification, however, in many cases, the appearance of the system or facility is one of the key functions of running a business. In this case, this category could lead to operational or economical consequences. If the answer to the final question yields a 'yes' the failure mode is placed in the appearance consequences list. A 'no' means that the considered failure mode has no consequences.

The fact that some of failure mode could hurt or effect the health and safety of the user, could lead to breach of the environment regulations, or could lead to economic, operational or non-operational effect does not necessary mean that they will do so every time they occur. Many failures occur quite often without doing so. Therefore, one of the most difficult aspects in determining the consequences is the extent to which beliefs about what is acceptable vary from one system to another and from one circumstance to another.

RCM Decision Diagram

System/subsystem/item _____

Indenture Level _____

Page No. _____ of _____

Date _____ Revision _____

Prepared By _____

Id. No	Item/Eleme	Failure Mode	Failure Consequence									Maintenance Task				Remark
			HF	EF	HC	SC	EC	OC	AC	NC	FBM	TBM	IBM	FF	RD	

Element Id. No:	Failure Mode Id. Code	Maintenance Task code	Maintenance Task Description

Failure Consequences: HF= Hidden Failure, EF= Evident Failure HC= Health, SC= Safety, EC= Economic, OC= Operational, AC= Appearance, NC= No consequence. **Maintenance Task:** FBM= Failure-Based Maintenance, TBM= Time-Based Maintenance, CBM= Condition-Based Maintenance, FF= Failure Finding and RD= Redesign.

Figure 6.5 RCM process worksheet

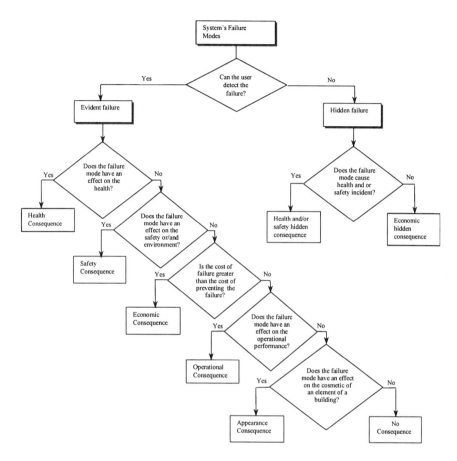

Figure 6.6. Failure Consequences: Decision Logic

Selection of Maintenance Tasks

Once the consequences of each failure mode are identified based on the decision logic shown in Figure 6.6 the second part of RCM decision process is to select the most applicable and cost-effective maintenance task or combination of tasks which prevent each failure consequence [Anderson 1990]. Generally speaking, all types of maintenance tasks are described in Chapter 5 could be applied to every failure mode of each item in the system, but only one task or combination of tasks will yield optimal results. The RCM task evaluation requires that a task meet both the applicability and effectiveness criteria to be acceptable.

Two criteria for selecting maintenance task used in RCM are:

1. *Applicability* - The applicability of the task depends on the failure characteristics of an item. Therefore, the applicable task must satisfy the requirements of the type of failure mode. These requirements are different for each type of maintenance task. After the applicable task is selected, the effectiveness of that task in preventing the failure consequences must be determined. A maintenance task is applicable in relation to consequences of failure and it should satisfy the applicability criterion. For example a preventive maintenance task is applicable if it can eliminate or avoid the failure, or at less reduce the probability of occurrence to an acceptable level.

2. *Cost- Effectiveness* - A maintenance task is effective in relation to economical consequences, which means that the task does not cost more than the failure it is intend to prevent. The effectiveness could be evaluated by balancing the cost of performing the maintenance with the cost of not performing it. The direct and indirect maintenance costs are described in Chapter5. The effectiveness criteria vary by failure consequences. Therefore, each type of task must meet the effectiveness criteria under the consequences of failure.

Effectiveness criteria for health, safety and environment consequences

For health, safety and environment failure consequences, the effectiveness criteria requires that the task reduce the probability of failure to an acceptable level. For hidden failure consequences, the task must reduce the probability of multiple failure to an acceptable level. In order to assess the effectiveness of preventive maintenance task, it is necessary to define the values of acceptable probability of failure, actual probability of failure and probability of multiple failure. If a task proves not to be cost effective, no preventive maintenance is required. However, in these cases redesign is required.

Effectiveness criteria for economic and operational consequences

For economic and operational consequences, the effectiveness criteria is cost related. For purely economic consequences, a task is effective if it costs less than the cost of the failure it prevents. For operational consequences, a task is effective if its cost is less than the combined cost of operational loss and the failure it prevents. If cost effectiveness can not be determined form evaluating the failure rate, operational consequences, repair and operating costs, an economic trade-off analysis must be performed. This analysis determines whether a task is cost effective and identifies the optimal interval at which to perform the task. If a task proves not to be cost effective, no preventive maintenance is required. However, in some cases redesign may be desirable.

The RCM team will use the decision logic process, which is shown in Figure 6.7a and b, to select the most applicable and cost-effective maintenance task or combination of tasks which will be one of the following:

Condition-based maintenance

This is an on-condition task designed to monitor the condition of an item in order to detect incipient failure modes with identifiable condition parameter(s). It is a task, which could be in a form of a scheduled inspection or examination that is designed to monitor the performance and/or the condition of an item in order to detect incipient failure. This task can vary from visual inspection to more advance inspections using a variety of condition monitoring tools. For health and safety consequences this task must reduce the risk of failure to ensure the safe use of system and its surrounding. For economic and operational this task must be cost-effective, which means, that the condition-based maintenance cost must be less than the cost of the failure.

Time-based maintenance

This is a scheduled replacement or reconditioning task in order to retain an item to satisfactory conditions before a functional failure takes place. Functional failure is the inability of any system to fulfil a function to a standard of performance, which is acceptable to the user. This task performed in accordance with a predetermined plan at regular, fixed interval. For health and safety consequences replacement at a specified age must reduce the risk of failure to ensure the safe use of system. For economic and operational consequences replacement at fixed frequency must be cost-effective, which means, that the replacement cost must be less than the cost of the failure it prevents.

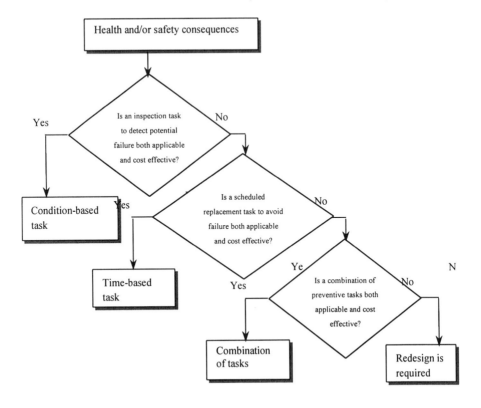

Figure 6.7a. RCM decision logic for maintenance task selection for health and safety consequences

Failure-based maintenance

This is a corrective task to restore an item following functional failure. It is reactive maintenance task performs on item which has ceased to meet an acceptable level of operational and functional requirements. This task usually takes place in ad hoc manner response to the breakdown of the item. This task is most cost-effective for failures, which have no health, safety, and economic or operational consequences. No pre-determined action is taken to prevent failure modes, which have an affect on appearance and failure modes, which effect neither health, safety, environment nor economics and operations consequences.

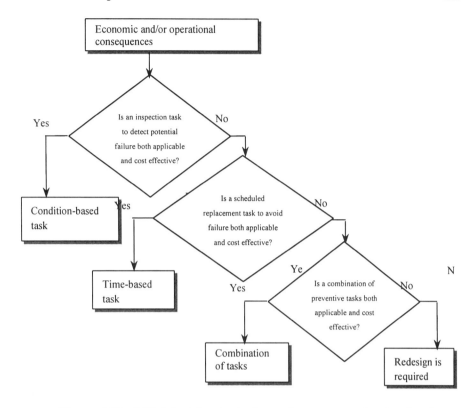

Figure 6.7b. RCM decision logic for maintenance task selection for economic and operational consequences

Failure finding task

This is a task used to locate hidden failures, which can not be otherwise detected. Its purpose is to prevent them or at least reduce the risk of the associated second failure. It is an inspection of a hidden function to identify any potential failure. Failure finding task is applicable to items, which are subject to a functional failure that, is not evident to the user.

Redesign

This is an engineering action where no applicable and cost-effective maintenance task was found. For health and safety consequences, a design change is required to eliminate the failure mode. For the economic and operational consequences, a design change may be desirable to reduce the economic losses. If the design changes is needed for reasons other than health and safety, a cost and benefits analysis is required, in order to show the expected cost saving.

6.4.3 Potential Failure and P-F Curve

In order to determine the frequency of condition-based maintenance task, it is necessary to identify the potential failure (P) and functional failure (F) points, which both define the P-F interval. The potential failure point is defined as the point where the deterioration in condition or performance can be detected. The functional failure is defined, as the point at which an item fails to perform it required function. Once these points have been defined, the interval can be determined by examining the change in the trend of operational performance or the change in the deterioration mechanisms of condition parameter. The P-F curve, which is illustrated in Figure 6.8 shows how a failure starts to deteriorate from the P point, if it is not corrected, it continues to deteriorate usually at an accelerating rate- until reaches the F point. The P-F interval can be known as the "lead time to failure" [Rausand, 1998], which is the time between potential failure and functional failure. The longer the P-F interval the more time one has to make a good decision and plan actions. Having identified the potential failure point, then two actions can take place:

1. To prevent the functional failure, depending on the nature of failure mechanism it is sometimes possible to intervene to repair the existing item before it fails completely.

2. To avoid the consequences of the failure, in most cases, detecting a potential failure does not actually prevent the item from failing, but it makes it possible to avoid or reduce the consequences of the failure.

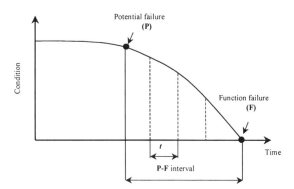

Figure 6.8 The P-F Curve

P-F curves have considerable variation in length from minutes to months or more which depends on the types of failure mechanism and the speed of deterioration. Identifying P-F curve will lead to determine the inspection or examination interval which is shown as *t* in Figure 6.8, adjust the original inspection or examination frequency, organise the logistic resources needed to correct the potential failure without disturbing operation or mission. For condition-based maintenance task to be applicable, the P-F interval and the interval *t* must be reasonably consistent.

RCM -The team

RCM is cross-discipline exercise, which requires the combined skill from several disciplines to carry out the analysis. An effective RCM implementation requires a multi-discipline team effort involving active participation from the various disciplines such as design engineers, system analyst, reliability and safety engineers, maintainability and maintenance engineers, production and process engineers, operators or users and other people with a specific expertise, for example representative from the manufacturer. The collaborative team work effort will implement the RCM analysis steps, which will identify the system functions, the system performance standards, functional requirements, the failure modes, the causes and consequences of failures, and finally the most applicable and effective maintenance tasks to mitigate failure consequences. RCM can be carried out manually, however this could be time consuming and not cost beneficial. There are a number of computer software packages available to carry out RCM. These packages vary according to their ease of use and the logic that is used.

6.5 AGE RELATED MAINTENANCE

The ideal maintenance plan would be to replace the component just before it is about to fail. This can only be done if there is a high probability of being able to detect that the component has started to fail. For a mechanical component, this requires that there is a high probability that it will be inspected between the time when a crack first becomes visible and when the component breaks and, that the inspection process will actually identify a crack if one is present. Under ideal conditions, i.e. bright new metal with no oil or dirt contamination, a crack first becomes visible, to the naked eye, when it is 0.1 mm long. Normally, unless the aircraft engine is stripped down to part level, inspection has to be done using an *intrascope* or *boroscope*, which can often only see a part of the surface and then may be at a very oblique angle. The surface being inspected is usually contaminated and the picture seen through one of these instruments is difficult to interpret. The conditions under which the inspector has to work may be anything but ideal; cold, wet, dark, windy, contorted or, even blinded by sunlight.

RCM is defined as part of the LSA exercise and, by implication, should use the data held within the LSA record (LSAR) database. This database holds just one piece of information relating to the time to failure for each failure mode of each component. This item of data is the "MTBF" - mean [operating] time between failures. The only (continuous) failure distribution that can be defined by a single parameter is the exponential distribution. The unique property of the exponential distribution is that replacing an old, but still functional, component with a new one does not improve, in any way, the probability that it will survive the next hour, day or year. To attempt to overcome this, it has been recognised (by the Department of Defence) that many components crack and that, if the crack propagation time is reasonably long, and the components are inspected sufficiently frequently, there is a high probability of detecting a crack before the component actually fails. In practice, very little is usually known about the crack propagation times, neither with respect to their duration or the amount of variance, so it is almost impossible to determine the probability of detecting a crack given a routine inspection probability. The effectiveness of inspection - the probability of detecting a crack given one is present is usually unknown.

The second deficiency is that the exercise is supposed to be done on each component in total isolation. It is assumed that when the system fails it is the result of one, and only one, component failing and when it is recovered only that component which failed is repaired or replaced. Whilst this may be true for some systems (or subsystems) such as electronic equipment, it is rarely the case for mechanical ones. Typically, for military gas turbine engines, over 50% of the modules, which comprise the engine, will be replaced (known as *opportunistic maintenance* or *on-condition maintenance*). The failure of one component can often cause significant damage to several other components within the engine. When an engine is disassembled, it becomes possible to inspect many of the components, which are otherwise inaccessible. These may be damaged, worn or corroded so will need to be repaired or replaced. Because it is expensive to remove and strip an engine, the opportunity will also be taken to replace safety-critical components, which are nearing their hard life. With aero-engines, it is quite possible for failed components to go undetected for some time, often until the engine is removed. Such failures may cause small increases in vibration, reduction in thrust or specific fuel consumption. These factors may lead to the engine being run hotter (at higher throttle settings) to achieve the required performance and hence could lead to more rapid wear/deterioration of some other components. This effect is difficult to quantify and has not been considered in this chapter.

The Department of Defence has, however recognised that when engines and/or modules are reconditioned (usually at Depot level or by the contractor), unnecessary work may be done and, parts may be replaced prematurely. The RCM process attempts to reduce this by requiring that parts which are unlikely to have failed (based, of course on MTBF!) should not be inspected for anything other than obvious damage. In particular, parts that have a protective coating should not be stripped (of that coating) unless there is evidence to suggest that the coating has been damaged or compromised. This is based on the engineering maxim "unless it's broken don't fix it" principle.

Hard life (hard time) and *soft life* are two maintenance concepts used in aircraft maintenance. *Hard life* is defined as the age of the component, at or by which the component has to be replaced. Upon achieving this age, the system or sub-system containing the given component will be rejected for subsequent recovery (by part exchange). It is, therefore, *age based preventive replacement*. This concept is already in common use with safety-critical parts such as discs, which can cause the loss of the aircraft if they burst. Associated with a hard life is usually a minimum issue life (*MISL*) which specifies how many flying hours the (safety-critical) part must have remaining for it to be re-issued - i.e. re-fitted into an engine. The purpose of the MISL is to reduce the number of unnecessary engine removals and recoveries that are expensive and, as such, is a purely economical device.

Soft life is the age of the component after which it will be rejected the next time the engine or one of its modules, containing it, is recovered (*age based opportunistic replacement*). It is effectively the same as the minimum issue life except that it can apply to any part (not just those with a hard life) and it is the age (from new) not the hours remaining to the hard life. Thus the fact that a component has exceeded its soft life would not be sufficient reason to ground the aircraft in order to remove the engine whereas this would be cause for rejection if it had exceeded its hard life.

The cost of a planned arising, one done to replace a component which has achieved its hard life, is likely to be considerably less than that of an unplanned arising. Firstly, it can be scheduled at the operator's convenience so minimising the disruption to the operation. Secondly, because the component has not actually failed, there will be no *caused or secondary damage*. Offset against this, however, is the fact that the component will have been replaced prematurely, i.e. it is likely to have lasted for a number of hours more before it actually failed. This means that, over the life of a fleet of aircraft, there could be more engine removals and recoveries than would otherwise have been the case. Given that the cost of a planned arising is less than that of an unplanned one and, that the probability of an unplanned arising can be reduced by replacing a given component before it fails, there may be an optimum age at which the given component should be replaced. If the cost of a component is relatively small, compared to the cost of a Line Replaceable Item (LRI) removal there is likely to be an optimum value for the soft life. Note that the longer the LRI lasts, between removals, the more likely the soft-lifed part will fail before the soft-life policy has had the opportunity to come into effect.

Let

$C_{u,i} =$ cost of an unplanned LRI rejection due to component i

$C_{p,i} =$ cost of a planned LRI rejection due to component i

$C_{s,i} =$ cost of replacing component i at time t

$f_i(t) =$ probability density function of time to failure (TTF) for component i

$f_i(t) =$ probability density function of TTF for LRI (excluding component i)

$T_{s,i} =$ soft life for component i

$T_{h,i} =$ hard life for component i

Using simple probability arguments, one can derive the following mathematical expressions. The expected costs associated with unplanned engine removals caused by the given component, $E(C_{i,u})$, is given by:

$$E(C_{i,u}) = H(T_{s,i})C_{i,u}, \qquad 0 < t < T_{s,i} \qquad (6.10)$$

where, $H(t)$ is the cumulative hazard function given by:

$$H(t) = \int_0^t h(x)dx = \int_0^t \frac{f(x)}{R(x)}dx = \int_0^t \frac{f(x)}{1-F(x)}dx \qquad (6.11)$$

For the case when the distribution of the times to failure for component i are given by a Weibull distribution, $W[\beta_i, \eta_i]$

$$H(t) = (\frac{t}{\eta_i})^{\beta_i} \qquad (6.12)$$

Cumulative hazard function, $H(t)$, is used here rather than the cumulative distribution function, $F(t)$, as it is assumed that a component which fails before it reaches its soft life will be repaired to a "same-as-old" state and hence can fail several times before eventually reaching its soft life. If the repair restores the component to a " same-as-new" state then $H(t)$ should be replaced by the renewal function with the cumulative distribution function of TTF given by $F(t)$.

The expected cost for the period when the component's age is greater than its soft life but less than its hard life can be derived in two parts: the first is when the LRI is rejected before the component and; the second when the component fails before the LRI. In both cases, the component would be replaced with a new one so there would be no opportunity of it failing two, or more times, within this period. The corresponding expression is given by:

$$E(C_{i,s}) = C_{s,i} \int_{T_{s,i}}^{T_{h,i}} f_l(t)(1 - F_i(t))dt$$
$$+ C_{u,i} \int_{T_{s,i}}^{T_{h,i}} (1 - F_l(t)) f_i(t)dt \qquad \text{for } T_{s,i} < t < T_{h,i} \quad (6.13)$$

In equation (613), $f_l(t)$ is the convolution of $f_j(t)$ for $j = 1, n$ and $i \neq j$ where n is the number of components which can cause an LRI failure. Similarly for $F_l(t)$. The expected cost of a planned LRI removal due to the component reaching its hard life is given by:

$$E(C_{i,p}) = R(T_{h,i})C_{p,i}, \quad \text{for } t > T_{h,i} \qquad (6.14)$$

where $R(T_{h,i})$ is the reliability function for component i. If the component reaches its hard life, the LRI is removed and the component is replaced with a new one.

Now the total expected cost of maintenance is given by:

$$E(C_i) = E(C_{i,u}) + E(C_{i,s}) + E(C_{i,p}) \qquad (6.15)$$

It will be noted that $E(C_{i,u})$ is a function of $T_{s,i}$, $E(C_{i,s})$ is a function of $T_{s,i}$ and $T_{h,i}$ and $E(C_{i,p})$ is a function of $T_{h,i}$.

If component i causes an LRI removal (fails or reaches its hard life), it will create an opportunity for the other components that have soft lifing policies, thus the costs $C_{u,i}$ and $C_{p,i}$ will depend on these other component soft lives (and, of course, vice versa). Similarly, there will be an opportunity to inspect other components for unexpected damage, wear or corrosion, that may have occurred before the component has reached its own soft or hard life and hence may avert a failure.

For safety critical components, the hard life is determined by its failure distribution(s) and is not subject to economic considerations, in the same way as non-safety-critical ones. However, the soft life, usually referred to as the minimum issue life or MISL, is based purely on economic considerations and is subject to the above analysis.

If a component has several failure modes the whole process gets somewhat more complicated. It is possible that if the component fails due to failure mode i it will be repaired to the same-as-old whereas if the cause was failure mode j ($i \neq j$) then the component is replaced or restored to same-as-new. The component, which failed due to mode i could fail again for the same reason or fail due to a different (competing) failure mode. The amount of damage (to other components) may also be significantly different if the component fails in different ways. As an example of this, a blade can, melt, if its cooling holes become blocked or, it can break off at its root. If it melts, the amount of damage to other components is minimal but if it breaks, the damage can be extensive.

A further complication is that different soft lives (and MISLs) may be applied at different echelons in the maintenance environment. Typically, components held (in storage) at the deeper echelons (3rd and 4th lines) will be required to have potentially more life remaining (before causing a planned LRI removal) than those which are held at 2nd line. A typical 2nd line MISL might be 100 hours whilst the 3rd or 4th line MISL might be 400 hours. This is generally due to the fact that the 2nd line MISL would normally only apply to modules which have not had to be recovered, i.e. they have simply been removed for access to other modules which have had to be recovered. Modules held at 3rd or 4th line would normally only be there if they have been rejected and hence would have needed to be recovered. To put it another way, if a module contains rejected components and hence has to be stripped and re-built then the marginal cost of replacing a hard or soft lifed component is relatively low compared to the case when the module is rejected purely to replace such a component.

6.5.1 Age related maintenance - Case Study on Aircraft Engine

A simulation program was coded to consider a very simple case in which the LRI (Engine) arisings are modelled by an MTBF and just one part is considered for soft/hard life optimisation. The MTBF for the engine should be adjusted to exclude the failures resulting from this part which is modelled using a Weibull distribution. No attempt has been made to model either the maintenance or supply activities - recovery of the LRI is instantaneous and the component in question is also instantly replaced by a new one every time it is rejected (for whatever reason). Suppose the MTBF for an engine is 1000 Flying Hours excluding component 1 which has a Weibull time to failure distribution with scale parameter 5000 hours and shape parameter 3 (i.e. W[3,5000]).

Let the cost of an engine failure (and recovery) due to:

an unplanned failure of component 1 $(C_{u,i})$ = 100,

a planned rejection of component 1 $(C_{p,i})$ = 50

Let the cost of replacing Part 1 (soft-lifed) $(C_{s,1})$ = 10

[Note: the (expected) cost of an "unplanned failure" would include the cost of repairing the failed component and any others that were either secondary or found damaged or that had exceeded their soft lives or MISL. The (expected) cost of a "planned rejection" would include the cost of replacing the component and any others that were found damaged or that had exceeded their soft lives or MISL. It would not include secondary damage because the component had not failed so could not have caused any]

Using Monte-Carlo simulation with 1000 replications of one engine flying 10,000 Flying hours (FHrs) the following 3 graphs were produced. The Figure 6.9 shows how the costs vary with hard and soft life. This indicates that there is no benefit in setting a hard life - all curves are decreasing monotonically (allowing for random variations) as the hard life increases.

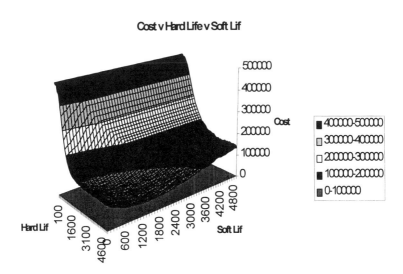

Figure 6.9 Cost Vs Hard life Vs Soft life

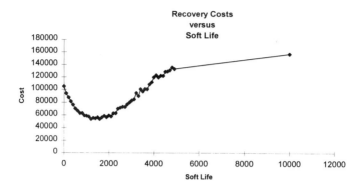

Figure 6.10 Recovery cost Vs soft life

The Figure 6.10 shows how the recovery cost varies as the soft life increases for an infinite hard life. It appears to become asymptotic to a value of approximately 157,000. Due to run times and the fact there was considered to be little benefit, soft lives between 4900 and 10000 were not simulated (hence the straight line).

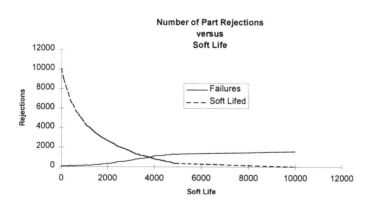

Figure 6.11 Number of unplanned engine rejections versus soft life

The Figure 6.11 shows how the numbers of unplanned engine rejections due to component 1 "Failures" and the number of soft-lifed removals of component 1 "Soft Lifed" vary with the soft life.

It should be noted that if the engine fails for reasons other than the failure of component 1, it has been assumed that component 1 is not replaced unless it has exceeded its soft life. In practice, there is a certain probability that the part would be damaged as a result of the primary cause of failure. There is also a certain probability that the component will be found damaged during inspection while the engine is being stripped. This may actually be age-related, unfortunately, we have no data to be able to test this hypothesis.

In this particular exercise, the failed engine was recovered instantly and continued to operate until it either failed again or achieved 10,000 FHrs. No attempt was made to model spares or the recovery procedure. In practice, when an engine fails, it is replaced by a spare (as soon as one becomes available). The failed engine is stripped to its modules. The rejected modules are replaced with spares and are sent for part exchange. The fact that the parts will not need to be inspected, which often involves removing their coatings, the use of dye-penetration and being re-coated (if found satisfactory) all by relatively skilled personnel, means additional savings may be made. Strictly, this only applies to parts that have exceeded their soft life; as there is no point in inspecting a part if is going to be replaced regardless. However, if the part has not yet reached its soft life, which from the graphs is around 1/3 the expected life (1500 versus 4500) there is unlikely to be any signs of sub-surface damage.

6.6 TOTAL PRODUCTIVE MAINTENANCE

The demand for high quality products at lower costs is driving manufacturers to shift focus toward equipment management programs. Leading the way is the Japanese theory known as *Total Productive Maintenance*, *TPM*, which is a proactive equipment maintenance strategy designed to improve Overall Equipment Effectiveness, OEE. TPM is an integrated approach to maintenance and production which is developed and introduced by Japan Institute of Plant Maintenance during 1970's. Nakajima (1986) defined TPM as: *productive maintenance carried out by all employees participating through small group activities*. According to Nakajima the concept of TPM includes the following elements:

1. TPM aims to maximise overall equipment effectiveness by eliminating the major six equipment losses, which are equipment failure, set-up and adjustment, idling and minor stoppages, reduced speed, process defects and reduced yield.
2. TPM is implemented by various departments of a company such as maintenance, operation and production.
3. TPM involves every all employees, from top management to the shop floor workers.

4. TPM is based on motivation management through involvement of small-group activities.

Thus, TPM introduce measures to maximise the overall equipment effectiveness which is a function of the equipment availability, its performance efficiency, and the corresponding quality rate taking into consideration the equipment losses. The OEE, is given as [Nakajima, (1989)]:

$$OEE = \text{Avaliability} \times \text{Performance efficiency} \times \text{Quality rate} \quad (6.16)$$

where; availability can be expressed as a ratio of actual operating time to loading time. Thus,

$$\text{Availability} = \frac{\text{Loading time} - \text{Downtime}}{\text{Loading time}} \qquad (6.17)$$

where; Loading time is the planned time available per a period of time say day or month for production operations, and downtime is the total time that the plant or part of the plant is not operating due to equipment failure or/and set-up and adjustment requirements.

Performance efficiency can be expressed as the product of operating speed rate to net operating rate, thus.

$$\text{Performance Efficiency} = \text{Net operating time} \times \text{Operating speed rate}$$

The operating speed rate refers to the discrepancy between the ideal speed (based on equipment capacity as designed) and its actual speed.

$$\text{Operating speed rate} = \frac{\text{Theoretical cycle time}}{\text{Actual cycle time}}$$

The net operating rate measures the maintenance of a given speed over a given period. This calculates losses resulting from minor recorded stoppages, as well as those that go unrecorded on the daily log sheets.

$$\text{Net operating time} = \frac{\text{Actual processing time}}{\text{Operating time}}$$

$$= \frac{\text{Processed amount} \times \text{Actual cycle time}}{\text{Operating time}} \tag{6.18}$$

where; processed amount presents the number of items processed per a given period of time (day or month), and operating time is difference between loading time and downtime. Thus

$$\text{Performance Efficiency} = \text{Net operating time} \times \text{Operating speed rate}$$

$$= \frac{\text{Processed amount} \times \text{Actual cycle time}}{\text{Operating time}} \times \frac{\text{Theoretical cycle time}}{\text{Actual cycle time}}$$

$$= \frac{\text{Processed amount} \times \text{Actual cycle time} \times \text{Theoretical cycle time}}{\text{Operating time}}$$

Quality rate can be expressed as a ratio of non-defect amount produced to total amount produced over a given period. Thus,

$$\text{Quality rate} = \frac{\text{Processed amount} - \text{Defect amount}}{\text{Processed amount}} \tag{6.19}$$

where; the defect amount refers to the number of items rejected due to quality defects of one type or another. TPM is therefore a philosophy aim to maximise OEE through the optimisation of equipment availability, performance efficiency and quality rate.

TPM Achievement

The main contribution made by TPM to maintenance is that it destroys the barrier between the maintenance department and production department within a company (Williams et al, 1994). This means, that the operators have been given a new role, which is not only to operate equipment, but to monitor the condition of the equipment and prevent it from breaking down. It also encourages the operator, who is idle to provide first-line maintenance, to perform simple maintenance tasks. In many industries, the OEE ratio for equipment and processes is currently running at 50% to 60%; TPM can effect improvements to the level to 80% or 90% [Willmott, (1989)]. TPM has made excellent progress in many areas. This include:

- measure and eliminate much of the non-productive time
- measure and eliminate specific equipment performance problems and provide specific tools to use to improve equipment performance
- improve teamwork and less adversarial approach between production and maintenance
- help operators and maintenance staff to understand how they can improve the efficiency of the equipment with which they work.
- improve work areas around the equipment
- aim at zero defects and zero failures.

Noting the above benefits from the application of TPM to the Japanese manufacturing, many companies in USA, Europe and Asia are being active implementing of TPM.

6.7 COMPUTERISED MAINTENANCE MANAGEMENT SYSTEM

It is generally accepted that the main functions of management are planning, organising, staffing, leading and controlling. These functions therefore also apply to the maintenance management. Over the recent decades, the maintenance management, perhaps more so than many other management disciplines, has undergone significant change. Maintenance management refers to the application of the appropriate planning organisation and staffing programme implementation and control methods to maintenance task and its activities. Maintaining systems involve the collection of large amount of data and information to record historical system performance, identify spares, etc. Development of information systems for improving maintenance management has over the years focused upon improved means for optimising maintenance. Improving the task planning, scheduling and execution can enhance the effectiveness of maintenance work. This may be achieved by integrating control across all the maintenance tasks and by improving the control over work location, issue, execution and reporting.

The Key to maintenance work control is information in workload, on the available resources and plant running conditions. This information is dynamic, altering continuously due to ongoing changes in the production requirement, plant performance and work force availability. It is therefore, difficult and labour intensive to mange this information manually. The dynamic nature of the situation requires a dynamic response in scheduling, allocation, issue and feedback [Paulsen et al. (1991)]. Historically this information has been held in paper based records. Large amounts of paper based records can become difficult and expensive to store and analyse, errors and omissions can then easily take place.

The development of the Personnel Computer has been followed by an increasing choice of Computerised Maintenance Management Systems (CMMS) which replaced the paper-based records. The objectives of CMMS are to provide timely, accurate information that will assist management in planning, organising, staffing, budgeting and controlling for maintenance. It also provides a systematic, automated procedure for standardising maintenance of a system. The structure of CMMS is shown in Figure 6.12. CMMS generally includes elements such as:

- Work orders system
- Maintenance task selection (FBM, TBM, and CBM)
- Maintenance resources (manpower management, tools and facility management)
- Spare parts inventory management (Purchasing functions such as ordering, requisition of materials, etc.)
- Data and information management (e.g. equipment history)
- Finance and budgeting system
- Reporting and documentation

Most of the above mentioned elements of CMMS offer the important modules that maintenance managers are seeking to assist with the effective and efficient maintenance management activities. CMMS is designed to help maintenance departments reduce costly downtime, control expenses, increase maintenance staff productivity, track spare parts inventories and costs, effectively deploy available personnel and support equipment, improve the efficiency of purchasing parts and maintain data required for reporting and control.

There are many maintenance management systems available from various vendors. These range from the simple work planning and control systems to very comprehensive systems with on-line, real time, multi-access, relational database computer systems which can run on stand alone or main frame systems. Many companies now own and use a CMMS to good effect with, in many cases, substantial improvements in maintenance department efficiency as results. CMMS will administer a preventive maintenance programme, stock inventory etc more efficiently, more reliably and more cheaply than a comparable manual system. Companies that have implemented more advanced maintenance systems have on average achieved a return on investment of 11 times the programme cost. Maintenance costs having reduced up to 27%, productivity gains up by 21%, unscheduled downtime reduced by 40% and a 74% reduction in system breakdown has been achieved.

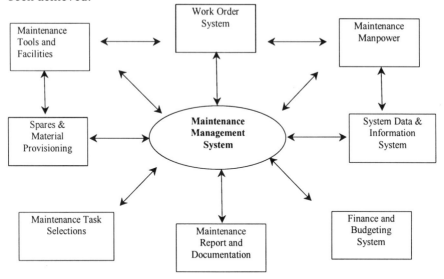

Figure 6.12 Structure of Maintenance Management System

Benefits of CMMS

CMMS can improve the effectiveness of maintenance management by prioritising day-to-day maintenance activities and maintenance scheduling, procurement and material management along with integration of these functions with human resources planning and financial management solution. CMMS could contribute to cost saving through cost effective management of system and maintenance resources, optimisation of workflow, improving of integration between various application such as accounting, planning and document management systems and finally it could provide information, which is presented in a comprehensive manner, to maintenance management to make the right decisions at the right time.

Chapter 7

Supportability and Logistics

Think of the end before the beginning

Leonardo da Vinci

Very few systems remain functional throughout their life: aircraft, buses, cars, ships require repair, replacement; manufacturing plants require supplies of raw materials maintenance and replacement of worn tools. When any part of the system changes from state of functioning (*SoFu*) to a state of failure (*SoFa*), the system loses a certain amount of functionality. Restoration of the system's functionality is invariably achieved through maintenance. And, all maintenance activities require support from facilities, equipment and resources. System failures may be anything from inconvenient to downright dangerous. Running out of fuel, in a car, may mean a long walk, in an aircraft, you will be exceptionally lucky if you can walk. Knowing in advance when a system will require maintenance can save embarrassment and even lives but, above all, it can save money. From relatively simple devices such as fuel gauges, oil pressure warning lights, magnetic oil filter plugs through to highly complex equipment such as engine health and usage monitoring systems (*HUMS*), these all play their part in allowing the operator to decide the best time to perform preventive maintenance.

Reliability can predict (with varying levels of confidence) when a system will enter a state of failure. Having done so, maintainability will predict how long the maintenance tasks, to recover the system, will take. But, supportability will determine whether it is worth recovering and, if so, where it should be done and what will be needed to do it. *Supportability engineering is concerned with designing the system so it can be supported at minimum (life cycle) cost.* This will need to take into account what facilities, equipment and resources will be required to enable the system to be supported in the most cost effective manner. Support or logistics is the process of determining what (facilities, equipment and resources) will be needed when and where and making sure these requirements are satisfied to ensure that the maintenance task can proceed with minimal delay. Nowadays supportability engineering plays a leading role in the life cycle considerations of a product because it is recognised as making a considerable contribution towards the shape of the functionality profile and, as a consequence, the operational cost. Figure 7.1 illustrates where these fit within system operational effectiveness.

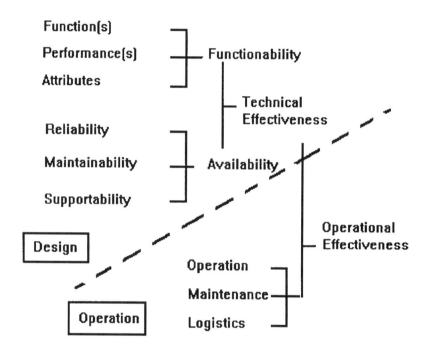

Figure 7.1 System operational effectiveness

The most appropriate time to perform supportability analysis is at the beginning of the life cycle, i.e. the early design stage. At the early design stage there is a wider choice of option for selecting the best support alternative and also can make full use of the existing resources. A change at a later stage of the development cycle might be very expensive. A good example for successful consideration of supportability comes from the Boeing Company, which decided to design the cockpits of their latest aircraft to have the same design and layout. Thus, once a pilot was trained to fly one of them he or she would not have to visit a flight simulator for any of the other similar planes. This makes great savings for the customer as the pilots will be available more often, and also reduce the number of flight simulators that would have to be purchased or hired (M Turner, 1999).

Boeing showed supportability considerations quite clearly from the development of their new 777 aircraft. The Boeing 777 is the largest aircraft and has 8 doors, it was decided where possible to make them all common. Boeing achieved about 95% commonality, which will reduce the number of different spares to be held by the customers.

7.1 SUPPORTABILITY – TERMS AND DEFINITIONS

In this section we introduce various terms and definitions used in supportability engineering.

7.1.1 Supportability

Knezevic (1993) gives the following definition of supportability:

Supportability is the inherent characteristics of an item related to its ability to be supported by the required resources for the execution of the specified maintenance task.

The first important point in the definition of supportability is that *supportability* is inherent, i.e. it is a consequence of design whether deliberate or accidental. It is basically how well the item has been designed for support. In order to explain the physical meaning of supportability; let us establish the link between the maintenance process and the additional length of time during which the item is in *SoFa*. Thus, supportability can be graphically presented as shown by Figure 7.2, where *T* represents the instant of time when the required support resources have been made available and the specified maintenance task can be performed.

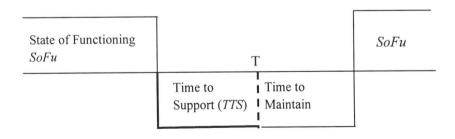

Figure 7.2 Concept of supportability

The additional time spent in SoFa, due to the performance of the support task, that is, the time to support, TTS, is a random variable.

The randomness in support time is caused due to the factors listed below:

Maintenance factors: these are related to the management of the maintenance process, in particular its concept, policy and strategy.

Location factors: the influence of the geographical location of the items, communication systems, or transport.

Investment factors: these influence the provision of support resources (spares, tools, equipment, and facilities).

Organisational factors: these determine the flow of information and support elements.

Thus, the time-to-support random variable depends on the above mentioned factors, that is:

$TTS = f$ (maintenance, location, investment, organisational factors)

Taking into account the analysis performed so far, it could be concluded that *TTS* has an unpredictable nature, being the result of the variability and complexity of all the influential factors to the restoration process, together with the provision of support resources. It is therefore reasonable to say that it is impossible to give a deterministic answer regarding the additional length of time during which any specific item will spend in the state of failure. It is only possible to assign a probability that it will happen at a given instant of time, or that a certain percentage of trials will, or will not, be completed during a specific time interval.

7.1.2 Supportability Engineering

Supportability engineering can be defined as (Knezevic, 1993):

A scientific discipline which studies the processes, activities and factors related to the support of a product with required resources for the execution of specified operation and maintenance tasks, and works out methods for their quantification, assessment, prediction and improvement.

Military and aerospace companies have recognised the importance of information regarding the supportability of their product.

7.1.3 Logistic Delay Time (LDT) or Time to Support (TTS)

The *time to support or logistic delay time* is defined as the time taken to restore a system excluding the time taken to perform the maintenance tasks. Essentially, it is the time spent waiting for facilities, equipment, manpower and spares. In practice, this time may be made up of number of elements as the system recovery may require several maintenance tasks, each of which possibly requiring different facilities, equipment and resources.

7.1.4 Support Resources

The resources needed for the successful completion of every operation and maintenance task, could be grouped into the following categories:

- Supply Support
- Test and Support Equipment
- Transportation and Handling
- Personnel and Training
- Facilities
- Data
- Computer Resources

Each of these categories identified, are briefly described below.

Supply Support

Supply support is the generic name, which includes all spares, repair parts, consumables, special supplies, and related inventories needed to support the operation and maintenance processes. Considerations include each operation and maintenance task and each geographical location where spare/repair parts are distributed and stocked; spares demand rates and inventory levels; the distances between stocking points; procurement lead times; and the methods of material distribution. Supply support factors will largely be determined by the maintenance policy which addresses such as depth to which maintenance will be carried out, where this will be done and what level of system availability is to be achieved (Walsh, 1999).

Test and Support Equipment

Any equipment that is required to support operation and maintenance tasks can be classified as support equipment. This category includes all tools, special condition monitoring equipment, diagnostic and checkout equipment, metrology and calibration equipment, maintenance stands, and servicing and handling equipment required supporting scheduled and unscheduled maintenance actions associated with the system or product. Most maintenance tasks will require certain types of equipment. These may include hoists, cranes, general purpose tools (e.g. hammers, screw drivers, spanners) special tools (e.g. jigs, plug spanners, valve-spring compressors, star-headed screwdrivers). Test and support equipment may be classified as "peculiar" (newly designed and/or off-the-shelf items peculiar to the system under development) or "common" (existing items already in the inventory). M Turner (1999) mentions that in the 1960's the United States Department of Defence discovered that millions of dollars had been spent on various support equipment that was not required. This became the catalyst for the beginnings of Logistic Support Analysis (LSA). Through such analysis, the inclusion of every piece of support equipment in the inventory has to be justified.

Transportation and Handling

This element of support includes all provision, containers (reusable and disposable), and supplies necessary to support packaging, preservation, storage, handling, and/or transportation of system, test and support equipment, spares and repair parts, personnel, technical data, and mobile facilities. In essence, this category basically covers the initial distribution of products and the transportation of personnel and materials for operation and maintenance purposes.

In some cases the failed unit will be already at the facility, in others it will need to be moved. If the aircraft has lost all of its power, a tug will be needed to move it off the runway/taxiway to the hanger. Similarly, if the ship's engine have failed then tugs will be required to tow it to a safe haven, harbour or dry dock. Aircraft engines are highly prone to salt water corrosion. If they are likely to be carried by ship then they will need to be protected from sea spray. They are also quite delicate, in so far as the external pipework can easily be damaged if knocked. To overcome these problems, special containers have been designed for some engines (e.g. Rolls Royce EJ200). If designed correctly, this also has the advantage that it allows stacking and handling by standard dockyard equipment.

Personnel and Training

The Personnel required for the installation, checkout, operation, handling, and sustaining the maintenance of the system (or product) and it's associated test and support equipment are included in this category. Maintenance personnel required for each operation and maintenance are considered. Personnel requirements are identified in terms of quantity and skill levels for each operation and maintenance function by level and geographical location. Formal training includes both *initial* training for system/product familiarisation and *replenishment* training to cover attrition and replacement personnel. Training is designed to upgrade assigned personnel to the skill levels defined for the system. Training data and equipment (e.g. simulators, mock-ups, and special devices) are developed as required, to support personnel training operations.

At the heart of every maintenance task is the mechanic. This person will have certain skills but may need special ones for certain tasks. They will need training both general and specific. For example, in the British armed forces, three skill levels are identified. Ideally, all the tasks should be designed to be within the capability of the lowest level mechanic to allow maximum flexibility, however, this is not always possible. The use of boroscope, intrascope or endoscopes allows the opportunity to look inside an aircraft engine or wherever, but it does require skilled personnel (inspectors/mechanics) to interpret the pictures. Use of video could play important role in training personnel.

Facilities

This category refers to all special facilities needed for completion of operation and maintenance tasks. Physical plant, real estate, portable buildings, housing, intermediate maintenance ships, calibration laboratories, and special depot repair and overhaul facilities must be considered. Once the failure has been registered the first maintenance task can start, however, certain resources will be needed. Firstly, there will be a need for somewhere to do the work, i.e. a maintenance facility. A facility is a physical location where maintenance activities can be performed. Specifically, it is a location which protects both the system and the maintainers from whatever elements (e.g. wind, sun, rain, snow, sea, sand, nuclear, biological and chemical contamination, dust or smoke) are considered likely to be detrimental. Capital equipment and utilities (heat, power, energy requirements, environmental controls, communications, etc.) are generally included as part of the facilities. Often, the first level of maintenance, (e.g. removing an aircraft engine, radar set or car wheel) can be done in the open, by the side of runway or motorway, so no facilities, as such, are required.

Technical Data

Technical data includes all the documented technical procedures in either electronic or hard copy, system installation and checkout procedures, operation and maintenance instructions, inspection and calibration procedures, overhaul procedures, modification instructions, facilities information, drawings, and specifications that are necessary in the performance of system operation and maintenance functions are included herein. Such data not only covers the system but also the test and support equipment, transportation and handling equipment, training equipment, and facilities.

Computer Resources

This facet of support refers to all computer equipment and accessories, software, program tapes/disks, data bases, and so on, necessary for the performance of system operation and maintenance functions. This includes both condition monitoring and maintenance diagnostic aids.

7.1.5 Arising

An *arising* is any non-trivial event that causes the state of the system to change from functioning to failure. An arising may be routine or non-routine, planned or unplanned, predictable or unpredictable, age-related or non-age related.

Planned Arising

Planned arising refers to an event when an item with age related failure is replaced to avoid any deterioration in system characteristics. Turbine discs in aircraft engines are given 'predicted safe cycle lives', which defines the maximum number of stress cycles a disc can be subjected to before it has to be replaced. Typically, this is about a 1/4 of its expected time to failure (MTTF). By knowing the age at which the disc has to be replaced, it should be possible to predict, with a reasonable level of confidence when the engine will be needed to be removed so resources can be made available in a timely manner.

Unplanned Arising

Components may fail due to non-age related external factors such as a stone being thrown up from the road and cracking the windscreen, a nail puncturing a tyre, a bird being ingested by an engine etc. These events might result in an unplanned arising. Age-related failures also can result in unplanned arising, however this can be reduced by planned maintenance.

On Condition Arising

A lot of maintenance actions are done as a result of inspection/examination which may, may not, be routine. Some of these will cause the system to enter a state of failure whilst many are likely to be identified whilst the system is already in a state of failure. The former becomes unplanned arisings the latter come under the category of opportunistic maintenance.

7.2 SUPPORTABILITY MEASURES

The support task, whose main objective is provision of the support resources required for the performance of the specified maintenance task, can be considered as a random variable, called *Duration of Support Task, DST,* as (*or time to support, TTS*). Since it is readily accepted that a population of supposedly identical items experience states of failure for different lengths of time, it follows that the ability of the system to be supported can only be described in probabilistic terms. Hence, supportability is fully defined by the random variable *DST* and its probability distribution.

The most frequently used supportability measures are:

1. Supportability function,

2. DST_P Time,

3 Expected time to support,

A brief definition and description of these characteristics follows.

Supportability Function

The cumulative distribution function of the random variable, *DST*, which represents the probability that it will have a value equal to or less than some particular value, say a, $F(a) = P(X \leq a)$, is called the *supportability function*. At any instant of time *t* the supportability function presents the probability that the required support resources will be provided before or at the specified instant of time, *t*, thus:

$S(t)$ = P [support resources will be provided before time t]

$$S(t) = \int_0^t s(t)dt \qquad (7.1)$$

where, *s(t)* is the probability density function of support process.

DST$_p$ Time

This is the length of time by which required support resources will be provided for a given percentage of demands. DST_p time can be mathematically represented as:

$$DST_p = t \rightarrow for\ which\ S(t) = \int_0^t s(t)dt = p \qquad (7.2)$$

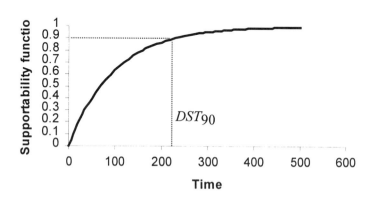

Figure 7.3. *DSTp* Time

The most frequently used *DSTp* is *DST90*, which presents the length of time during which 90% of support tasks will be completed, as shown in Figure 7.3.

$$DST_p = t \rightarrow \text{ for which } S(t) = \int_0^t s(t)dt = p = 0.90$$

Expected time to support

The expected value of the random variable DST can be used as another measure of supportability, thus:

$$E(DST) = \int_0^\infty t \times s(t) \; dt \qquad (7.3)$$

This characteristic is also known as Mean Time to Support, *MTTS*.

Example 7.1

Assume that the duration of support task for restoring a weather radar follows Weibull distribution with scale parameter $\eta = 24$ hours and $\beta = 2.7$.

1. What is the probability of providing support resources within the first 18 hours?
2. What is the length of support time by which required resources will be provided in 90% of the cases?
3. What is the Mean Time to Support?

SOLUTION

1. P (DST ≤18) = S(18)

For Weibull distribution,

$$S(18) = 1 - \exp(-(\frac{18}{24})^{2.7}) = 0.3686$$

That is, less than 40% will be satisfied within 18 hours.

2. *S(t)* = 0.9

That is, $1 - \exp(-(\frac{t}{24})^{2.7}) = 0.9 \Rightarrow t = 24 \times \{\ln[1/0.1]\}^{1/2.7} = 32.68$

That is, to be 90% confident one should allow nearly 33 hours.

3. MTTS for Weibull distribution is given by (using Gamma Table, or using the function EXP(GAMMALN(1+1/β)) in EXCEL)

$$MTTS = \eta \times \Gamma(1 + 1/B) = 24 \times 0.889 = 21.33 \text{ hours}$$

7.3 RECONDITION

Whereas repair may be considered to be minimum needed to restore a component to a state of functioning, *reconditioning,* is defined as that which is necessary to restore the system to an as-good-as-new. Short of replacing a component with a new one, reconditioning will seldom restore it to a truly as-good-as-new state.

Typically, reconditioning is applied to assemblies rather than to piece parts. In this case, the restoration is usually achieved by replacing rejected parts with new ones. A possible exception would be if the windscreen of a car has been broken in an accident. In this case, an undamaged windscreen taken from a similar car that has been scrapped will have very nearly the same life expectancy as a new one. This is because the vast majority of the failures of windscreens are unrelated to its age.

Aircraft engines are usually recovered by module exchange. The engine is disassembled to a depth necessary to access (and replace) any rejected modules. The rejected modules are sent to a workshop for recovery (by part exchange). In the meantime, the engine will be rebuilt with spare modules. These will either be new or as a fleet of aircraft ages, with modules previously recovered by part exchange. Unless new ones replace all of the modules, the engine cannot really considered as reconditioned (according to the definition above). Similarly, unless all the parts have been replaced with new ones in any given module, that module has also not been reconditioned. In practice, the term recondition is used to define a module recovery in which one, or more, of the lifed parts has been replaced because they have been due for replacement. The implication is that none of the other parts have age-related failures, or at least, these failures are not the primary causes of rejection. In practice this is often not the case.

7.4 OBSOLESCENCE

Obsolescence plays a crucial role in logistic support, specially, when it comes to forecasting spare parts requirement. Any overstock with potential obsolescence would result in heavy loss of capital investment. In the early days of television, the broadcasting was 405 lines in black and white, today it is 625 in colour, tomorrow it will be digital (already sold in the market) and one day it will probably be in 3-D. The original phonographs used cylinders as their recording medium. These were later replaced with circular disks. The early version operated at a speed of 78 rpm and discs were generally 10 inches in diameter and normally played about 2 minutes per side. Later the speed was reduced to 45 rpm and the size to 7 inches in diameter but the duration remained approximately same. Next extended play (EP) and long play (LP) records were introduced. The former squeezed more grooves onto the same 7 inch 45 rpm disc to give up to 10 minutes per side. The LP however, increased the size to 12 inches (with some 10 inches) and reduced the speed to 33 1/3 rpm to give around 30 minutes per side. At the same time the record player has been developing, tape recorders, cassette players and now compact disc (CD) players have also been developed. A major disadvantage of the record player was that it was bulky and very sensitive to movement. The discs were also susceptible to damage (scratching and warping). The tape recorder had the major advantage that users could not only play pre-recorded tapes but could make their own recordings from radio, other tapes, records and of their own sounds. With the advent of the transistor, it became possible to produce easily transportable tape recorders, indeed, today's players will fit easily into a pocket. Next came the compact disc that uses laser technology and record digitally. The discs are approximately 5 inches diameter (although smaller ones are produced), single sided and will hold up to 75 minutes of music.

These are just some examples. At the turn of the 20th century no heavier than air machine had flown. At the start of World War II, no one in the British government, at least, could see any benefits in developing the jet engine, now almost every aircraft is powered by them. At the end of the same war, it was thought the market for computers would be two or three per country, now millions of people use them everyday. In 1970, the British Steel Corporation owned one of the first 'portable' computers, it filled the whole of a ten ton lorry and had a tiny fraction of the memory power of one of today's computers that will fit into the palm of the hand.

At a somewhat more mundane level, it has been said no two RB199 engines (which power the Panavia Tornado) were built the same due to the speed of developments and modifications to parts. This is perhaps not too critical during the production phase but, when it comes to deciding how many spares to hold, it becomes a major consideration.

7.5 LEVEL OF REPAIR ANALYSIS

Level of Repair Analysis, *LORA*, is a systematic procedure to determine the cost of alternative maintenance options and maintenance levels, taking into account spare parts, support equipment and manpower cost [Blanchard, 1992]. *LORA* determines the alternative maintenance tasks at different maintenance levels. Maintenance task complexity, manpower skills-level requirements, frequency of occurrence, special facility needs, economic criteria etc., dictate to a great extent the selection of specific repair task to be accomplished at each level. Each option reflects the characteristics of system design which is evaluated in terms of effectiveness criteria such as availability, reliability, maintainability, supportability, Life cycle costing etc. The maximum benefits in implementing *LORA* is obtained by performing it at the early stages of system design and use the analysis to change the design accordingly, to prepare maintenance plans and to determine logistics resources allocations.

Level of repair analysis technique can be used to decide whether an item should be repaired or discarded and if repaired, to find the location where repair or discard will be performed. Whenever a system fails, the faulty *LRU* is isolated and replaced with a spare *LRU* if available. The removed *LRU* may be discarded or repaired. If it is decided to discard a LRU, then all the *SRUs* within the *LRU* are also discarded. If the *LRU* is repaired, each *LRU* or *SRU* may be repaired at intermediate or depot level. The repair of *LRU* is carried out by replacing or repairing the consisting *SRUs*. Maintenance or repair levels in *LORA* are determined for each item to be organisational level (1st line), intermediate level (2nd line), depot level (3rd line) or 4th line (contractor). Obviously, each of these decisions has a different economic impact. LORA attempts to find the best combination of repair/discard decisions and the maintenance level that minimise the total support cost. For example, if an item is recommended to be repaired by the manufacturer, then there is no need to procure logistic support resources such as test and support equipment, facilities etc. However if the repair is to be carried out by the operator, then the operator needs to procure all the support resources required.

The following optimisation model is a refinement of the model presented by *Barros* (1996), by considering the impact of expected number of failures on *LORA* optimisation.

The optimal repair/discard decisions for a system can be derived using 0-1 non-linear programming formulations. Let,

N   =   The total number of indenture level for the system under consideration.

n_j =   Number of components at the indenture level j.

K   =   Total number of repair options.

$M_i(T)$ =   Expected number of failures for item i during the cumulative
Life (T) of the fleet.

$FC_{r,i,j}$ =   Fixed cost associated with repair option 'r' for item i at the indenture level j.

$VC_{r,i,j}$ =   Variable cost associated with repair option 'r' for item i at the Indenture level j.

$$z_{r,i,j} = \begin{cases} 1, & \text{if repair option } r \text{ is selected for item } i \text{ at} \\ & \text{indenture level } j \\ 0, & \text{otherwise} \end{cases}$$

$$y_{r,i,j} = \begin{cases} 1, & \text{if repair option } r \text{ is for item } i \text{ at} \\ & \text{indenture level } j \text{ requires fixed cost} \\ 0, & \text{otherwise} \end{cases}$$

The objective function for the optimisation model is given by:

Minimise:

$$\sum_{j=1}^{N} \sum_{r=1}^{k} \sum_{i=1}^{n_j} z_{r,i,j} [VC_{r,i,j} \times M_i(T) + FC_{r,i,j} y_{r,i,j}] \tag{7.4}$$

Subject to the constraints:

$$\sum_{r=1}^{k} z_{r,i,j} = 1, \, i = 1, 2, \ldots, n_j \, ; j = 1, 2, \ldots, N \tag{7.5}$$

For $i = 1, 2, ..., n_j$, $j = 1, 2, ..., N$, $r = 1, 2, ..., k$

$$z_{r,i,j} \leq z_{r,i,j-1} \tag{7.6}$$

$$z_{r,i,j} \geq z_{r,i,j-1} \tag{7.7}$$

$$z_{r,i,j} = 0,1 \tag{7.8}$$

The objective function (7.4) tries to minimise the support cost associated with different repair options over the expected life of the system. Constraint (7.5) makes sure that only one repair option is selected for each component in the system. Constraint (7.6) and (7.7) controls the decisions at the higher and lower indenture levels. That is, if an LRU is discarded upon failure, then all the consisting SRUs are discarded. If an SRU is repaired at lower maintenance level, then the LRU in which the SRU is enclosed is also subject to repair. The above optimisation problem can be carried out by variety of special purpose software or using general purpose software such as *SOLVER of EXCEL*.

7.6 TESTABILITY, INSPECTION AND DIAGNOSTICS

Good troubleshooting is nothing more than good deductive reasoning. At the centre of that reasoning is a careful collection and evaluation of physical evidence. Unfortunately, many aircraft devices use computer chips to provide a function formerly fulfilled by substantial mechanical parts or subsystems. Consequently, troubleshooting in the traditional sense of searching for physical evidence of failure is hindered. You can't troubleshoot a computer chip by looking for physical evidence of failure. A broken chip does not look any different than a healthy one. Although it can be argued that broken chips occasionally make smoke, evidence of malfunction is seldom readily apparent. Broken chips do not leak, vibrate, or make noise. Bad software within them does not leave puddles or stains as evidence of its misbehaviour. Ones and zeros falling off the end of connector pin are difficult to see.

(J Hessburg, Chief Mechanic New Airlines, Boeing)

Testability is the characteristics of a system that governs the extent to which the system supports fault detection and, once found, fault isolation down to a particular component of the system. Before a system can be recovered we must first be able to determine that it does, indeed, need to be recovered. The problem is to decide when a system's state changes or better still, is about to change from a state of functioning to state of failure. Most maintenance is actually done before the system stops working. It is more likely to be done at the convenience of the operator or following a warning. In this section we look at some of the ways system designers can help the operators prevent failures by forewarning the users that the system is about to enter a state of failure unless some form of preventive maintenance is carried out. These ranges from simple gauges found in almost all vehicles through to complex built-in test equipment to the need for data entry checking in software. To provide overall system operational effectiveness, consideration must be given to providing the proper test and inspection capabilities in the basic equipment design. Testability must be established within the early design stage for effective support and minimum life cycle cost.

7.6.1 Built in Test (BIT)

The objective of any system designer should be to design-in most cost-effective approach to support for his product. This can be achieved by including a self-test mechanism in the design of the product. The requirement for built-in test is included in many avionics systems. Built in test (BIT) provides monitoring and fault detection capabilities as an integral feature of system design. Advanced BIT sub-systems are processor controlled and are fully isolated by hardware interlocks from the safety critical parts of the system. All faults are identified to the BIT subsystem and the LRU (line replaceable unit) responsible identified. Fault isolation can be accomplished locally or remotely following menu driven software prompts. As technology advances to increase the capability and complexity of modern systems, there is a necessity for the use of automatic failure detection capability. The need for BIT is driven by operational availability requirements, which do not permit the lengthy duration of maintenance activities associated with detecting and isolating failures in electronic equipment. Also, a well-designed BIT system can substantially reduce the need for highly trained maintenance personnel. The performance measures for BIT are:

1. Percentage detection, *BPDp*, the percentage of all faults or failures that the BIT system detects.

2. Percentage isolation, *BPIp*, the percentage of detected faults or failures that the system will isolate to a specific level of assembly.
3. Percentage of false alarms, *BPFRp*, is the percentage of BIT indicated faults where no fault is found to exist.
4. Percentage of false removals, *BPFRp*, is the percentage of units that removed as indicated by BIT whose condition was found to be satisfactory at the higher maintenance level.
5. Automatic fault isolation capability, *AFIC*, is the percentage isolation and detection, that is

$$AFIC = BPD_p \times BPI_p \tag{7.9}$$

The percentage of false alarms is a difficult parameter to measure accurately because initial fault detection followed by an analysis indicating that no fault exists can be due to several possible events such as

1   The BIT system erroneously detected a fault.
2   An intermittent out-of-tolerance condition exists somewhere.
3   A failure exists but cannot be readily reproduced in a maintenance environment.

The percentage of false removals can be a more difficult problem to address, because it may be caused by the following events.

1   Incorrect BIT logic.
2   Wiring or connection problems which manifest themselves as faulty equipment.
3   Improper match of tolerances between the BIT and the test equipment at the next maintenance level.

The resolution of each type of false alarm and false removal requires a substantially different response. From a supportability point of view, false alarms often lead to false removals creating unnecessary demands on supply and maintenance. Another concern is the fact that false alarms and removals create a lack of confidence in the BIT system to the point where maintenance or operations personnel may ignore certain fault detection indications. The specification of BIT performance must be tailored to the specific system under consideration as well as the available funds and, most importantly, the mission requirement. This tailoring activity must include a comprehensive definition of BIT capability based upon the figures of merit presented above.

The design of BIT is based upon two assumptions regarding the reliability of the basic system, accurate identification of failure modes and correct estimation of the frequency of occurrence of the failure mode. However if either of the assumptions is proven incorrect by test or operational experience, the resultant BIT performance is likely to be inadequate or at least less effective than anticipated. The development of BIT and diagnostics has traditionally been an activity that has chronologically followed basic system efforts.

Example 7.2

A system has five line replaceable units (LRU) with the BIT and system performance characteristics as mention below:

1   Mean Time Between Failure of the system: 50 flying hours.
2   Total mission duration: 5000 flying hours.
3   Percentage detection: 90%
4   Percentage isolation: 90% (to LRU level)
5   False alarm rate: 5% (of all BIT indications)
6   MTTR (with BIT): 2 hours (includes all failures which have been both detected and isolated).
7   MTTR (non BIT): 5 hours (includes failures which have not been isolated but may have been detected)

Making use of the above information determine:

1   The expected number of failures during 5000 flying hours.
2   The expected number of failures detected by BIT
3   The expected number of failures isolated to an LRU.
4   The automatic fault isolation capability (*AFIC*).

SOLUTION

5   The expected number of failures, *E[N(t)]*, during 5000 hours is given by (since we have only the MTBF value, we will assume exponential time-to-failure distribution)

$$E[N(t)] = T/MTBF = 5000/50 = 100 \text{ failures}$$

6   Expected number of failures detected by BIT is

$$E[N(5000)] \times BPD_p = 100 \times 0.9 = 90$$

7   Expected number of BIT isolation is the product of isolation percentage and the expected number of failures detected by BIT. That is, $90 \times 0.9 = 81$ failures.

8   $AFIC = BPD_p \times BPI_p (at\ LRU\ level) = 0.9 \times 0.9 = 0.81$

7.6.2   Built-in-Test-Equipment (BITE)

Built in test equipment, commonly known as *BITE*, refers to the part of the system which performs the built-in test function. In most digital avionics the equipment part of BITE includes some hardware and much software. Typically, the system to be tested is connected to BITE through an interface unit. This is basically a routing system so that the stimulus and measurement devices can be connected to the system under test. Once the system has been connected, system test is selected and the test sequence is started. In response, the central control then sets up the first test in the sequence such that appropriate stimulus and measurement devices are connected to the system under test. The measurement is taken and the result is compared to critical levels and a 'pass' or 'fail' determined and displayed to the operator.

The BITE fault record stored inside an LRU is often the only useful data to assist in failure investigation. It is therefore important that shop staff record and track BITE records of all LRUs that are received. BITE data must be credible if maintenance work is to be based on it. Avionics suppliers must keep records of equipment repairs and BITE data. Memory for BITE data recording is now at low cost and big enough to record much useful data, such as phase of flight, aircraft and route, location of fault, state of the system at that time and so on. The event recording frequency of BITE systems is programmable by users, a useful feature if learnt.

7.6.3   Health and Usage Monitoring Systems (HUMS)

Health and usage monitoring systems, commonly known as *HUMS*, are used to improve the airworthiness of aircraft. *HUMS* use the maintenance data collected through several health and usage monitoring techniques to improve the safety and supportability of aircraft. *HUMS* are expensive and they must be extremely reliable and accurate with diagnosis whilst operating in difficult conditions. Before they can be fitted to an aircraft, a *FMECA* has to be conducted on all aircraft systems. *HUMS* have two main objectives, first, to give the air crew an assurance of airworthiness and give a well advanced warning of a potential critical failure. The second objective is to provide the maintainer with detailed maintenance information on usage and health measurements through the ground station. P Brain (1995) mentions that information generated by HUMS and collated at the ground station can be fed back to fleet management for performance analysis and back to the design authority for continuous product improvement. HUMS can be broadly partitioned into three categories:

1. **Health** – relates to detection of either malfunctions or impending failures.
2. **Usage** – relates to the detection of the usage spectrum and the lifing of critical components.
3. **Status Monitoring** – relates to the availability and status of aircraft subsystems. This may include mechanical systems and avionics. For the latter it is performed by harnessing the LRU Built in Test.

Health monitoring functions should be able to monitor component degradation, abnormal vibrations, performance degradation etc. Where appropriate, warnings are issued to the crew when a problem is hazardous to flight. Engine health monitoring includes calculation of engine power index, torque available, engine vibration analysis, temperature monitoring, debris monitoring. Usage monitoring is used when component failures are age related. Its objective is to measure true-life consumption in an appropriate unit of measurement (e.g. stress cycles).

Effective and reliable HUMS can contribute greatly to safety. Fleet management can use HUMS data for effective maintenance. Currently several aircraft and helicopters are fitted with HUMS. For example HUMS within the Westland Augusta EH101 helicopters have the following features (P Brian 1995)

- HUMS are fully integrated into the central management computer.
- HUMS monitors engines, transmission, rotors, structures, electrical systems, hydraulic systems, fuel system, anti-ice system and the avionics.
- Extensive processing and filtering of HUMS data takes place in real time, providing operational information and warnings to the aircrew.
- Detailed HUMS data is captured in non-volatile RAM and written to a data transfer cassette for by ground based data system, *GBDS*.

7.7 NO FAULT FOUND (NFF)

An all too common problem is when a part of the system malfunction but no matter what tests, checks, examinations are performed no cause of the fault is found (also known as *phantom failures)*. It is basically a fault arising that is subsequently found to be erroneous at a deeper level of maintenance. Isolating the true cause of failure of a complex system naturally demands a greater level of analytical skill, particularly where there is a fault ambiguity present. If the technical skills cannot resolve a failure to a single unit then the probability of making errors of judgement will increase, dependent on the level of ambiguity (E Chorley, 1998). This problem is not unique to the support of military hardware. Data presented by Knotts (1994) quotes a Boeing figure of 40% for incorrect part removals from airframes, and British Airways estimate that NFF cost them in the order of twenty million pounds per annum.

Very often the problem only happens in certain circumstances which cannot be reproduced by the examiner, e.g., turning right at high G-forces, flying inverted for more than so many seconds/minutes. In electronic equipment, it is not uncommon for two conductor strips separated by a few micrometers to be bridged by a dust particle or drop of condensation and having bridged it to burn out or evaporate as soon as the current flows leaving no visible evidence behind. Such an incident would result in a no fault found recording. For some systems these are highly significant and extremely time consuming. On a particular type of personal radio (used by armed infantry forces), over one third of reported failures were subsequently diagnosed as NFF.

Before on-board engine monitoring systems were fitted pilots often reported, on landing, that they noticed the turbine temperature falling from above a certain critical level but could not be sure as to how long it had been above this value. Such an incident had to be recorded as a potential 'over temperature excursion' and as such, would need the engine to be stripped so the blades could be properly examined for burning. Most times, nothing untoward would be found but the engine would be out of service for several days. Table 7.1 shows the percentage of no fault found recordings in several parts used in defence and aerospace industry reported by Morgan (1999). The data covers over 15 years and the cumulative usage of up to half million flying hours. He also noted that a NFF can cost between $ 1000 to $ 3000 for a note reading NFF and wait as long as 60 days, depending on the contractor or procedure.

Table 7.1. Percentage no fault found records in aircraft systems

System	Arising's	No Fault Found	Percentage NFF
Communication	12853	4184	33 %
Auto Flight Control	7607	2438	33%
Navigation	17298	4808	28%
Duct Sensor	261	162	62%
Temperature Controller	502	260	52%
Water extractor	130	126	97 %
Plenum	130	110	85 %
Condenser	195	145	79 %
Surveillance / Search	1346	453	33 %
Temp. controller valve	33	17	52 %

Inadequate troubleshooting procedures, human factors and limited training might all lead to no fault founds. NFF can reduce the availability of the system by a significant factor. Chorley (1998) list the following external factors that may cause the level of no fault found:

1. Quality and depth of training.
2. Quality of technical data.
3. Test equipment suitability, accuracy and calibration.
4. Design of BIT, its resolution of ambiguities and clarity.
5. Intermittent faults.
6. Human stress and fatigue induced error.
7. Wilful intent.
8. The increasing impact of software.

In due course technology may provide truly intelligent products that not only detect a fault but are capable of analysing the actual cause and perform self repair. In fact, this is already available in many computers.

7.8 COMMERCIAL OFF THE SHELF (COTS) AND SUPPORTABILITY

Commercial off-the-shelf (COTS) are items that are used for non-governmental purposes that have been offered for sale or lease to the general public. In order to reduce the development times and resources, NATO countries led by the US Department of Defence have encouraged an extensive use of '*commercial off the shelf (COTS)*' items, also known as 'Non-Developmental Items (NDI)'.

There is an increasing emphasis on systems and products from a perspective of total cost of ownership. This further mandates explicit inclusion of supportability and ILS (integrated logistic support) issues and concerns into the design and development process. This requirement becomes even more urgent given the emphasis on using COTS elements, both hardware and software, in the development of complex, distributed and multifunctional systems. This emphasis on commercially available technology and system elements has already resulted in significant, relatively immediate, reductions in system development and production costs on selected programs. These benefits and cost reductions notwithstanding, an emphasis on utilising COTS system elements results in unique challenges from a systems and supportability engineering standpoint. (Verma *et al*, 1999).

One of the main advantages of COTS to producer is potential large market size for the product. Depending on the type of the system, there may be the possibility of future technological upgrades to the equipment. To customers, he cost of the product will be low in comparison to a bespoke system. The reason for relative low cost of COTS product is that the market size of COTS products will be much higher than that of bespoke system. However, one of the main disadvantages of COTS product is the possible need for a customer to compromise on its original system specification.

7.9 CONTINUOUS ACQUISITION AND LIFE CYCLE SUPPORT (CALS)

CALS originated around September 1985 as a joint initiative of the United States Department of Defence (DoD) and US Defence Industry and at that time it was known as Computer Aided Logistics Support. The main objective of CALS is to make the transition from the traditional paper based organisation to an integrated computer aided enterprise. In 1986 it became known as Computer Aided Acquisition and Logistic Support (and the impact that the data can have on the life cycle of the system). By 1989 there was an International interest on CALS and by 1991 it had become the Department of Defence acquisition policy. In 1990, the UK MoD developed its own strategy to implement CALS under the banner of Computer Integration of Requirements, Procurement and Logistic Support (CIRPLS). In 1991 a NATO Government working group on CALS was formed. The CALS working group was tasked to formulate and implement a NATO CALS programme. By the end of 1993 the NATO CALS office was established with a mission to develop CALS policies and standards. Since then, CALS has become known as Continuous Acquisition and Life Cycle Support.

In August 1988, the US DoD issued a directive stating that plans for new weapon systems and related major equipment purchases should include the use of CALS standards. The objectives of CALS are:

a. To modernise the infrastructure of customers and suppliers to eliminate paper and create a communication network for automatic interchange of data.
b. To aid the improvement of new product designs in terms of supportability, by increasing the control, management and availability of data throughout the product life cycle.
c. To improve quality, eliminate duplication and error; reduce information storage requirements, invoice and payment etc.
d. To develop an almost paperless design, manufacturing and support process.
e. To integrate the data used by various departments and in the various phases of life cycle.

The benefits of CALS concept are obvious, however, implementation of CALS is proving difficult and expensive at present. Due to long product life many defence systems, the legacy of data and data systems gathered over many years also hinder the transition to the electronic age.

7.10   CASE STUDY: SUPPORTABILITY AND LOGISTICS AT BRITISH AIRWAYS

Today, British Airways is rated as one of the world's biggest carriers of international passengers and has a scheduled rate network covering around 170 destinations in almost 80 countries. On average, a British Airways flight departs every 90 seconds, contributing to a total of over a quarter of a million flights per annum. Along with such an outstanding statistics in its curriculum vita, *BA* has an equally enormous responsibility towards its passengers, which amounts to over 30 millions per annum. It is not an easy task to support, service and maintain a fleet of over 250 aircraft, where a single mistake or a simple overlook can cause a disaster claiming hundreds of lives and the aircraft. So the fact that British Airways employs a staff of 48,000 to support the logistics of its aircraft is not surprising.

The Fleet

British Airways operates one of the largest and the most modern fleets of any airline in the world. It consists of over 250 aircraft serving routes as diverse as short-haul regional services in the Scottish Highlands and intercontinental long-haul flights around the world: from high-density, short-range shuttle flights to rarely visited distant cities. In order to meet such conflicting requirements of a worldwide route network, British Airways flies several aircraft types (Table 7.2). The following table gives brief details of all aircraft, which are currently used by British Airways and its subsidiaries.

Table 7. 2. Fleet size at British Airways

Aircraft Type	Number in the Fleet
Airbus A320	10
Boeing 737-200	33
Boeing 737-400	33
Boeing 747 -200/100	31
Boeing 747-400	32
Boeing 757	44
Boeing 767	24
Boeing 777	13
Concorde	7
McDonnell Douglas DC 10-30	7

Cabin Crews

To the average passenger the face of British Airways is represented by the flight attendant, who looks after them during the flight. The primary purpose of the cabin crew is to safeguard the aircraft passengers. The airline employs almost 11,000 cabin staff who are all trained at the Heathrow Cabin Services Training Centre, where in addition to learning how to look after passengers and their requirements, they are also taught the vital procedures to be used in the event of any emergency.

Flight Simulator Centre

Apart from a fleet of over 250 aircraft which carry passengers all over the world, British Airways also owns another fleet, which is in constant operation. This fleet can fly anywhere and do anything but does not carry a single passenger and never leaves the ground. This is, of course, the 17 strong fleet of aircraft simulators housed in the British Airways Heathrow facilities. These amazing electronic devices are realistic reproductions of the flight deck of a specific aircraft and in most cases they have a six-axis motion system to give absolute realism during the course of simulated flights. In addition, the more modern simulators are fitted with advanced computer generated visual systems which give a realistic wide-angled view over areas of terrain and airports for use during simulated landings and take-offs. The fleet includes module simulators for all modern types of aircraft owned by British Airways, as well as for older types such as the early versions of the 747(100 and 200 series), the 737-200, BAC-111 and Lockheed Tri-Star.

British Airways Engineering

Each aircraft of British Airways is equipped with a variety of complex electrical, hydraulic and pneumatic systems as well as increasingly sophisticated avionics including navigation and communications equipment. To keep these aircraft flying reliably and safely, and to maintain a total of almost 700 engines (excluding spares), British Airways has a workforce of 9,700 staff that performs the required maintenance and overhaul activities. This engineering division also earns a valuable 70 million a year through the overhaul of aircraft belonging to other airlines such as Canadian Airlines International, Continental Airlines and Cathy Pacific. Indeed such is the size and complexity of the engineering task, that in April 1995 British Airways Engineering became an operating division and profit centre run as a separate business within the British Airways Group and has its own Board of Directors and marketing department. Its projected turnover in the first five years is predicted to reach 1,000 million pounds.

Since its new establishment, the British Airways Engineering has started reorganising its staff and facilities in order to improve efficiency, reduce cost and attract new business. Basically this involves, the establishment of three 'Fleet Streams' each dedicated to support specific aircraft types. Fleets 1 and 2 utilise the massive engineering base occupying a 220-acre site at Hatton Cross, Heathrow, where fleet 1 looks after the Airbus A320, Boeing757 and Boeing 767 fleets, while Fleet 2 takes care of Boeing 747, Boeing 777 and Concorde fleets. And fleet 3, which is situated at Gatwick, looks after the entire 737-200/400, McDonnell Douglas DC-10s and the British Aerospace ATPs (based at Glasgow).

In addition to these three fleets, there is British Airways Maintenance Cardiff (BAMC) which is solely dedicated to the overhaul and maintenance of all variants of Boeing 747. And also, British Airways Avionics Engineering Ltd (BAAE), Wales, whose activities include servicing and repairs of a vast range of avionics equipment from radio and radar through air data computers and navigation equipment and taking care of the increasingly sophisticated in-flight entertainment and communication facilities.

Maintenance Schedules

Due to the enormity and complexity of modern aircraft, the work involved in maintaining the airliner to its full potential is enormous. The following inspection and maintenance schedule for a long-haul Boeing 747-400 provides a fascinating insight. This schedule consists of a series of increasingly complex checks and maintenance procedures as the aircraft passes various milestones based on accumulated flying hours, which include transit check, ramp checks, service checks, inter checks and major service.

Transit Check

Transit check is performed before every flight, with two engineers and one flight crewmember. It mainly consists of exterior check of the aircraft and engines for damage or leakage, as well as specific checks on listed items such as brake and tyre wear. In addition, on the new aircraft this will include interrogation of the on-board diagnostic computers and downloading of HUMS data to the ground service station.

Ramp Checks

- Ramp 1 check is performed on daily basis and it requires 4 engineers. This consists of transit check plus additional checks on engine oil levels, tyre pressures, aircraft external lighting, cabin emergency equipment, engine health monitoring systems and assessment of technical log entries.
- Ramp 2 check is performed every 190 flying hours with four engineers, where, in addition to transit check and ramp 1 check, checks on component oil levels, engine component oil levels, cabin interior condition and windows are performed.

- Ramp 3 check is performed every 570 flying hours with the help of 6 engineers. This check consists of transit check, ramp 1 and 2 checks plus replacement of hydraulic-system filters, checks on cockpit, cabin seats and attachments, sterilisation of water system and detailed inspection of system filters. This also includes more detailed inspections on items covered in previous checks including avionics systems and standby power systems and change of batteries.

Service Checks

- Service 1 check is done every 1060 flying hours or 85 days, which requires 50 engineers and is performed during overnight stopovers at a maintenance base. This involves all previous checks plus partial strip-down of structure and engines for detailed inspection, replacement of worn components and soiled or damaged cabin equipment and furnishings and servicing of undercarriage struts. Total service check takes around two shifts (approximately 16 hours) to complete.
- Service 2 check is performed every 2120 flying hours or 160 calendar days or 320 landings and requires 50 engineers. This includes all the above checks plus additional and more detailed inspection of specific areas, external wash of aircraft, system clarification function checks and deep cleaning of cabin water and waste systems. Requires three shifts to complete.
- Service 3 check is performed every 3875 flying hours, or 300 calendar days or 500 landings with the help of 50 engineers. All the above checks plus detailed inspection of flying controls, structure and engines, fluids are drained and refilled in major mechanical components, aircraft is washed, integrated checks on avionic systems are performed and cabin condition is assessed and repaired in depth. This requires four shifts.

Inter checks

- Inter check 1 is done every 6360 hours, or 500 calendar days or 900 landings and its completion requires 160 engineers. Detailed inspection and repair of aircraft, engines, components, systems and cabin including operating mechanisms, flight controls and structural tolerances are carried out. Typical duration of this maintenance task is 7-8 days.
- Inter check 2 is performed every 12720 hours, or 900 calendar days or 1600 landings, which requires 160 engineers to complete. This consists of Inter check 1 plus additional system function checks and takes between 8-9 days.

Major Service

In case of a Boeing 747-400, the major service is performed every 24000 flying hours, or 1825 calendar days (5 years) or 3800 landings. These intervals are 'never to exceed' values. A major service requires 180 engineers and takes 20-25 days for the successful completion.

Plans for a major service takes place two months in advance with a general inspection of the aircraft to assess the work to be done. The planning group then prepares a list of work requirements according to the Approved Maintenance Schedule (AMS) and presents it to the Production Control Group (PCG), who ensures the availability of manpower and equipment for the required tasks. Three weeks before the Major service support groups and workshops are briefed. Seven days before the Major, the planning group meets again with PCG to agree on an informal contract with proposed downtime estimate (cost of downtime is £ 100,000 per day) and task list.

The PCG then converts each task into job cards using a computer based work control system. The first day the aircraft arrives at outside hanger and flaps are lowered, all wings are recessed and aircraft bays are washed. Functional tests are done on pneumatic, fuel and electrical systems in order to establish a baseline when the aircraft is returned to service. The aircraft then enters the hangar and is stabilised on jacks, depanelled where required (2 days), and stripped of certain components and most of the interior including seats, carpets, soundproofing, floor-boards, galleys and toilets (4 days).

Inspection of aircraft takes 8 days, which may reveal the need for additional work. At this time the group meets for a Post Initial Review (PIR) and details of the check are finalised. If additional work is needed the manpower may be increased to ensure that the Scheduled Time to Serviceability (STS) is still achieved. Sometimes this is not possible due to the unusual content of the additional work, for example, a cracked keel beam requires an additional 10 days to repair. All the equipment that needs lubrication receives a routine service now, finite life components are replaced and instrumentation is inspected and tested.

The rebuild begins 8 days before the end of the Major with flying control rigging, cabin rebuild and functions. Getting the aircraft off the jacks is critical since flying controls can not be tested on jacks, which requires about 5 days to test. One day before the completion, the aircraft goes to the test pen for engine runs. Note that engine overhaul is governed by different operating constraints and does not form part of a Major check. Once checks are complete, the aircraft returns to the service. A mechanic and avionics engineer accompanies the first post-check flight and a post-check review assesses the work progress and considers possible improvements.

The main aim is to reduce downtime for Majors by 50 percent through better planning. In order to maximise the availability of the aircraft at peak periods, full use of the time limits for the check cycle may be beneficial. A helpful factor to note is that civil aircraft have much redundancy built into the systems and structure so that a single failure is rarely catastrophic, but this redundancy and integrity has to be maintained by test or inspection.

Chapter 8

Spares Parts Provisioning and Management

We produce defence stocks and inventories for just in case,
rather than just-in-time.

Lincoln, H

Spares forecasting and inventory management is one of the most challenging problems in the whole integrated logistic support process. On the one hand the operators want replacement parts to be in stock when required but on the other hand they cannot afford to have capital tied up in inventory. Every pound spent on spares is a pound less to pay for fuel, wages, new systems or to gain interest. In addition, every spare being held is incurring on-going costs in the form of rates, handling charges and possibly deterioration costs through a limited shelf life or obsolescence. The cost of spares for an operator of a fleet of aircraft, whether civil or military, will over the life of the system far exceed the cost of the original aircraft. By how much will depend on how the fleet is operated, maintained and supported.

The Royal Air Force, for example, operates a single echelon, centrally controlled inventory system with approximately 855,000 line items; of which 680,000 are consumable (Kendrick et al, 1998). Some of these items may stay on the shelf from a few days to 30 years. Similarly, their value may vary between a few pence and several million pounds sterling. At any given time, the value of the total stock held will run into hundreds of millions of pounds.

Many of these parts may never get used. Some will pass their "sell by" dates, others will become obsolete and be superseded by new "improved" standards and, others will be surplus to requirement. And, with so many parts to keep track of, some may simply become "lost". For such large quantities of items, even a very small error in forecasting the demand for spares can make a huge difference in the support cost.

A number of factors help contribute to the difficulty of forecasting the demand. Determining the time-to-failure distribution at the design stage even for the perfectly manufactured component is as yet not exact science. The way the system is used may also have profound effects on the times to failure. These factors are further exacerbated by the fact that some of these parts may take months or even years to manufacture, especially after the production line has been closed down. (The last Concorde was built nearly 20 years ago but the aircraft are expected to remain in service for at least another 10-20 years).

Initially, spares forecasting and optimisation is done as part of the life cycle cost exercise. This value has taken on an extremely important role in the tendering and decision-making processes for determining which system will be bought. Having made the decision, the chosen contractor(s) will, almost certainly, be put under pressure to make improvements to further reduce the estimated life-cycle cost. Ultimately, under leasing type contracts, the life-cycle cost analysis may be used to determine hire charges, for example in the form of a "power-by-the-hour" agreement.

In many cases, spares are overstocked, resulting in high inventory costs. The longer parts stay on the shelf, the higher the risk that they will become obsolete or become unserviceable through deterioration or become "lost". It is therefore important to predict the demand for spares as accurately as possible to avoid unnecessary inventory costs and to provide cost effective support. In the fast changing technological world it does not take much time for assets (spare parts) to become liabilities (disposal cost). Spare provisioning plays a much more crucial role in defence and aircraft industries because of the long lead-times and huge budget spent on spares every year.

Usually, the manufacturer/supplier provides the information on the required number of spares of each component of the system for a stated period of time (initial provisioning). Unfortunately, as mentioned by Pironet (1998), demand prediction for spare parts as well as maintenance requirements is the weakest aspect of stock management today in all armed forces and industries alike. Since 1990, spares worth $ 7 billion have been sent for disposal by Ministry of Defence, UK (Bateman, 1999). The same figure for the Department of Defence, USA is over $ 34 billion. In 1997, It was reported that the commercial aviation industry holds an inventory of spare parts worth more than $52 billion (Aircraft Economics, 1997). Approximately $23 billion of this is repairable spare parts. There are $29 billion worth of non-repairable items in commercial airline inventory. One estimate shows that the amount of surplus spare parts in commercial airline industry is between 25% and 40%. Between 1993 and 1997, the commercial aviation stockpile grown at annual compound rate of between 7% and 10%, out pacing both traffic and fleet growth (Aircraft Economics,

1999). Disposal of unused spares, risk of running out of spare parts and overstocking are undesirable events for any organisation. Any successful model used for stock management should be able to predict the demand as closely as possible.

In this chapter we discuss a few spares forecasting models with their applications and limitations. One of the traditional approaches used to predict spares is queuing theory. Cox (1962) carried out a comprehensive work on the subject. We will also look at a method based on Palm's Theorem (1938), renewal theory and finally, simulation.

In looking at these methods we will also consider what data is required and how this might conflict with the current Military Standards. In particular, we note that the Military Standards have no space in their logistic support analysis (LSA) database (MIL-STD-1388) for anything other than MTTF or MTBF. Since the exponential distribution is the only failure distribution with a single parameter, one is forced to use this when only the MTTF or MTBF value is known. This means, one cannot model age-related failures that are a very sensitive factor in spare parts prediction.

8.1   BASIC CONCEPTS AND DEFINITIONS

In this section, we discuss some terms and definitions that are used in the spare parts management literature.

Line replaceable units (LRU) – A system can be considered as comprising of repairable assemblies that can be replaced in order to recover the system. Replacements are normally carried out at an organisational or first level maintenance facility, that is, where the systems operate. The main aim of such a replacement is to minimise the downtime in recovering the system.

Shop replaceable Unit (SRU) – Each LRU will normally comprise of sub-assemblies or modules that can be replaced to facilitate recovery of the LRU within the second (intermediate) or third (depot) level maintenance units (or shops).

Consumables – Consumable parts are those non-repairable items that may be used to recover a system, LRU or module. The categorisation of an item as consumable does not necessarily imply that the item has no recovery potential.

Turn Around Time (TAT) - The time elapsed between the removal of a repairable item from the system till the time it is returned as fully serviceable is called turn around time (some times *turn round time,* TRT).

Fill rate represents the percentage of demands that are met from the stock on hand for a stated period of time. For example, 95% fill rate requirement stands for the stock level for which the demand can be met 95 percentage of times or 95% chance that there will be no stockout. Also known as *'probability of no stockout'* (*PNS*) or *spares adequacy.* The most desirable is fill rate would be 100%, but one has to invest a lot to achieve 100% fill rate. Most commercial airlines prefer 85% fill rate. The difference between 100% fill rate and the achieved fill rate is called *provisioning gap.*

Expected Back Order (EBO) represents the expected number of demands that cannot be met for a given stock level for a given mission period. In inventory literature this is called *lost sales.*

Pipeline for a site denotes a random variable that represents the number of items that are under repair or being resupplied to the site from higher echelons. Average pipeline for a site is the average number of units under repair or resupply.

Multi-Indenture – Indenture level refers to the hierarchy of engineering parts. Systems can have two or more indenture levels (multi-indenture). In general, *first indenture* refers to the items that are directly assembled in to the system, *second indenture* refers to the modules within the first indenture items. In case of aircraft, *Line replaceable Unit (LRU)* refers to the items that are directly fitted into the aircraft and *Shop replaceable Unit (SRU)* refers to the items that are removed from LRU at maintenance shops. In weapon systems the term *Weapon Replaceable Items (WRI)* is used to refer to first indenture items. Figure 8.1 shows the basic indenture configuration relating to the aircraft engines.

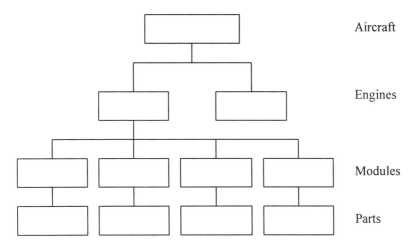

Figure 81. Multi-Indenture Environment

Multi-Echelon – The hierarchy of operating locations and supporting depots is referred to as multi-echelon. For example, aircraft may be dispersed in squadrons (1st echelon) across a number of bases (2nd echelon) such that each base may have one or more operational squadrons as illustrated in Figure 8.2. The bases are supported by a 3rd line depot or maintenance unit and these are supported by the contractors (4th line).

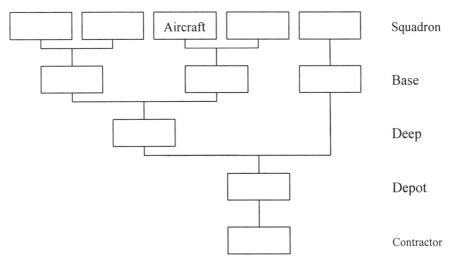

Figure 8.2 Multi-Echelon Environment

8.2   MODELLING DEMAND RATE

There are several factors that can influence the demand for a spare (A detailed study on the factors that influence the demand is carried out by I Robinson (1998).   Apart from the functional failure, maintenance and incorrect removals play a major role in spare parts demand. For this reason, many models use removal rate instead of failure rate or renewal rate. In this section, we discuss some of the factors that can influence the demand.

For many parts, failure accounts for the majority of the removals. The demand due to failure can be modelled either using a Poisson process (if the time to failure is exponential) or renewal process (when the time to failure is non-exponential). In most of the cases, one would expect a non-exponential time-to-failure distribution. However, under steady state conditions one can use constant failure rate models to predict the demand for spares at higher indenture (LRU level).

The second major factor that influences the removal of an item is maintenance. In general during planned maintenance and overhaul most systems are removed and during recovery some of the parts within the system may be replaced with new parts. Safety critical items are always replaced once they reach their hard life (age based replacement policy), thus requiring a spare. Parts whose ages are close to their hard lives or have exceeded their soft lives will also be replaced (to minimise the number of system removals). In addition, the opportunity is taken to inspect parts, which are normally not visible, and this, naturally, can lead to further parts being replaced.

No faults found (NFF) forms another major influencing factor for removal of an item from a system. Following a reported fault or failure, items will be removed for testing. Sometimes no faults will be found during these tests.   This is a particularly common problem with electronic equipment in which dampness may cause temporary short-circuiting, stresses to the boards may open dry joints or inconsistent electrical supply may cause temporary malfunction of sensitive components.   Similarly, faulty built-in test equipment (BITE) may generate erroneous warning messages resulting from errors in the software which only occur under a particular combination of inputs, say. It is important to note that this is a fictitious demand in the case of repairable spares and not the actual demand. However, if the turn around time is long, then this can affect the system parameters such as availability, expected backorders, fill rates etc. Also, some administrative and maintenance cost will be incurred whenever there is a no fault found. Thus one has to consider the effect of no fault found while calculating spares related costs.

Software-related demands could be due to software failure or software upgrade. As the software is upgraded, it usually creates demand for new hardware configuration. Another factor that can cause a demand for spare parts is *technological obsolescence.* This further complicates the modelling problem as one has to know how long a particular technology is going to last.

For modelling demand rate or removal rate we consider the influence of failure, maintenance and software related removal. Other factors such as no fault found, incorrect removal by maintainers, false failure indication by built in test equipment (BITE) etc create a fictitious demand for repairable spares that can affect availability of spares, backorders, inventory costs etc. While forecasting spares, we look at two measures, the expected number of failures and the expected number of removals. For any item, the following inequality is valid,

$$NF(t) \leq ND(t) \leq NR(t) \tag{8.1}$$

where NF(t), ND(t) and NR(t) denote the number of failures, number of demands and number of removals by time t, respectively. In the following sections, we study models for predicting demand for spares for consumable (discardable) and repairable items.

8.3 MATHEMATICAL MODELS FOR DEMAND PREDICTION

Two popular mathematical models that are used in spare parts provisioning are based on Poisson processes and renewal theory. The Poisson process can be used whenever the demand rate is constant (this means each failure mode and other factors which influence the demand should follow the exponential distribution). Whenever the demand rate is not constant we use renewal theory to forecast demands for spares.

8.3.1 Homogeneous Poisson Process Models for Forecasting Spares

Homogeneous Poisson process (HPP), $\{N(t), t \geq 0\}$, is a counting process that satisfies the following conditions.

1. $N(0) = 0$
2. The process has stationary and independent increments.
3. The number of demands in any interval of length t is a Poisson distributed with mean demand λt. That is, for all h, $t \geq 0$

$$P[N(t+h)-N(h)=n] = \frac{\exp(-\lambda t) \times (\lambda t)^n}{n!}, \qquad n = 0,1,2... \qquad (8.2)$$

The expected number of demands during the duration of length t is given by:

$$E[N(t)] = \lambda t \qquad (8.3)$$

To examine whether an arbitrary process is actually a Poisson process, one should show that conditions (1), (2) and (3) are satisfied. Condition (1) states that the counting starts at time t = 0. A stationary increment means that the distribution of number demands (or removals) that occur in any interval depends only on the length of the interval. That is, in a stationary process, the number events in the first 100 hours are the same as the number of events occurring between 500 and 600 hours or any other 100-hour interval. A process is said to have *independent increments* if the numbers of demands that occur in any two disjoint intervals are independent. The *time between demands* in a Poisson process follows an exponential distribution with mean time between demands (removal) $(1/\lambda)$.

Expression for Fill Rate using HPP Model

First we derive the expression for fill rate using HPP for simple cases. Assume that initially N spare items are stocked for an item. Also assume that the stocks are not renewed and the failed parts are not repaired. Under these assumptions, one will run out of spares only when the number of demands exceeds the initial stock level N during the stated operating period. Using the HPP model, the expressions for fill rate for a mission length of t is given by:

$$Fill\ rate = \sum_{k=0}^{N} \frac{\exp(-\lambda t) \times (\lambda t)^k}{k!} \qquad (8.4)$$

The above expression is simply obtained by adding the probabilities for the demand being equal to 0, 1, 2, etc up to N. Since $\lambda = 1 / MTBR$, the equation (8.4) can be written as:

$$Fill\ rate = \sum_{k=0}^{N} \frac{\exp(-\frac{t}{MTBR}) \times (\frac{t}{MTBR})^k}{k!} \qquad (8.5)$$

where MTBR stands for mean time between removals. It is important to note that the above expression is valid only for non-repairable spares that are not renewed. In most cases, MTBR could be replaced with MTBF. In fact, MTBR is used in the above expression because a demand could be caused by reasons other than failure as discussed above. If the stock is regularly replenished, say, as soon as the stock level reaches the reorder level 's', then the approximate expression for fill rate is given by

$$Fill\ rate \approx 1 - \sum_{k=0}^{s} \frac{\exp(-\frac{LD}{MTBR}) \times (\frac{LD}{MTBR})^k}{k!}] \times F(LD) \qquad (8.6)$$

where, LD stands for *lead time,* that is, the time between placing an order and receiving the stock. F(LD) is the probability that the replenishment stock arrives by the lead time LD. The above expression is only an approximate expression, as we do not consider the probabilities such as receiving the order before the lead-time and running of the stock.

Expression for Expected Backorder using HPP Model

Assume that initially N spare items are stocked and the stocks are neither repaired nor replenished. During a mission length of t, a backorder will occur only when the number of demands exceed N. The expression for an expected backorder (EBO) using HPP model for a mission length of t is given by:

$$EBO(N) = \sum_{k=N+1}^{\infty} (k - N) \times \frac{\exp(-\lambda t) \times (\lambda t)^k}{k!} \qquad (8.7)$$

The above expression is obtained by using the logic that there will a backorder for one spare, when the number of demands is N+1 and there will be a backorder for two spares whenever the number of demand is N+2 and so on. Since $\lambda = 1 / MTBR$, the above expression can be written as:

$$EBO(N) = \sum_{k=N+1}^{\infty} (k - N) \times \frac{\exp(-\frac{t}{MTBR}) \times (\frac{t}{MTBR})^k}{k!} \qquad (8.8)$$

Example 8.1
Time between failures of an item can be modelled using an exponential distribution with mean time between removal 128 hours. If 7 spares are stocked for this item, find the fill rate and expected backorders for mission duration of 800 hours.

SOLUTION:

The distribution of number of demands is given by:

$$P[N(t) = n] = \frac{\exp(-\frac{t}{MTBR}) \times (\frac{t}{MTBR})^n}{n!}$$

Substituting MTBR = 128 hours and t = 800 hours in the above equation, we have:

$$P[N(800) = n] = \frac{\exp(-\frac{800}{128}) \times (\frac{800}{128})^n}{n!}$$

The distribution of number of failures is plotted in Figure 8.3

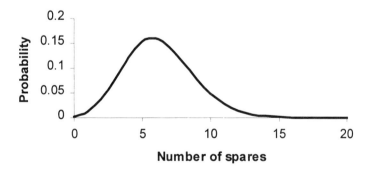

Figure 8.3. The distribution of demands for the example problem

The fill rate is given by:

$$Fill \ rate = \sum_{k=0}^{7} \frac{\exp(-800/128) \times (800/128)^k}{k!} = 0.7089$$

That is, there is a 70% chance that one will be able to meet the demand during 800 hours of mission length with 7 spares. Figure 8.4 shows how the fill rate varies as the stock level changes. From the graph it is easy to find the stock level to meet the required fill rate.

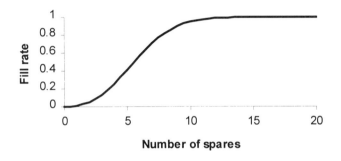

Figure 8.4. Fill rate versus the stock level.

The expected backorder can be found using the following expression.

$$EBO(7) = \sum_{k=7+1}^{\infty} (k-7) \times \frac{\exp(-800/128) \times (800/128)^k}{k!} = 0.673$$

That is, the expected backorder for an initial stock of 7 is less than one.

Example 8.2

The time to failure of an item follows an exponential distribution with mean life 100 hours. 1. Plot the probability for the number of demands for 200 hours of operation. 2. Find the expected number of demands for 200 hours of operation. Find the fill rate for the corresponding expected number of demands.

SOLUTION:

Since we have only the time-to-failure information, we have to assume that all the demands are caused due to failure (that is there is no incorrect removals). Under these assumptions we can compute the probability distribution of demands and fill rate as follows.

Given that the mean life $(1/\lambda) = 100$, that is $\lambda = 0.01$. Since the time to failure is an exponential distribution, the demand process follows Poisson process, thus the probability for number of demands during 200 hours of operation is given by:

$$P[N(t) = n] = \frac{(\lambda t)^n \exp(-\lambda t)}{n!}$$

Thus for $\lambda = 0.01$ and $t = 200$ hours, the above equation can be written as:

$$P(N(200) = n] = \frac{(0.01 \times 200)^n \exp(-0.01 \times 200)}{n!} = \frac{2^n \times \exp(-2)}{n!}$$

The values for various values of n are given in the following table 8.1. Figure 8.5 shows the distribution of number of demands for 200 hours of operation. The expected number of demands for 200 hours of operation is $\lambda t = 0.01 \times 200 = 2$. The fill rate corresponding to the expected demand value of 2 is given by:

$$Fill\ rate = \sum_{k=0}^{2} \frac{(\lambda t)^k \exp(-\lambda t)}{k!} = \sum_{k=0}^{2} \frac{2^k \times \exp(-2)}{k!} = 0.6766$$

That is, if 2 spares are stocked, then the fill rate is 0.6766.

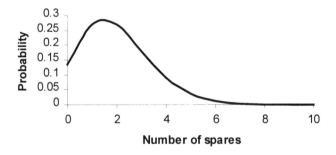

Figure 8.5. The distribution of number of demands during 200 hours

Table 8.1 Distribution of demand

N	P[N(t)] = n	Probability
0	$2^0 \times \exp(-2) / 0!$	0.1353
1	$2^1 \times \exp(-2) / 1!$	0.2706
2	$2^2 \times \exp(-2) / 2!$	0.2706
3	$2^3 \times \exp(-2) / 3!$	0.1804
4	$2^4 \times \exp(-2) / 4!$	0.0902
5	$2^5 \times \exp(-2) / 5!$	0.0360
6	$2^6 \times \exp(-2) / 6!$	0.0120
7	$2^7 \times \exp(-2) / 7!$	0.0034

Applications and Limitations of the Homogeneous Poisson Process Model

In general Poisson process models can be used only when the time between demands (removal) follows an exponential distribution. That is, the removal rate is constant. This means that, items with age related failure mechanisms cannot be modelled using Poisson processes. Strictly speaking, Poisson process models can be used only when the removals are caused by non-age related causes such as accidental damage and incorrect removal due to maintenance inefficiency. However, Poisson processes can be used to model higher indenture spares such as LRU in steady state.

In an LRU with a large number of components, where each component can be modelled using a independent renewal process, Theorem's by Palm and Drenick state that in steady state the time between removals at the LRU level follows an exponential distribution. That is the demand follows a Poisson process. The statement of Drenick's theorem (1961) is given by:

Assume a piece of equipment made up of n sockets, each the bearer of a renewal process that is statistically independent of all others.   The equilibrium states of these processes are characterised by the component mean lives m_i (i = 1, 2, ..., n; $0 < m_i < \infty$) and the residual survival probabilities $G_{ci}(t)$. Assume further that

(i) $lim_{n \to \infty} sup_{1 \le n \le \infty} M / m_i = 0$
(ii) $1 - F_{ci}(t) \le a_i t^\delta$ *as* $t \to 0$
(iii) $a_i < A$

where, (1/M) is the replacement rate for all components of the equipment in equilibrium. The residual survival probability of the equipment is given in the limit by the relation

$$\lim_{n \to \infty} G_{ci}(Mt) = \exp(-t) \text{ for fixed } t > 0.$$

The proof of the theorem is strongly based on the central limit theorem. The result is true also for initial distribution but with different constant (different from M). The theorem provides a good insight about the behaviour of the complex systems. Without any doubt, the theorem has an important application in forecasting demands for spares. However, there are many other issues that should be addressed before one can use exponential time between removals. The main problem in using Drenick's limit theorem is that it is proved under steady-state conditions, that is, by setting $t \to \infty$. This is the major drawback of any model that tries to argue that all complex systems have constant failure rates.

It is very easy to show that the systems with relatively few parts and consumable spare parts do not follow constant failure rates during a finite time horizon. A simple analysis of hazard function is sufficient to prove this. For example, consider a complex system with N items. Assume that all the items have constant hazard function except one.   Without loss of generality, assume that the item N has an increasing hazard function. Now, the hazard function of the system can be written as:

$$h_s(t) = \sum_{i=1}^{N-1} h_i(t) + h_N(t)$$

$$h_s(t) = K + h_N(t) \tag{8.9}$$

where, K is a constant which is equal to the sum of all the constant hazard function values. Equation (8.9) is an increasing function between any two failures of item N. This is an example where Drenick's Theorem fails. In fact, it is almost impossible to find a part or component with few items that can be modelled using Drenick's Theorem (1961) in a reasonable time frame so that the system will reach steady state. As an example, consider a system with the number of items N = 100. Let the time-to-failure of the item N can be modelled using Weibull distribution with scale parameter η = 100 and β = 3. Assuming K = 0.01 in equation (1), the hazard function of the system is given by:

$$h_s(t) = 0.01 + (\frac{\beta}{\eta})(\frac{t}{\eta})^{\beta-1} \text{ , that is, } h_s(t) = 0.01 + (\frac{3}{100})(\frac{t}{100})^2$$

Figure 8.6 shows how the hazard function varies over time. It is clear that the failure rate is not constant. It is easy to see that the item with non-constant failure rate dominates the failure pattern of the system. When the system is subject to preventive maintenance and repair, the hazard function will have slightly different shape as shown in Figure 8.7. Simulating the failures of the item N produces figure 8.7. It is easy to notice that the hazard function is not constant. Thus, a careful analysis of time-to-failure data is necessary before using any commercial model for forecasting spares, in particular during the initial provisioning.

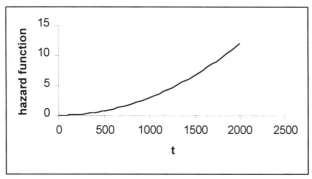

Figure 8.6 Hazard function of consumable spare parts

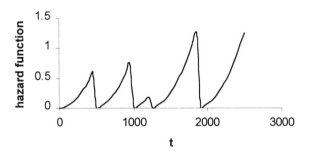

Figure 8.7 Hazard function for repairable SRU.

8.3.2 Renewal Process Models for forecasting Spares

If the sequence of time between removal random variables $\{X_1, X_2, ...\}$ is independent and identically distributed, then the counting process $\{N(t), t \geq 0\}$ forms a renewal process. $N(t)$ represents the number of renewals (in our case the number of demands) that occur by time t. Assuming that the time between removal random variables X_i, $i \geq 1$, are independent and have common distribution $F(t)$, then the probability distribution of number of removals is given by:

$$P[N(t) = n] = F^n(t) - F^{n+1}(t) \tag{8.10}$$

where $F^n(t)$ is the n-fold convolution of $F(t)$ and is given by:

$$F^n(t) = \int_0^t F^{n-1}(t-x)dF(x) \tag{8.11}$$

$F^n(t)$ denotes the probability that the n-th removal occurs by time t. The expected number of removals, $M(t)$, during a length of t is given by:

$$M(t) = \sum_{n=1}^{\infty} F^n(t) \tag{8.12}$$

The equation (8.12) is known as the *Renewal Function*. Computing $M(t)$ from the above equation is usually difficult for most of the time to failure distributions. One distribution for which one can compute the above function easily is the normal distribution. For the normal distribution the expression for expected number of demands, M(t), during t hours of operation is given by

$$M(t) = \sum_{n=1}^{\infty} \Phi(\frac{t - n \times \mu}{\sqrt{n \times \sigma}}) \tag{8.13}$$

The expected number of renewals, M(t), also satisfies the following equation (known as *renewal equation*)

$$M(t) = F(t) + \int_0^t M(t - x)dF(x) \tag{8.14}$$

Equation (8.14) is easier to compute compared to equation (8.12). Computing the expected number of removals from equation (8.12) involves successive calculation of $F^n(t) = \int_0^t F^{n-1}(t - x)dF(x)$, which is usually complex. One can easily calculate the integral $\int_0^t M(t - x)dF(x)$ in equation (8.14), for example by using simple numerical integration. By substituting t = n × h, and using the trapezoidal equation, the expression (8.14) can be written as:

$$M(nh) = \frac{1}{1 - (h/2)f(0)}[f(nh) + \frac{h}{2}M(0)f(nh)$$

$$+ h\sum_{k=1}^{n-1} M[(n - k)h]f(kh)] \tag{8.15}$$

Another approach for computing the equation (8.14) is using recursive computation suggested by Xie (1989). For t > 0, we partition the time interval [0,t] according to $0 = t_0 < t_1 < t_2 < ... < t_n = t$ where $t_i = ih$ for a given grid size h > 0. For mathematical simplification put,

$M_i = M$ (ih) and $F_i = F[$ (i - 1/2) h] and $A_i = F(ih)$, $1 \le i \le n$.

The recursion scheme for computing the M_i's is as follows:

$$M_i = \frac{1}{1 - F_1}[A_i + \sum_{j=1}^{i-1}(M_j - M_{j-1})F_{i-j+1} - M_{i-1}F_1] \qquad (8.16)$$

Starting with $M_0 = 0$. The recursion scheme is easy to program and gives accurate results. The recursive computation is able to resist the accumulation of round off error as t gets larger. How to choose the grid size h depends on the desired accuracy in the answers, but also on the shape of the distribution F(t) and the length of the time interval [0,t]. The usual way to compare the answers for grid sizes h and h/2. If these are significantly different then it may be necessary to halve the interval again to h/4 and repeat the comparison. In many cases of practical interest a four-digit accuracy has been obtained with a grid size h in the range of 0.05-0.01. The discretization method has been applied to compute the exact values of the renewal function for the Weibull distribution.

Example 8.3

Time to removal of an item can be modelled using normal distribution with mean $\mu = 200$ hours and the standard deviation $\sigma = 40$ hours. Find the expected number of removals during 800 hours operation. Also find the fill rate if 4 spares are stocked initially.

SOLUTION:

The expected number of removals during 800 hours of operation is given by:

$$M(t) = \sum_{n=1}^{\infty} F^n(t) = \sum_{n=1}^{\infty} \Phi(\frac{800 - n \times 200}{\sqrt{n} \times 40}) \approx 3.51$$

Table 8.3 gives values of $\Phi(\dfrac{800 - n \times 200}{\sqrt{n} \times 40})$ for different values of n.

Table 8.2. $F^n(t)$ values for different n

n	$F^n(t)$
1	1
2	1
3	0.99805
4	0.5
5	0.0126
6	0.00002

The expected demand during 800 hours of operation is 3.51. The fill rate for an initial stock of four spares is given by:

$$Fill\ rate = \sum_{n=0}^{4} F^n(800) - F^{n+1}(800) = 0.9873$$

Example 8.4

The time-to-failure distribution of discs in a jet engine follows a Weibull distribution with scale parameter $\eta = 20$ flying hours and the shape parameter $\beta = 4$. Find the expected number of demand for spares for 100 hours of operation.

SOLUTION:

To calculate the expected demand we use the equation (8.14), which needs recursive computation of $M(i*h)$. For mission duration of 100 hours we get the expected number of demands as 5.0556. Table 8.3 shows the expected demand for different values of t for an item with Weibull time to failure with scale parameter 20 and shape parameter 4.

Table 8.3. Expected demand for different t values

t	M(t)
20	0.6446
40	1.7466
60	2.8490
80	3.9523
100	5.0556

8.4   DEMAND FOR SPARES – CONSTANT VS INCREASING FAILURE RATE

In this section we look at the sensitivity of making a wrong assumption regarding the failure rate of an item on forecasting. A good question would be to ask what happens if someone assumes constant failure rate for the demand when it is not true. The answer to this question is given by the inequality stated in equation (8.17). If mean life or Mean Time to Failure (MTTF) of the item is known, then for an item with *increasing hazard function* the following inequality is valid (Gnedenko, 1961).

$$\frac{t}{MTTF} - 1 \le M(t) \le \frac{t}{MTTF} \qquad (8.17)$$

Where, M(t) is the renewal function that gives the expected number of demands for spares during t hours of operation. The right hand side of equation (8.17) is in fact the expected number of demands if one assumes a constant failure rate.

Equation (8.17) can be interpreted as follows. The expected number of demands for an item with an increasing hazard function will be less than (or equal to) that of an item with a constant hazard function having the same expected life by at most one. Now one might think that it is not worth bothering with non-constant failure rates, if the difference in the expected number of demands under constant and increasing hazard function is going to be less than one. The answer to this question really lies in the fleet size. The difference in the number of demands under constant and increasing hazard increases with fleet size. For a fleet of 1000 items, one might store as many as 1000 more spares than actually required. This automatically risks a high inventory cost.

Consider two items A and B. Assume that the time-to-failure of item A is exponential with mean life (MTTF) $1/\lambda = 100$ hours and the time-to-failure of item B is Normal with mean $\mu = 100$ hours and standard variation $\sigma = 10$ hours. The expected number of demands for item A, E[N(t,A)] is given by:

$$E[N(t, A)] = \lambda t \qquad (8.18)$$

The expected number of demands for item B, E[N(t,A)], can be derived using renewal theory and is given by:

$$E[N(t,B)] = \sum_{n=1}^{\infty} \Phi(\frac{t - n \times \mu}{\sqrt{n} \times \sigma}) \qquad (8.19)$$

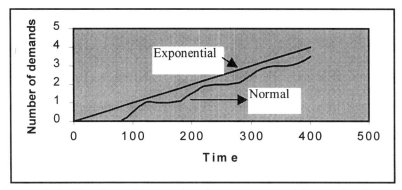

Figure 8.8. Number of demand under constant and increasing hazard

Figure 8.8 shows the difference in the number of demands between the constant and non-constant demand rate. When the demand rate is non-constant with increasing hazard, the number of demands shows a cyclic pattern. The Figure 8.8 also indicates that all the spares might be used at some stage or other, but the main concern is that the part might be become obsolete before the demand occurs.

8.5 FORECASTING MODELS FOR CONSUMABLE (DISCARDABLE) SPARE PARTS

We use the generic term consumable (discardable) spare parts to denote those parts that are not repaired up on failure/removal. As the parts are discarded, it is reasonable to assume that the number of demands is equal to the number of removals. We consider the following factors for predicting the number of demands.

1. Time to failure distribution of the item.
2. Time between incorrect removal (TBIR) of the item.

Time-to-failure

In case of a consumable spare part, the time-to-failure can have constant demand if and only if all the failure modes involved follow an exponential distribution. This is a very unlikely event, thus it is reasonable to assume in most cases the time-to-failure will not be exponential. In fact, one would expect an increasing demand rate due to wear out characteristics. We would recommend the use of the renewal process model for all consumable spare parts. However, if there is strong evidence to suggest that the time to failure is exponential, then one can use Poisson process model.

Time To Incorrect Removal

Time to incorrect removal of a consumable spare part refers to the factors such as incorrect removal by the maintenance crew, false indication by the Built-in Test Equipment etc. Since incorrect removal is a *purely random* event (i.e. unrelated to the age of the component), one can use the exponential distribution to model the demand due to incorrect removal. The mean time to incorrect removal (MTTIR) can be used to calculate the number of such removals over a period of time.

The demand for a consumable spare part is a function of time-to-failure and time to incorrect removal.

Demand = f(time to failure, time to incorrect removal)

The expected demand during t hours of operation can be found as follows. Let T_1 be the largest integer less than (t / MTTIR) and let T_2 be t – $T_1 \times$ MTTIR. The expected demand during t hours of operation is given by:

$$T_1 \times M(MTTIR) + t / MTTIR + M(T_2) \qquad (8.20)$$

M(.) is the renewal function. In many cases, finding MTTIR may be difficult for consumable spares, as data may not be available. Also there are no NFF recorded. Consumable items are usually non-complex and thus the percentage of incorrect removal is minimum. Surprisingly, many organisations seem to take consumables lightly even though they account for bulk of the money. Commercial aviation industry has consumable worth $29 billion out of the total of $52 billion (Aircraft Economics, 1997). Since most of the consumable parts are non complex fault isolation percentage is very high and it will most appropriate to use the renewal function M(t) to find the expected demand.

8.6   MODELS FOR REPAIRABLE SPARE PARTS

Classifying a spare as repairable in most cases is based on economical factors. If an item can be repaired upon failure and, if it is less costly to repair the item compared to buying a new spare, then it is classified as repairable. As a rule of thumb, many airlines replace the item if the repair cost is more than 65% of the replacement cost. There are about 2000 repairable in Boeing 737. Boeing 757 and 767 have about 1800 repairable, the number of repairable in 777 is about 2100 and the 747-400 has about 2300. The maintenance level to which the components will be moved for repair or recondition will depend on the capabilities and capacities of the various sites and these will depend on the user.

Let us consider an example of an aircraft engine. The bases (2nd line maintenance) will normally be able to strip, rebuild and test engines and may also be able to repair modules (by striping and replacing the parts). They will also have limited storage capacity for spare engines and modules. These bases will be supported by maintenance units or depots which have the capacity of doing any work on the engine, modules or part and should have, effectively, unlimited storage capacity. In this case, the engine and modules will, almost certainly, be considered as repairable.

8.6.1 Birth and Death Process Model for Repairable Spare Parts

An item with exponential failure and repair time (or turn around time) can be modelled using birth and death process. A birth and death process is a continuous time Markov chain with states $\{0, 1, 2, ..., \}$ for which transitions from state i may go only to either state i -1 or state i + 1. Assume that λ_i and μ_i are the transition rates from i to i + 1 and from i to i -1. In repairable spares provisioning, the states represent the number of items under repair or waiting for repair. The relationship between the birth and death rates and the state transition rates and probabilities are

$$P_{i,i+1} = \frac{\lambda_i}{\lambda_i + \mu_i}, \qquad i > 0 \tag{8.21}$$

$$P_{i,i-1} = \frac{\mu_i}{\lambda_i + \mu_i}, \qquad i > 0 \tag{8.22}$$

Let $P_i(t)$ represent the probability that the system is in state i at time t. The system of differential equations describing the birth and death process is given by:

$$\frac{dP_0(t)}{dt} = -\lambda_0 P_0(t) + \mu_1 P_1(t) \tag{8.23}$$

$$\frac{dP_i(t)}{dt} = -(\lambda_i + \mu_i)P_i(t) + \lambda_{i-1}P_{i-1}(t) + \mu_{i+1}P_{i+1}(t) \tag{8.24}$$

$$\frac{dP_n(t)}{dt} = \lambda_{n-1}P_{n-1} - \mu_n P_n \tag{8.25}$$

The *limiting probabilities* ($\lim_{t \to \infty} P_i(t) = p_i$) exists when $\lambda_i > 0$ and $\mu_i > 0$. It is easy to solve the above system of equations, and it can be shown that the probabilities p_i are given by

$$p_i = \frac{\lambda_0 \lambda_1 ... \lambda_{i-1}}{\mu_1 \mu_2 ... \mu_i} p_0, \qquad i = 1,2,....n \tag{8.26}$$

$$p_0 = \frac{1}{1 + \sum_{k=1}^{n} \prod_{j=0}^{k-1} \frac{\lambda_j}{\mu_{j+1}}} \tag{8.27}$$

Where p_i is the steady-state probability that there are i items under repair or waiting for repair. The above equations can be used to find the fill rate for different types of problems. Some of those are discussed below.

Case 1. Single repair facility

Consider an item with a single repair facility. Assume that (n-1) spares are stocked initially, and whenever an item fails, it is sent for repair. Assume that λ and μ represent the failure and repair rate of the item respectively. The transition rates λ_i and μ_i are given by

$$\lambda_i = \lambda, \qquad i = 0, 1, 2, \ldots n\text{-}1$$

$$\mu_i = \mu, \qquad i = 1, 2, \ldots, n$$

Steady-state probability that i items are under repair (or waiting for repair) is given by:

$$p_i = \left(\frac{\lambda}{\mu}\right)^i p_0$$

$$p_0 = \frac{1}{1 + \sum_{k=1}^{n}\left(\frac{\lambda}{\mu}\right)^k}$$

As there would be any spares when the system is in state n, the fill rate is given by:

$$Fill\ \ rate = 1 - \frac{(\lambda/\mu)^n}{1 + \sum_{k=1}^{n}(\lambda/\mu)^k} \qquad (8.28)$$

Example 8.5

Time to failure of a gearbox in an armoured vehicle can be modelled using an exponential distribution with mean distance between failures 2500 miles. In calendar time this is approximately equal to 20 weeks. It takes about four weeks to repair a failed gearbox. An ILS manager in charge of this armoured vehicle stocks 2 spare gearboxes. Find the steady state fill rate. Also find the probabilities of 0, 1 and 2 gearboxes under repair or waiting for repair. Assume that there is only one repair facility.

SOLUTION:

The mean distance between failure (in calendar days) = 20 Weeks, this means the failure rate $\lambda = 0.05$ per week. The failure rate $\mu = 1/4 = 0.25$ per week. There will be a stock out when the process is in state 3, that is there are 3 gearboxes waiting for repair (2 spares and one in the vehicle). The corresponding probability is given by

$$p_3 = \frac{(\lambda/\mu)^3}{1 + \sum_{k=1}^{3}(\lambda/\mu)^k} = \frac{(0.05/0.25)}{1 + \sum_{k=1}^{3}(0.05/0.24)^k} = 0.00641$$

Now the fill rate is given by:

Fill rate $= 1 - p_3 = 0.99359$

The probabilities, p_0, p_1 and p_2, that is, there will be 0, 1 and 2 gearboxes under repair or waiting for repair is given by

$$p_0 = \frac{1}{1 + \sum_{k=1}^{3}(\lambda/\mu)^k} = \frac{1}{1 + \sum_{k=1}^{3}(0.05/0.25)^k} = 0.8012$$

$$p_1 = (\lambda/\mu)^1 p_0 = (0.05/0.25)^1 \times 0.8012 = 0.16025$$

$$p_2 = (\lambda/\mu)^2 p_0 = (0.05/0.25)^2 \times 0.8021 = 0.03205$$

Case 2. Model for a fleet of k units with r repair facilities and (n-k) spares

Now let us consider a fleet of k items with (n-k) spare parts. Also assume that there are r repair facilities. Here again we assume that the failure and repair times are exponential with mean failure time $(1/\lambda)$ and mean repair time $(1/\mu)$. This situation can be modelled using birth and death process with state space $\{0, 1, 2, \ldots, n\}$, where state i denotes the number of items under repair or waiting for repair. The birth and death rates of the process are as given below:

$$\lambda_i = \begin{cases} k\lambda, & i = 0,1,\ldots,n-k \\ (n-i)\lambda, & i = n-k+1, n-k+2,\ldots,n-1 \end{cases}$$

$$\mu_i = \begin{cases} i\mu, & i = 1,2,\ldots,r \\ r\mu, & i = r+1,\ldots,n \end{cases}$$

Using equations (8.26) - (8.27), one can find fill rate for a given fleet of size k with (n-k) spare parts and r repair facilities.

Example 8.6

In the previous example, assume that 5 spare gearboxes are stored for a fleet of 4 armoured vehicles. The repair is carried out by two repair facilities dedicated for gearbox failures. Find the fill rate by assuming same failure and repair rate as discussed in previous example. Also find the probability that the whole fleet of armoured vehicles will be grounded due to gearbox failure.

SOLUTION:

The state space of the birth and death process is {0, 1, ..., 9}. We have, fleet size k = 4 and the number of spare gear boxes n-k = 5. The birth and death rates are given by:

$$\lambda_i = \begin{cases} 4\lambda, & i = 0,1,\ldots,5 \\ (9-i)\lambda, & i = 6,\ldots,8 \end{cases}$$

$$\mu_i = \begin{cases} i\mu, & i = 1,2 \\ 2\mu, & i = 3,\ldots,9 \end{cases}$$

Probability that there will be no gearboxes under repair, p_0, is given by:

$$p_0 = \frac{1}{1 + \sum\limits_{k=1}^{9} \prod\limits_{j=0}^{k-1} \left(\lambda_j / \mu_{j+1}\right)} = 0.4290$$

As long as the process is in states 0 to 5, there will be no spare stock out, Thus fill rate can be calculated by subtracting the summation of probabilities p_6, p_7, p_8 and p_9 from 1.

$$p_6 = \frac{\lambda_0...\lambda_5}{\mu_1...\mu_6} p_0 = 0.00351 \quad, \quad p_7 = \frac{\lambda_0...\lambda_6}{\mu_1...\mu_7} p_0 = 0.00105$$

$$p_8 = \frac{\lambda_0...\lambda_7}{\mu_1...\mu_8} p_0 = 0.00021, \quad p_9 = \frac{\lambda_0...\lambda_8}{\mu_1...\mu_9} p_0 = 0.000021$$

Fill rate $= 1 - (p_6 + p_7 + p_8 + p_9) = 0.9951$

Probability that the whole fleet is grounded is given by $p_9 = 0.000021$.

8.6.2 Palm's Theorem and its Application in Spares provisioning for repairable items

Palm a Swedish statistician, working for Ericsson's the Telephone Company, proved that the probability that a telephone call would be lost depended in a simple way, on the mean time between calls arriving at the exchange and the mean duration of these calls. From this he was able to determine how large a switchboard should be and how many lines would be needed for any two switchboards. For the purpose of spare parts provisioning, Palm's theorem is defined as follows

If the failure distribution for an item is given by exponential distribution with mean time between demand $(1/\lambda)$ and if the repair time for each failed item is independent and identically distributed with mean time to repair MTTR, then the steady-state probability distribution for the number of items in repair has a Poisson distribution with mean $(\lambda \times MTTR)$.

That is, regardless of the repair time distribution, the distribution of the number of items in repair is Poisson with the following probability mass function

P[number of items in repair = n] $= \dfrac{\exp(-\lambda \times MTTR) \times (\lambda \times MTTR)^n}{n!}$

(8.29)

In Kendall-Lee notation, Palm's theorem describes a M/G/∞/GD/∞/∞ queue, and for this reason, is sometimes referred as '*infinite channel queuing assumption*'. Although the Poisson distribution is discrete and is quite heavily skewed for low means, it is quite common to use a formula that is based on the normal approximation. Because the mean and variance of the Poisson distribution are identical the formula is simply:

$$S_\alpha = \left[\frac{MTTR}{\lambda} + z_\alpha \sqrt{\frac{MTTR}{\lambda}} \right] \uparrow \qquad (8.30)$$

where z_α is the normal variate for the α percentile. The $\uparrow$ indicates that the value should be rounded up to the next nearest integer. S_α is then the recommended number of spares required. This is often referred to as the α% fill rate. Strictly speaking the Poisson distribution approaches the normal distribution asymptotically as the mean increases.

Combining Palm's and Drenick's theorems we can obtain an expression for the number of spares required of a system or complex LRU. Suppose an LRU consists of n components whose time to failure distributions are known, or more particularly, their mean times to failure are known and are given by $1/\lambda_i$ (for i = 1 to n). Suppose, also that the mean time to recover the LRU, given the cause of the LRU rejection is component i, is μ_i then the steady-state probability distribution for the number of items in repair has a Poisson distribution with mean

$$E[\text{No O-o-S}] = MOOS = \sum_{i=1,n} \lambda_i \mu_i$$

and

$$P[\text{number of items O-o-S} = n] = \frac{\exp(-MOOS) * MOOS^n}{n!}$$

where No. O-o-S is the number out-of-service.

Note that these expressions are only valid once the system/LRU reaches steady-state and, it assumes that all failures and recovery times are independent.

Occasions when Palm's Theorem may not be appropriate

Palm's theorem applies when the times between failures are exponentially distributed and the recovery/repair times are independent and identically distributed. At the LRU level, due to Drenick's theorem, the assumption that the times between failures will tend to become exponentially distributed as the complexity (of the LRU) and (operational) time increases.

In cases where the number of systems operating is relatively small and most the of causes of LRU rejection are heavily age-related, as is very often the case with aero-engines, Drenick's theorem may not start to apply until well into or even beyond the life of the fleet. This will be particularly true for systems in which most of the maintenance is preventive due to lifing (i.e. when a component is rejected because it has exceeded its age limit or hard life).

For a system which consists of a number of components whose mean times to failure are given by $1/\lambda_i$ Drenick's theorem states that the mean time to failure for the system is given by $1/\lambda$ where $\lambda = \Sigma\lambda_i$. However, this is only valid if the failures are independent.

Example 8.7

Suppose that there are 10 parts in an engine that each has a hard life of 1000 hours then they will each have a "failure rate" $(\lambda_i) = 0.001$. Using the formula, $\lambda = 0.01$ giving $1/\lambda = 100$ so, if the engine is used for 1000 hours we would expect to see 10 engine rejections (on average), or would we? If each part is new at the start of the 1000 hours and the 1000 hours hard lives is based on a probability of failure of 1 in 1000 (the $B_{0.1}$ percentile) then it is extremely unlikely that any of parts will have failed therefore all of them will be rejected simultaneously after exactly 1000 hours. This means that rather than there being 10 engine rejections, we would only expect 1 but, we would expect to replace all 10 of the parts at the same time.

Taking this a stage further, suppose the mean time to recover the LRU is the same for each of these parts at 10 days. Just as we would not expect the number of engine removals to be 10 so, we would not expect the time to recover the engine to be $10*10 = 100$ days but, much closer to just 10 days. The recovery of an engine is usually achieved by stripping the engine to its modules, replacing each rejected module with a serviceable spare and rebuilding the engine from this new set.

This example may sound particularly contrived but, in fact, it is actually quite common. True, there will often be other causes of engine rejection. These will be less strongly age-related but, when the hard lives are very similar, or, as in this case, identical it is quite common to re-build the engines using as much as possible of the original engine. This keeps the ages of the lifed parts closely aligned so when one of them reaches its hard life, most, if not all of the others, will be due for replacement.

When a component in an engine fails (as anyone who has suffered a broken cam or timing belt will know) damage is quite likely to be caused to other parts of the engine. This is even more common with gas turbines than it is with internal combustion engines. Very often, it is this *secondary damage* which actually causes the engine to malfunction. Many parts also suffer minor damage from various factors ranging from ingested runway debris to hard carbon (soot) deposits breaking off the lining of the combustion chamber to sand erosion or salt corrosion. When the engine or its modules are stripped, this damage may be found during inspection and the affected parts will then be rejected. If the times to failure of these parts are age-related (e.g. described by a Weibull with a shape greater than 1) then replacing them before they have actually failed will reduce the number of engine removals. Similarly, if several components are replaced at the same time then the total recovery time is also likely to be significantly less than if each recovery involves one and only one component.

Another factor, which may cause Palm's theorem to be invalid, is that, in most cases, the number of spares is not infinite. This means the recovery time may increase significantly if a spare is not available when required. The effect of shortages will naturally depend the "fill rate".

If we are dealing with systems whose failures are mainly age-related then, the number of spares will affect the elapsed time to failure. This is sometimes referred to as the *dilution effect* and is perhaps best illustrated by a simple example.

The average family car has five wheels. If it is a front-wheel-drive car then the two front tyres will tend to wear out quicker than the two rear tyres which, in turn, will wear out rather quicker than the spare. Now, if you rotate these tyres so that the front off-side becomes the rear off-side which becomes the spare which becomes the rear near-side which becomes the front near-side every month, say, then each tyre will receive roughly equal usage. The result of this is that the time between replacements will be increased but, it is likely that all five tyres will need to be replaced simultaneously.

Now consider a fleet of aircraft. Each aircraft has an engine which is an assembly of modules, some of which containing parts with hard lives. If such a module is replaced due to the failure/rejection of a non-lifed part then the lifed parts in the removed module will stop accumulating usage (hours). If such occurrences are relatively frequent then the more spare modules there are available, the less time each will tend to spend in service so it will tend to take longer for the lifed parts to achieve these lives. This will then tend to reduce the number of engine removals (due to lifing) and hence the out-of-service time leading ultimately to a reduction in the need for spare modules and possibly spare engines. We will return to these complications later.

8.7   MULTI-INDENTURE SPARE MODEL

If a system consists of a number of LRUs which, in turn, consist of a number of SRUs then the number of spare SRUs will affect the availability of the LRUs which will affect the availability of the system. In practice, we would expect the demand on each of the different types of LRU to be different. We would also expect the cost of a spare LRU to be dependent on the type of LRU. The problem we now look at is how to decide how many spares of each LRU we should hold.

In a multi-indenture system, recovery of the system is usually achieved by replacement of an LRU which, in turn will usually be recovered by the replacement of one, or more, SRUs. The time the system will be out-of-service will depend on how long it takes to identify and remove the offending LRU, acquire a suitable replacement (LRU) and install this replacement. Of this time that the system is out-of-service, the only time that is relevant to determining the number of spare LRUs needed is the time it takes to "acquire a suitable replacement".

The actual acquisition time will depend on how long it takes to move a spare from where it is being held to where the system is being recovered. It will also, of course, depend on how long it takes for a spare to become available. If the operator decides not hold any spare LRUs then the acquisition time will be the time to move the LRU to a workshop, identify the failed SRU(s) within it, remove these SRUs, acquire replacements, re-build the LRU, carry out any necessary testing and "adjustments" and, finally, move the recovered LRU back to the system. Having one, or more, spare LRUs will only reduce this acquisition time if the time to (find and) move the LRU to the system is less than the time it takes to recover the removed LRU.

If the time to remove the LRU is significant, then the system recovery time could be reduced if the acquisition of the replacement LRU is initiated as soon as the cause of the system failure has been identified (rather than waiting for the LRU to be removed). In some cases, this may not be possible because it may be necessary to remove the LRU to test it in order to determine whether it is, in fact, the cause of the system failure. Indeed, for systems with poor, or non-existent, diagnostic capabilities, it may be necessary to remove several LRUs before finding the offending one.

In other systems, it may only be possible to determine whether an LRU has failed by running the system. In the early days of televisions, it was quite common for the maintenance person to remove and replace each of the "cards" in the set in turn until a faulty one had been identified in much the same way as one might trace the failed light bulb in a set of Christmas tree lights. If in the unfortunate event, two, or more of the boards happened to have failed (or, if one, or more, of the replacements was faulty) then the set would inevitably be taken away.

Whenever a demand for a first indenture (LRU) occurs at the base, a spare LRU is used if available at the site or a backorder is placed. The failure of an LRU could have been caused by any of the comprising SRUs. The failed SRU is identified and sent for repair. The following results on addition of Poisson processes can be used to analyse the multi-indenture problems. If $N_1(t)$ and $N_2(t)$ are two Poisson processes with rates λ_1 and λ_2 respectively, then the addition of these two processes, $N_1(t) + N_2(t)$, is also a Poisson process with rate $\lambda_1 + \lambda_2$. This is a very important result, which is very useful in predicting demand at system level. The following Figure illustrates the relation between individual Poisson processes and the addition of Poisson processes.

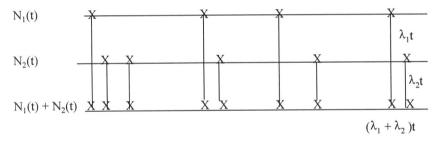

Figure 8.9. Addition of two Poisson processes

The above result means that, for a system with n items with demand rate λ_i for item i, the demand rate for the system is given by the addition of the demand rates of the items, $\lambda_1 + \lambda_2 + \dots + \lambda_n$.

Example 8.8

In this example, we consider demand for a system with many items connected in series for reliability purpose. For example, consider a system with four items connected in series. Let the time to failure of all the four items follow exponential distribution with rate λ_i. Where $\lambda_1 = 0.05$, $\lambda_2 = 0.002$, $\lambda_3 = 0.008$ and $\lambda_4 = 0.005$. Find the expected number of demand for the system for t = 400 and t = 1000.

SOLUTION:

Since the times to failure of all the items are exponential, the demand process for each of these items follows a Poisson process. Since the addition of Poisson processes is again a Poisson process, the demand for the system is also a Poisson process with rate $\sum_{i=1} \lambda_i$.

The expected number of demands for the system for t = 400 and t = 1000 is given by:

$$E[\,N(400)\,] = [\,\sum_{i=1}^{4} \lambda_i\,] \times 400 = 0.065 \times 400 = 26$$

$$E[\,N(1000)\,] = [\,\sum_{i=1}^{4} \lambda_i\,] \times 1000 = 0.065 \times 1000 = 65$$

The fact that the system is multi-indentured really has little effect on the method for estimating the number of spares required. If a component at a given level of indenture is made up of a large number of parts, the failure of any of which will cause the failure of the parent (component) and hence the system then we can use Palm's and Drenick's theorems to give us an approximation of the number parent components needed to support the system to a given level of availability.

As we have seen earlier, if some of these failures are age-related and the maintenance policy includes the provision for preventive and/or opportunistic maintenance then we can no longer add the failure rates as this will tend to give a system/LRU/parent failure rate which is too high.

8.8   MULTI-ECHELON SPARE MODELS

In multi-echelon spare parts management, the main objective is to optimally distribute the spares across different locations. In a typical multi-echelon inventory of spares, the first echelon maintains inventories with line replaceable units. In this book, we use the term organisational level to represent the first echelon. The second echelon (intermediate or 2nd echelon) supports each of the first echelon. Both first and second echelon is supported by third echelon called depot. The establishment of different echelons for sparing rather than single echelon has proven to be cost effective. In this section we discuss a simple two-echelon model.

Two-echelon models

In the two-echelon model, the bulk of spares are usually held at a depot, which supplies the higher level (base) facilities. The number of spares held at the bases will then be kept to a minimum. Normally, such a policy would only be used if the component was recovered at the deeper echelon (depot) as few base managers would be very enthusiastic about sending a serviceable component away to a deeper echelon so that it could be used by another base. There is also the added disadvantage that once the serviceable component has left the first echelon (base), it is not available whilst it is in transit. In extreme circumstances, if components are transported by ship, the transit time could be several days, weeks or even months.

In a two-echelon model, whenever there is a failure (or removal) at the first echelon, it would be replaced with a first echelon spare, if one is available and the failed item would then be sent for recovery at the second echelon. At the same time, the first echelon would issue a demand on the second echelon for a spare to replace the one used. An alternative policy, sometimes referred to as *hole-in-the-wall,* is when the second echelon delivers a spare at the same time as collecting the failed item. This policy is particularly common when, the second echelon is run by a contractor, rather than the operator.

Given that none of the first echelon sites have shortages (or outstanding demands), the failed item, once it had been recovered, would be held at the second echelon. Provided there are sufficient spares in the system, the first echelon sites should never have to wait for spares to recover the operational units. However, in determining how many spares are needed, in total, it is important to recognise that when a component is in transit, it is not "available" whether it is being moved to the second echelon for recovery or back to first as a serviceable spare.

The calculations of how many spares are required and where is done in two parts. We need to determine how many spares should be held at the first echelon sites and then how many at the depot or second echelon. If we assume that there are an infinite number of spares at the depot then we need only hold sufficient spares at each first echelon site to cover the transit time from the depot to the given site.

If we assume that the times between failures are exponentially distributed with mean $1/\lambda_i$ and the mean transit time from the depot to the given (first echelon) site i is μ_i then the expected number of demands during the transit time is $\lambda_i\mu_i$ using Palm's theorem. Thus, the number that needs to be held at site i to give a fill rate of α can be determined using the Poisson distribution with parameter $\lambda_i\mu_i$. That is, the required number of spares to meet a fill rate of α, is given by the minimum n+1 such that

$$\sum_{k=0}^{n} \frac{\exp(-\lambda_i\mu_i)\times(\lambda_i\mu_i)^k}{k!} \geq \alpha \qquad (8.31)$$

Similarly, assume that λ denotes the constant arrival rate of items to the second echelon, where $\lambda = \Sigma\lambda_i$. Let μ now denote the mean time to repair each item. Now using the Poisson process model, we can find the probability that a spare is available at the depot when requested.

Using Palm's theorem, the mean number of items that are under repair or waiting for repair (i.e. out of service) is given by $(\lambda\mu)$. If the required fill rate at the depot is α, then the number of spares required to meet the fill rate is calculated by computing the least n for which the cumulative Poisson distribution with parameter $(\lambda\mu)$ is greater than or equal to α. That is, the required number of spares to meet the fill rate of α, is given by minimum n+1 for which

$$\sum_{k=0}^{n} \frac{\exp(-\lambda\mu) \times (\lambda\mu)^k}{k!} \geq \alpha$$

For example, assume that the removal rate for site i $(\lambda_i) = 0.005$ for i=1, 10). Assume also that the depot to base i transit time $(\mu_i) = 2$ for i=1, 10 then the expected number of arisings during the time a spare is in transit is $(\lambda_i\mu_i) = 0.01$. If we want a fill rate $(\alpha) = 0.995$ then using the above formula gives n+1 = 2. This means that we need to hold 2 spares at each base (or 20 spares in total).

Now, at the depot, the failure rate $\lambda = 0.05$ and the mean time to repair including the base to depot transit time $\mu = 50$. Then the mean or expected number of components in recovery is $(\lambda\mu) = 2.5$. Assume that the required fill rate is again 0.995. The cumulative Poisson probabilities for $(\lambda\mu) = 2.5$ are tabulated in Table 8.4. For n = 7, the cumulative Poisson probability is greater than the required fill rate 0.995 for the first time. Thus, the required number of spares is 7 + 1 = 8 spares. The logic used in this analysis is that, even if there are 7 spares already at the repair facility (as calculated using Palm's theorem), there will be a spare when a demand occurs.

From these two calculations, we find that we need a total of 28 spares – 2 at each base and 8 at the depot.

Now let us consider the same scenario except this time, the repairs are done at the bases such that they still take a mean time of 50 days but now there is no transit between the bases and the depot. For base i, the expected number out of service is given by $\lambda_i\mu = 0.005 \times 50 = 0.25$. Using the Poisson distribution again gives n = 2 for $\alpha = 0.995$ so we would need to hold 3 spares at each base or 30.

Table 8.4. Cumulative Poisson probability

n	$\sum_{k=0}^{n} \exp(-2.5) \times (2.5)^{k} / k!$
0	0.0821
1	0.2872
2	0.5438
3	0.7575
4	0.8911
5	0.9579
6	0.9858
7	0.9957
8	0.9988
9	0.9997

This is strictly speaking not a fair comparison, as it will be recalled that the number of spares in the first case at the base level was calculated assuming an infinite number of spares at the depot. The fact that there was a chance of there not being a spare at the depot might increase the numbers needed at the bases. This is because the transit time would be increased slightly to take account of the delay while waiting for a spare to become available on those few occasions when there was not one.

Provided all of the times to failure at every base are exponentially distributed and the recovery times are independent then the rates do not need to be the same. They were so chosen in the example to make the arithmetic easier. If the times to failure are age-related or the recovery times are not independent then the above will only give an approximate result – just how approximate will depend on how age-related and dependent the times are.

8.9 SPARES PROVISIONING UNDER SPECIAL CONTRACTUAL REQUIREMENTS

No customer would like to pay for the spares that are not used within an acceptable period of time. Especially, if the parts are costly. There are several concepts being tried by many commercial and defence industries to overcome the problems faced in provisioning spare parts. The main objective of all these concepts is to reduce the investment in spares. One such concept is 'buy back excess spares', under this contractual requirement, the supplier is required to buy back the excess spares if difference between the number of spares supplied and consumed is more than about 5-10% of the actual demand.

Consider an item with fleet demand rate 0.05 per hour. If it is required achieve a fill rate of 85% during 500 hours of operation, then the minimum number of spares required is 31. However, the expected demand during 500 hours is $0.05 \times 500 = 25$. Adding 10% to 25 we get the maximum number of spares that the supplier can issue without any risk is 28. That is 3 less than the acceptable contractual requirement. That is the supplier may have to buy back 3 spares. Suppose if the supplier also has to pay penalty if he supplies fewer spares than the actual demand, then the problem becomes more complicated.

Power by the Hour

'Power by the hour' is a leasing contract for engines, under which the user pays a fixed price between $1000 to $2000 per hour (depending on the type of engine), and simply returns it if there is any fault and get a new serviceable engine. Ideally, the user is not expected to maintain any spares. The supplier maintains all the spares required supporting the engines. *'Power by the hour'* is particularly popular among many commercial airlines compared to defence aircraft. Table 8.5 gives the leasing costs for some of the popular engines used in commercial aviation as reported in *Aviation Economics*.

Table 8.5 Leasing cost of some popular engines.

Engine Type	Market Value	Daily lease rate	Lease Term	Maintenance cost per hour
CFM56-2	$ 2.4 M	$ 1,750	90 days	$ 125
CFM56-3	$ 2.8 M	$ 1,850	90 days	$ 125
JT9D-7A	$ 2.3 M	$ 1,850	90 days	$ 175
JT9D - 7	$ 3.0 M	$ 2,200	90 days	$ 225
JT9D - 7Q	$ 3.3 M	$ 2,400	90 days	$ 250

These lease rates are usually on the condition that the lease term is no less than 30 days, and most lessors prefer leases to run for as long as six months. The engine shop visit can take as long as 90 days. The economics of buying versus leasing is relatively easy to assess. Although the lessee has to pay both rental and flight hour maintenance for the engine, it also has to pay similar maintenance amount if he owns the engine. The difference in cost is therefore purely one of comparing the cost buying versus lease for the length of the shop visit.

8.10   INVENTORY MANAGEMENT

Managing spare parts inventory in the cost effective manner plays a crucial role in the support chain. Even though there have been many attempts to reduce the inventories by applying techniques such as *just-in-time* etc., the annual investment in inventories by manufacturers, retailers, and wholesalers continue to be about 16% of the US Gross National Product. Among defence industries the amount of spare parts inventory is very high, as all the armed forces must maintain high operational availability for their equipment. The important questions to be answered

1.   What to stock
2.   How much to stock
3.   When to reorder or replenish the inventory.

The objectives of the inventory management include:

1.   To achieve required operational availability for the equipment cost with minimum investment in inventory.
2.   To minimise the expected backorder to assure highest availability of spares when required.
3.   Efficient procurement and forecasting the spare parts requirement.
4.   Minimise warehouse costs.
5.   Maintain efficient replenishment of inventories to maintain the required operational availability and minimum cost.
6.   To maintain efficient transportation of inventories between different echelons.
7.   To maintain efficient IT (information technology) support for inventory management.

The most important objective would be to assure operational readiness of the equipment with minimum investment in inventory. The main cost drivers in inventory management are procurement cost (setup cost), unit purchasing cost, holding or carrying cost and shortage cost.

Procurement cost

Cost associated with procurement of items include cost of paper work and billing associated with the order, following up on the orders, receiving the items and updating the inventory records. Procurement cost exists even when the items those are ordered internally. Usually, the procurement costs do not depend on the size of the order.

Unit purchasing cost

Unit purchasing cost refers to the variable cost associated with purchasing a single unit. Purchasing cost includes unit production cost, overhead costs, profit and transportation cost. Sometimes, the suppliers might reduce the unit purchasing cost if larger order is placed. Such price reductions are refereed as *quantity discount.*

Carrying cost

Carrying cost or holding cost is the cost of carrying (or holding) one unit of inventory for unit time period. Carrying cost includes storage cost, insurance costs, taxes on inventory, interest, handling, obsolescence and depreciation costs. The most significant element of carrying cost is the opportunity cost incurred by tying up the capital in inventory. They could be invested in many ways such as new equipment or research and development.

Backorder Cost

When there is a demand for a product and the demand is not met on time, a backorder is set to occur affecting the operational availability of the system. It is estimated that the delay cost for large airlines is in the region of $1000 per minute. When an aircraft is grounded for a day, it can cost the airline up to $900,000. Usually, the cost of backorder is harder to measure compared to procurement cost and carrying cost.

8.10.1 Economic Order Quantity

Replenishment of spare parts is repeated in a regular fashion. A reorder can be placed at anytime (*continuous review models*) or periodically (*Periodic review models*). The Economic Order Quantity (EOQ) models try to find the optimal order quantity for which the total inventory cost is minimised. The total cost is sum of the procurement cost, carrying cost, unit cost and backorder cost. As shown in Figure 8.10, when the quantity ordered increases the procurement cost decreases but carrying cost increases. The optimal order quantity will be the value for which the total cost is minimum.

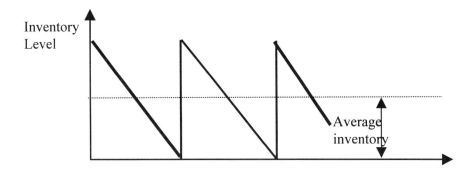

Figure 8.10 Inventory Pattern

Assume that:

1. Demand occurs at a constant rate.
2. Whenever an order of size 'q' is placed, procurement cost K is incurred.
3. The lead time for each order is negligible.
4. No backorders are allowed.
5. The cost of carrying the inventory is h.
6. Unit purchasing cost is p.

Assume that the λ is the demand rate per year. Then during any time interval t, the total demand will be λt. Let $TC(q)$ be the total annual cost incurred when q units are ordered.

$TC(q)$ = Annual procurement cost + Annual purchasing cost + Annual carrying cost

Since the order quantity is q units, (λ/q) orders per year will have to be placed. The annual procurement cost per year is given by

$$\text{Annual procurement cost} = \frac{K\lambda}{q} \tag{8.32}$$

Annual purchasing cost $= \lambda p$ (8.33)

The annual carrying cost = Carrying cost per cycle x Number of cycles per year

The average inventory is $q/2$ and the cycle length is q/λ. The annual carrying cost is given by

$$\text{Carrying cost per cycle} = \frac{q}{2}(\frac{q}{\lambda})h = \frac{q^2 h}{2\lambda}$$ (8.34)

$$\text{Annual carrying cost} = \frac{q^2 h}{2\lambda} \times \frac{\lambda}{q} = \frac{hq}{2}$$ (8.35)

Now the total inventory cost per annum is given by:

$$TC(q) = \frac{K\lambda}{q} + p\lambda + \frac{hq}{2}$$ (8.36)

To find the optimal q, we equate the first derivative of $TC(q)$ to zero. That is,

$$\frac{TC(q)}{dq} = -\frac{K\lambda}{q^2} + \frac{h}{2} = 0$$ (8.37)

Solving the above equation, the optimal economic order quantity is given by

$$q^* = \sqrt{(\frac{2k\lambda}{h})}$$ (8.38)

The optimal cycle length is given by

$$\textit{Optimal cycle length} = \frac{q^*}{\lambda}$$ (8.39)

Example 8. 9

Annual demand for a consumable spare for a fleet of five A310 is 2400. The airline orders the consumables periodically. It costs $1200 to initiate a purchase order. An estimated carrying cost for the consumable per year is $ 140. If the lead time is 30 days find the optimal order quantity.

SOLUTION

We have

$\lambda = 2400$ per year

$K = \$1200$ per order

$h = \$140$ per unit per year

Lead time $= 1/12$ years

The optimal order quantity is given by

$$q^* = \sqrt{\frac{2\lambda K}{h}} = \sqrt{\frac{2 \times 2400 \times 1200}{140}} = 202.83 \approx 203 \text{ units}$$

The associated cycle length is $= q^* / \lambda \approx 31$ days

8.11   OPTIMISATION METHODS

One of the important tasks in spare parts provisioning is to find the optimal spares allocation for different items within a system usually constraint on availability, volume, weight etc. The objective in many cases is to minimise the cost. Extensive research has been carried out in spare parts optimisation with the objective of minimising cost (or maximising availability) subject to constraints on minimum required availability (or maximum available cost). Traditional optimisation models such as dynamic programming and non-linear programming can be used to for optimising spares. However, due to the complexity of the problems due to the number of items (could be as high as a million), we may have to use either simulation or evolutionary programming techniques such as genetic algorithms.

8.11.1 Zero-one non-linear integer programming Model

Consider a reliability block diagram with n items connected in series. Assume that the time to failure distribution of the component *i* in the system follows an arbitrary distribution with cumulative distribution $F_i(t)$. Let c_i represent the cost of a unit spare and B be the maximum available budget. Also we assume that N_i is the maximum number of spare parts allowed for item i. The objective of the optimisation problem is to maximise the spare parts availability for the system satisfying the budget constraint. The mathematical programming formulation of the problem can be stated as follows. Let

$$z_{i,j} = \begin{cases} 1, & \text{if } j \text{ spares are allocated for component } i \\ 0, & \text{otherwise} \end{cases}$$

The availability of spares for item i, $A_i(T)$, when ever there is a demand for a spare under the assumption that the spares are not repaired, is given by

$$A_i(T) = \sum_{j=0}^{N_i} \sum_{k=0}^{j} [F_i^k(t) - F_i^{k+1}(t)]z_{i,j} \tag{8.40}$$

where, $F_i^k(t)$ is the k-fold convolution of the distribution function $F_i(t)$. The optimisation problem can be written as

$$\text{Maximise } A_s(T) = \prod_{i=1}^{n} A_i(T) \tag{8.41}$$

Subject to:

$$\sum_{i=1}^{n} \sum_{j=0}^{N_i} j \times c_i \times z_{i,j} \leq B \tag{8.42}$$

The above optimisation problem is a 0-1 non-linear integer programming problem. The objective function of equation (8.41) indicates that exactly j (= 0, 1, ..., N_i) spares are available for component *i*. Equation (8.42) guarantees that the total cost is less than or equal to the available budget B. The above optimisation problem can be solved using many general-purpose software optimisers such as SOLVER of EXCEL.

8.11.2 Marginal Analysis

Marginal analysis is a heuristic optimisation technique that is used in allocating the scarce resources to maximise the benefit. Marginal analysis can be defined as the analysis of the benefits and costs of the marginal unit. At each step of the algorithm in spare parts optimisation marginal analysis is used to determine the next item that should be bought that optimises the benefit, usually the expected backorder or availability. The first step in marginal analysis involves identifying a *control variable,* in our case, the number of different spare parts that should be bought for a system. Once the control variable is identified, then marginal analysis focuses on whether the control variable should be increased by one or not. Thus the main steps in marginal analysis involves

- Identify the control variables
- Determine what the benefit would be if the control variable is increased by one unit of each type. This is the *marginal benefit* of the added unit of each type.
- Determine what will be the increase in the total cost if one unit is added to the control variable (again for each type). This is the *marginal cost* of the added unit.

Note that in general, the marginal benefit will tend to decrease for each additional spare of a given type whereas the cost will normally remain constant (unless there is a bulk discount).

In marginal analysis, marginal benefit refers to the increase in total benefit per unit of control variable. Now consider a system with reliability block diagram represented by a series configuration. Our objective is to minimise the total expected backorder at the system level subject to a cost constraint (B). Assume that s_i represents the level of stock for item i, also the control variable in the marginal analysis. Let c_i denote the cost of each spare for item i. The mathematical programming formulation of the problem is given by

$$Minimise \quad \sum_{i=1}^{n} EBO_i(s_i) \tag{8.42}$$

Subject to:

$$\sum_{i=1}^{n} c_i s_i \leq B \qquad\qquad (8.43)$$

In equation (8.42), $EBO_I(s_i)$ represents the expected backorder for item i with stock level s_i. The above optimisation problem can be solved using the Lagrange multiplier method (see Sherbrooke, 1992). To use marginal analysis to solve the above optimisation problem, we use the following steps:

1. Starting from $s_i = 0, 1, 2, ...$ the marginal benefit for each item i (= 1, 2, ..., n) is calculated using the following equation.

$$\delta_i(s_i) = [EBO_i(s_i - 1) - EBO(s_i)] / c_i$$

   Once the $\delta_i(s_i)$ (i = 1, 2, ..., n, s_i = 1, 2, ...) values are calculated, they are rearranged in the increasing order.

2. With out loss of generality, assume that the set S = {[1], [2], ...} represent the set with the $\delta_i(s_i)$ (i = 1, 2, ..., n, s_i = 1, 2, ...) values arranged such that [1] > [2] > [3] etc. Now one spare is added to the item [1], next to item [2] etc. until the allocated budget is consumed.

   It should be noted that the marginal analysis assumes that the objective function to be convex.

Example 8.10

   An LRU consists of three modules A, B and C. The time to failure distribution of each of these modules are represented by exponential distributions with mean lives of 400, 500 and 800 hours. The total budget available for initial spares provisioning for approximately 2000 hours is $60,000. Each spare for modules A, B and C costs $ 4000, $ 8000 and $ 10,000 respectively. Find the optimal allocation of spares that minimises the total expected backorder at the system level.

SOLUTION:

The expected backorder and the $\delta_i(s_i)$ for items A, B and C are shown in Table 8.6.

Table 8.6. Spares allocation using marginal analysis.

s_i	$EBO_A(s)$	$\delta_A(s_I)$	$EBO_B(s_i)$	$\delta_B(s)$	$EBO_C(s_i)$	$\delta_C(s_i)$
0	5	----------	4	----------	2.5	----------
1	4.0067	0.000248	3.0183	0.000123	1.5820	0.000091
2	3.0471	0.00024	2.1098	0.000114	0.8693	7.1 E −5
3	2.1718	0.000219	1.3479	9.52 E -5	0.4131	4.5 E −5
4	1.4368	0.000184	0.7814	7.08 E -5	0.1707	2.4 E −5
5	0.8773	0.00014	0.4103	4.64 E -5	0.0619	1.08 E −5
6	0.4932	9.6 E -5	0.1954	2.69 E -5	0.0199	4.2 E −6
7	0.2554	5.96 E -5	0.0846	1.38 E -5	0.0057	1.4 E −6
8	0.1221	3.3 E -5	0.0336	6.38 E -6	0.0014	4.3 E −7
9	0.0540	1.7 E -5	0.0122	2.68 E -6	0.0003	1.1 E −7
10	0.0221	7.98 E - 6	0.0041	1.10 E -6	0.00007	2.3 E −8

Now arranging $\delta_i(s_i)$ in decreasing order, we have S = {[1], [2], [3], ...} where, [1] = $\delta_A(1)$, [2] = $\delta_A(2)$, [3] = $\delta_A(3)$, [4] = $\delta_A(4)$, [5] = $\delta_A(5)$, [6] = $\delta_B(1)$, [7] = $\delta_B (2)$, [8] = $\delta_A (6)$, [9] = $\delta_B (3)$, [10] = $\delta_C (1)$, ...

At the first step of the algorithm, we make a decision to buy one spare for item A, as $\delta_A(1)$ is the largest among the $\delta_i(s_i)$ values. At the next step, again we make a decision to buy one more spare for item A as $\delta_A(s)$ is the next largest among the remaining $\delta_i(s_i)$ values. This process is continued until the total available budget is consumed. In the present problem, since the total budget is \$60,000, the decision will be to buy 6 spares for item A, 3 spares for item B and one spare for item C at the total cost of \$58,000.

8.12 SIMULATION

For a system which is unlikely to reach steady-state conditions during its working life or for one for which we wish to investigate aspects of its behaviour during the period before it has reached steady-state, simulation is a particularly useful tool. In the above sections, we have seen a number of factors, anyone of which would be sufficient to invalidate the assumptions required for either Drenick's or Palm's theorems. This, however, would not necessarily invalidate their use in a practical sense if the accuracy of the forecasts they produced were within acceptable limits or were considered to be as good as could be expected.

The fact is that there is a large number of operators of a large number of systems who have very large inventories of spare parts that they will never use whilst, at the same time, they are experiencing shortages of other parts. In many cases they may be forced into buying spares at inflated prices or from suspect sources in order to keep their systems operational.

Simulation is not a panacea. It will not solve all of these problems. And, like any other scientific method, it requires good, clean data. All scientific methods are based on the assumption that by studying what has happened in the past will help us predict what will happen in the future and that the more factors considered the more accurate one can expect that forecast to be. The way most forecasting models are validated is to take data from a given period, split them into two (mutually exclusive and exhaustive) subsets. The first subset is used to calculate the input parameters and produce the forecasts. These forecasts are then compared with the second subset to determine the levels of accuracy and confidence.

8.12.1 Monte Carlo Simulation

Monte Carlo simulation was given its name by the nuclear physicists working on the Manhattan project in Los Alamos. It was recognised that if f(x) is a probability density function then

$$E[x] = \int_{-\infty}^{\infty} xf(x)dx$$

and that the sample mean of random samples from the given distribution would approach E[x] (the population mean) as the sample size is increased. In addition, from the central limit theorem is was possible to determine the level of confidence one had in the estimate of the mean. This meant that they could evaluate certain integrals numerically using far fewer function evaluations than would have been necessary had they used standard numerical integration (e.g. the trapezoidal rule or Simpson's rule). Since they did not have access to fast computers or even electronic calculators, this meant they could make much better use of the teams of human computers.

Although there is no clear definition of what constitutes a "Monte Carlo" simulation as opposed to any other type (e.g. discrete event-based or continuous event-based) we shall use the name to denote simulations that do not involve a "clock". In particular, this includes the use of random numbers to evaluate mathematical functions, numerically.

Example 8.11

Suppose the times to failure for a turbine blade can be fitted by a Weibull distribution W[3, 10000]. Now suppose there are 64 of these blades in a set (spaced equally round a disc). What is the mean time to the failure of the first blade in the set? If only this blade is replaced during a repair, what will the mean time to the next failure be?

SOLUTION:
The cumulative density function (cdf) of the Weibull distribution is given by:

$$F(t) = 1 - e^{-\left(\frac{t}{\eta}\right)^{\beta}} \quad \text{for } t \geq 0$$

Now, $0 \leq F(t) \leq 1$ so, if we sample a random number (which are also between 0 and 1) then we can re-arrange this equation to give a value of t (corresponding to the sampled random number p):

$$t = \eta(-\log_e(1-p))^{\left(\frac{1}{\beta}\right)}$$

If we sample a sufficiently large number of times to failure, in this way, we would see that they form a Weibull distribution W[β, η].

To find the mean time to first failure, we need to sample 64 random times to failure and sort them in ascending order (or simply record the minimum). Now repeat this at least 1000 times recording the minimum of the 64 times for each sample. Now find the average of these 1000 minima. The value should be close to 2230. In fact, it can be proved, mathematically, that the times to first failure are Weibull W[β, $\eta/N^{(1/\beta)}$] where N is the number of blades in a set. In general, to improve the accuracy of the result by a factor of 10, it is necessary to increase the sample size by a factor of 100.

To find the mean time to the second failure, use the 1000 sets of 64 random times to failure but this time, take the second smallest in each set and find the average of these. The sample mean should be approximately 2973. The difference between these two means (i.e. the mean time between the first and second failures for Weibull distribution W[β, η] in which the system is repaired to the same-as-old condition) is approximately MTTFF/β where MTTFF is the mean time to first failure.

Note: to be mathematically correct, a 65th time to failure should be sampled such that this time is the minimum of the first 64 plus a time to failure sampled from the original Weibull distribution. The time to first failure should then be removed from the set and the remaining 64 again sorted. The new minimum time should now be used as the time to second failure. In practice the time to second failure is very unlikely to be the new time to failure and a very reasonable approximation is to do as we did and simply take the second lowest of the 64 *TTFs*.

Example 8.12

Suppose we have a system made up of a large number of components. Suppose one of these components (P1) has times to failure that can be modelled by a Weibull $W[\beta, \eta]$ and the remainder can be approximated by an exponential distribution $E_x[\lambda]$ $(\equiv W[1, \lambda]$). What is the "best" maintenance policy for the system if it is to be operated for 10,000 hours?

SOLUTION:

Let the expected cost to repair Part P1 be C_{P1R}.

Potential damage to other parts within the system is an expected cost of C_{SD}.

Let the cost of recovering the system plus system downtime costs be C_{SR}.

Let the cost of the part be C_{P1}.

The cost of a failure of Part P1 (if repaired) $C_{P1FR} = C_{SR} + C_{SD} + C_{P1R}$

The cost of a failure of Part P1 (if replaced) $C_{P1FN} = C_{SR} + C_{SD} + C_{P1}$

The cost of replacing Part P1 as a planned arising $C_{P1P} = C_{SR} + C_{P1}$

The cost of replacing Part P1 opportunistically $C_{P1O} = C_{P1}$

(Note: the actual costs will not always be very easy to determine.)

Method:

Set SL to soft life limit
Set ROL to repair/overhaul limit
Set PF, SF, PST, NREP, NOH $= 0$
Set No.of.Passes $= 0$

INITIALISE:
 Set T $= 0$
  Sample a TTF for P1 (from its Weibull) as P_1
  Set $P_2 = P_1$

Sample a TTF for the (rest of the) system as S_1
START:

Advance the "clock" to the $\min\{P_1, S_1\}$ (= T)

If T > 10000 Go To ANALYSE
If P_1 = T then   'part failure'
    add 1 to PF
    If P_1 < ROL then   'repair'
      Add 1 to NREP
      Sample $P_2 \mid P_1$
      Set $P_1 = P_2$
    Else   'replace'
      Add 1 to NOH
      Sample P_2
      Set $P_1 = P_1 + P_2$
    End if
else   'system failure'
    add 1 to SF
    If $P_1 > SL_1$ then   'sec t/x part'
      Add 1 to PST
      Sample P_2
      Set $P_1 = T + P_2$
    End if
    Sample S_2
    Set $S_1 = S_1 + S_2$
End if
Go To START
ANALYSE:
Add 1 to No.of.Passes
If No.of.Passes < 1000 Go To INITIALISE
Find averages and variations of NREP, NOH, PST, PF and SF
STOP

Having arrived at estimates of the numbers of arisings of each type, we can now determine the expected cost for the given repair/overhaul limit and soft life.

The next stage is to change one or both of the parameters and repeat the exercise. Since there are only two parameters (ROL and SL) and both have an upper limit of 10,000, it would not be a massive task to evaluate a full grid with both variables increasing in steps of 1000. Note it does not make sense to set the value of ROL above that of SL. If there is a definite minimum at a value of SL less than 10,000 then you could reduce the step size to 100, say, within the interval containing the minimum.

Figure 8.11 shows the results of such a simulation. In this case, the minimum cost appears to be when the soft life is high but the repair/overhaul limit is low. This is due to the fact that the number of system caused failures was high compared to the potential number of part failures (given that they were replaced rather than repaired during recovery). The MTBF for the system was 1000 and the part TTF distribution was W[3, 5000]. The cost of a part C_{P1} = 10. The cost of a system recovery C_{SR} = 30. The cost of secondary damage C_{SD} = 20 and the cost of repairing the part C_{P1R} = 5.

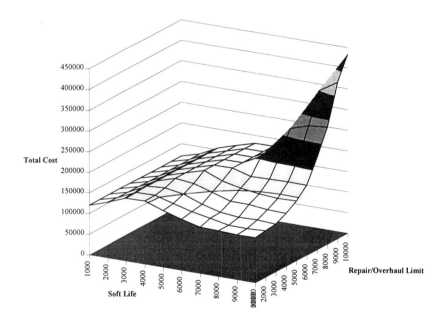

Figure 8.11 Simulation output

At this stage, if the "carpet plot" is not very smooth then it might be worth increasing the number of passes and re-running the cases in the immediate vicinity of the "minimum". In this particular case, the minimum was when the soft life was 9,000 and the repair/overhaul limit 1,000. It is likely that the actual minimum for these two sets of failure parameters would be when the soft life is 10,000 (i.e. no soft life limit) and the repair/overhaul limit 0 (i.e. always replace with new).

Having found the parameters which minimise the cost for one component, the exercise could be repeated for each of the other "prime drivers" that have age-related failures. To find the global optimum values, however, it would be necessary to write a model, which considers all such components simultaneously including any parts, which have hard lives (for safety-critical reasons).

In this example, the various costs were chosen quite arbitrarily. In practice, we are likely to know the cost of the part. We should be able to get a reasonable estimate of the likely amount of secondary damage and hence which other parts may need to be replaced and with what probability. This means our estimate of the secondary damage cost should also be reasonable. The cost of repairing a part (as opposed to replacing it) is more difficult as we would need to know what tools (special or common), manuals and procedures are needed, how long it will take and what skills and training the person will need.

When it comes to the cost of system recovery, the task becomes very much more difficult. One of the major factors affecting the time the system will be out of service will be the availability of a spare LRU to replace the one removed (to be recovered). This will, in turn, depend on how often the particular type of LRU has to be removed and how long it takes to recover each removal. As we have seen from the example, the frequency of LRU removals will depend on the maintenance policy. In cases where there are several components within the LRU that have age-related times to failure, the number of removals will also depend on the number of spare LRUs. The problem therefore becomes very much more difficult to solve.

8.12.2 Discrete Event Simulation

Whereas Monte Carlo simulation tends to be used to find numerical answers to often complex integrals, discrete event simulation is more likely to be used to model microcosms of the "real world". This technique has been used to design integrated iron and steel works, air and sea port design, hospital A&E facilities, equipment and resources, road layouts (including being responsible for the invention of mini-roundabouts) and aero-engine arisings forecasts and many more.

Although, it is generally considered by mathematicians as a method of last resort, many Operational Research scientists use it as a first line of attack. With discrete event simulation (hereafter referred to as "simulation") we are usually looking at a number of *entities* how they interact with each other and what calls they make on specific *resources*. The system being modelled is often described using *entity-cycle diagrams*. Typically, there will be a *source* or *generator*, which creates entities (e.g. customers to a bank, ships wishing to enter a dock, patients arriving at a hospital or engines requiring maintenance). These will then wait, in a queue, for one, or more, resources (e.g. a cashier, tug, pilot and berth, a triage nurse, doctor or consultant or, a workshop, mechanic and turnover stand, say). Once they have been allocated to the required resource(s) they are then *served*. At the end of the given time, the resources are *relinquished* and the entity enters another queue for the next stage in the process and so on until the particular entity leaves the system.

There may be many instances of "identical" entities, possibly entering the system at different times and requiring different services. For example, one ship may require a tug and a pilot, another may require two tugs and a third may be able to make its own way to the berth unaided by either tugs or pilots. Equally, there may be several different types of entity. We may have aircraft which are fitted with engines that comprise a number of different types of modules or sub-assemblies which in turn are assemblies of parts, say. We would keep track of each instance of each entity as it moves from one event to another through the system.

Such a system could, for example, be a fleet of military aircraft. Initially, each aircraft, in the fleet, will be delivered, probably according to a pre-defined (delivery) programme. Upon delivery, they will be allocated to a squadron, which is attached to a given base. They will also be allocated to a task performing a specific role. The task will define the number of hours each aircraft will be required to fly (on average) per month. The role will define the length of each sortie/mission and the type of flying involved (from which estimates of cyclic exchange rates may be made).

Once the task has started, each aircraft will be scheduled to fly sorties at certain times. Adjustments may be made to these schedules if any aircraft is unavailable at the time it is due to take-off. The detail of schedules will depend on the purpose of the simulation. If the main use is to determine the likely levels of availability, in terms, say, of numbers of delays and cancellations per 100 (scheduled) missions then the exact timings of the sorties may be appropriate. If on the other hand, the measure is in terms of the number of days aircraft spend on the ground due to logistic delay times then it is probably sufficient to schedule so many sorties a day.

Aircraft may be grounded following aborted sorties, pre and post flight checks and routine inspections. They may also be taken out of service for major and minor overhauls after a predefined number of flying hours or (calendar) days. Aircraft may also be grounded for maintenance prior to a deployment to a remote, forward base or during a "maintenance recovery period" prior to a "maintenance free operating period".

Some of these events may be purely random, i.e. totally unrelated to the age of the aircraft or any of its components, such as the ingestion of a foreign object (e.g. bird). Many may, however, be the result of ageing processes such as friction-caused wear, metal or thermal fatigue. A number of components will have life limiting restrictions such that the operator is obliged to ground the aircraft in order to replace such a component as soon as it reaches this limit (known to engine manufacturers as a *hard life*).

If the user only wants a rough approximation of the likely number of arisings during a given period then one could convert these age-related instances into "failure rates" and use exponential distributions. This actually would be rather pointless, as the mathematics is so simple that there would be very little benefit in using simulation.

The main benefit of using simulation techniques is to be able to try out various alternative scenarios which would be too expensive, impractical or even impossible to do in reality. We can try different time-to-failure distributions – it makes virtually no difference to the complexity of the model if we use Weibull, lognormal, exponential or any other type of distribution. We can see what happens if there are insufficient spares at one or more locations of one, or more, types of component. We can also see the effect on the overall life-cycle costs or availability, etc. of making design changes that improve the reliability of individual components or of changing maintenance and/or support policies.

Some 40 years ago, it was considered sensible to support military aircraft by performing routine preventative maintenance. Every so many days, the aircraft would be taken out of service and virtually stripped down to every nut and bolt to see if anything was in need of repair or replacement. Unfortunately, this policy was not only expensive and time-consuming but it did very little for the reliability of the aircraft. Indeed, it is generally considered that it made it significantly worse.

This policy has largely been replaced by *on-condition monitoring*. In this case the various components of the aircraft are monitored for detectable changes. This may be via analysis of debris in the oil using spectrometry and crystallography to identify which components are wearing and at what rate. Various types of trend monitoring are performed on the performance of engines measured against various parameters from within the engine to determine when it is advisable to change components, etc.

The latest idea is one of *maintenance-free operating periods* separated from each other by *maintenance recovery periods*. This is actually not dissimilar to the discredited routine preventative maintenance policies of 40 years ago. The main differences are that the manufacturers will be required to guarantee these periods with possible penalties if they are not achieved and only sufficient maintenance is performed during the MRP to be able to guarantee the next MFOP to the required level. The manufacturers are being encouraged to design their components so that they have a high level of fault-tolerance such that various parts within them can fail but, the system will still be able to continue not only for the given mission but right through until the end of the period. During the MRP, those parts, which have failed will be replaced along with any others which are considered unlikely to survive the next MFOP.

Simulation can be used to model each of these scenarios and determine which is likely to be the cheapest option. Unfortunately, it is very much more difficult to use this technique to determine by how much various parameters would need to be improved in order to make one policy "better" than another.

8.13 USE OF SIMULATION MODELS IN SPARES PROVISIONING

With most of what has gone before in this chapter, we have assumed that the times to failure are exponentially distributed. We have also assumed that every failure is the result of one, and only one, component and that the recovery of the system is by the replacement of just that component. In the rest of this chapter we consider systems in which these conditions do not apply.

Most mechanical systems are made up of a number of components, which tend to deteriorate with age. Bearings, piston rings, fan and cam belts, gears, seals, switches, shafts, discs and blades (both turbine and compressor), for example, all exhibit failure modes, which are due to an accumulation of stress through usage and hence are age-related. This phenomenon is by no means exclusive to mechanical parts. Spark/ignition plugs, contact breaker points, HT (high-tension) leads and even light bulbs (both fluorescent and incandescent) also display similar characteristics. Indeed, it is very likely that transistors, capacitors, resistors and even conductors have age-related failure modes – the only reason we do not recognise them is that other failure modes resulting from external factors (such as "dirty" electrical supplies, electro-magnetic interference, dry joints and dust contamination) tend to predominate.

It is often argued that because the system and most of its LRUs are complex (in so far as there is a large number of potential causes of failure/rejection) then steady-state conditions prevail resulting in the times between failures and/or rejections being (asymptotically) exponential. Whether this is true, or not, is largely irrelevant. If we can predict the times to failure or rejection more accurately and with greater confidence by looking at the individual causes (many of which will not be exponential) then our forecasts of spares demands (both repairable and non-repairable) should, in turn, be more accurate.

Repair Effectiveness

A factor, which will tend to increase the number of arisings, is that of *repair effectiveness*. Generally speaking, repairing a component will not be as effective as renewing or reconditioning it. Very often a repair will do little more than restore the component to a "same-as-old" condition whereas, renewing and reconditioning are more likely to restore it to an "as-good-as-new" condition. If the times to failure are (truly) exponentially distributed then these two conditions are identical (as far as the probability of failure is concerned). However, if the times to failure are age-related then these two conditions can be very different.

Example 8.13

1. Suppose your 10-year old car has failed its roadworthy test due to corrosion of a (structural) part of the body. If you get the body repaired (by welding in a new section) would you then expect the repair and the rest of the body to last another 10-years?.

2. Suppose all of your tyres have been on your car from new for 20,000 miles/km when one of them gets a puncture due to penetration by a nail. You get the tyre repaired by inserting and vulcanising a mushroom plug in the hole. Would you now expect this repaired tyre to last as long as if it had been replaced by a new one?

3. Suppose the time to failure distribution for a blade is W[3, 6000] (that is Weibull distribution with $\beta = 3$, $\eta = 6000$) and there are 64 identical blades in a set. If the first one of these fails after 1338 hours and is replaced with a new one would you expect the next blade, in the set, to fail in another 1338 hours?

SOLUTION:

1. Whilst it is possible the welded section will last another 10 years, it is unlikely that the area of body around it or, indeed, the rest of the body will last that long. In fact, there is a high probability that the car will need further welding the following year to get it through its annual test, if it lasts that long.

2. Here we are looking at two different failure modes (punctures and wear). As far as punctures are concerned, one would expect all four tyres to be equally likely to suffer a puncture. As regards wear, then one would expect the repaired tyre to last the same mileage as its

partner on the same axle; one would certainly not expect a plug, however well it had been fitted, to restore the tyre to "as-good-as-new".

3. It can be proved that the mean time to failure of the first blade in this set is ¼ ($=1/\sqrt[\beta]{N} = 1/\sqrt[3]{64}$) of the mean time to failure of each individual blade (where N is the number of blades in the set and β is the shape parameter). It can also be shown that the mean time between the first and second failures in such a set (given only the failed blade is replaced) is approximately $1/\beta=1/3$ of the mean time to first failure. In this case, the mean time to first failure would be 1338 and the mean time to second failure would be 1784 (hours from new) so, because the first failure was after 1338 hours, we would expect the next failure to be in 446 hours time on average or less than 660 hours with a 90% probability or between 185 and 720 hours again with a 90% probability or $\geq$1338 hours with a probability of 10^{-12}.

Dilution Effect

With many systems, their LRUs and often the sub-assemblies are repairable. In addition, they may also be *rotable* which means that they do not have to be refitted to the same system or LRU from which they were removed.

When such a component is removed from its parent, recovery of the parent is generally achieved by replacing these removed items with "identical" ones from the spares pool. The removed items will then be added to the same spares pool when they have been repaired.

Now, when an item is not installed in an operational system it is not being used. As the number of spare units increases so the amount of usage each item gets will tend to decrease. This means that the amount of elapsed or calendar time will tend to increase in order for a part to reach its time to rejection. If this time to rejection is age-related then the number of arisings will decrease as the number of spares increases (provided the items are rotated sufficiently often). This phenomenon is called the *dilution effect*.

Modifications

Military equipment, in particular, tends to remain in service for very long periods. It is not uncommon for aircraft, for example, to have lives of over 30 years. (It has been suggested that the B-52 bombers might stay in service for 100 years). Ships' lives tend to be even longer. During this time, technological developments do not stand still.

Many, if not most, components will undergo a number of modifications after they have entered service and before the system in which they are installed reaches the end of its useful life. Some of these changes will be to improve the performance of the system, others may be to make their manufacture cheaper or easier and others will be to improve the availability of the system. In practice, almost all modifications will have some impact on the reliability of the component and hence the system. Occasionally, this impact will be negative but, most times it will improve the reliability and may also improve the maintainability, testability, supportability and ultimately the availability.

Sometimes modifications can be achieved by making changes to an existing component possibly by re-machining. In other cases, it may be necessary to replace the component completely. It will then depend on how urgent or beneficial the modification is as to how quickly the new components are introduced (and hence the old ones discarded).

Life Extensions

Parts whose failure may cause the loss of the system and/or human life are given a hard life. This length of this is calculated using data collected from a number of similar parts usually tested to destruction or, at least, until they show signs of cracking. The age by which there is less than a given probability of "failure" is then determined. (Here "failure" is usually taken to be the point at which a crack reaches a visible size – known as an "engineering crack"). This age is normally measured in stress cycles and is called the *predicted safe cyclic life* (or PSCL).

Because test conditions are not always closely correlated with in-service operating conditions, it is common practice to give these parts a *release life*. This is initially set to half the PSCL, say. When the first few parts have achieved their release lives in service, they will be removed and put on test to determine how well the test simulates reality and re-calculate the "safe life". If this further testing proves satisfactory then the release life (of those still in service) may be increased to say 75% of PSCL. This procedure is repeated until the release life can be set to the original PSCL. (Note: if the original PSCL was unacceptably low then design changes would have been initiated to improve the reliability so that an acceptable PSCL could be achieved.)

Once a life extension has been agreed, the hard lives of all of the similar parts in service are "instantly" increased to the new release life. If the systems are fitted with usage monitoring equipment which can "measure" the amount of stress each part is experiencing throughout its life then these life extensions simply have the effect of changing the date on which the part is expected to reach its hard life. On the other hand, for systems not so equipped, the new release life will need to be converted from stress cycles to (usage) hours (usually using a straight-forward multiplying factor called the *cyclic-exchange rate*).

Cyclic-Exchange Rates

The main difference between the "reliability" of the engines on a commercial airliner and those on a military combat aircraft is the number of stress cycles per hour (or the cyclic-exchange rate) that each experiences. Typically, an airliner will use near to full throttle during take-off and until it has reached a certain altitude. Less throttle is needed while it climbs to cruise altitude, less again during cruise, still less as it descends towards its destination and then, as it touches down, a quick increase to apply reverse thrust to assist braking and finally right back while it taxis to the "gate". The total flight/sortie/mission accounts for between 1 and 2 stress cycles in total over a flight time which could last for over 12 hours giving a cyclic-exchange rate of considerably less than 1 (stress-cycle per hour).

A typical (training) mission for a fast-jet, combat aircraft only lasts for about an hour. However, during this time, the aircraft is likely to be climbing, on full (reheated) power, throttling right back to dive to treetop level, using full power again to pull up to +9G in turns possibly using thrust vectoring for increased manoeuvrability. The number of stress-cycles in any given mission can be as high as 30, or more, per hour.

The actual number of stress cycles will vary for different parts within the engine during any given mission. They are also likely to depend on the experience of the pilot, the trim of the aircraft, the performance of the engine, the ambient conditions and a number of other factors that may, or may not be known or measurable.

8.14 OPTIMISATION USING SIMULATION

Although simulation is an extremely powerful technique, which can help us understand what is likely to happen under different conditions, it is a technique which does not lend itself to optimisation. In the following we will look firstly at why it is so difficult then we will look at some techniques that can be tried to overcome or circumnavigate these problems.

One of the main tasks facing the Logistics Engineer is to minimise the life-cycle cost within the constraint of achieving a given level of availability. We will look at various measures of availability in Chapter 10 but, for now, let us assume that we can run our simulation and obtain a single value of the "availability" for each pass. Note that a "pass" is one run using one sequence of random numbers for one particular set of input parameters. Normally, we would run several passes, each using a different sequence of random numbers, for each scenario then average these to reduce the variance due to the random numbers (noise).

Example 8.14

Now, let us consider a typical scenario for a fleet of military aircraft. We will have a number of aircraft stationed at a number of bases, say 10 bases. Each of these "squadrons" may fly a different role, consist of a different number of aircraft and/or fly a different number of hours per month. These 10 bases will be able to remove and refit engines and strip, rebuild and test these engines (down to module level). The rejected modules will be sent to either a depot (3rd line) or to a contractor (4th line) for recovery by removing and replacing rejected parts. The recovered modules will then either be held at the point of recovery or sent to a main spares holding echelon (invariably the 3rd line depot) or to one of the bases to be immediately fitted to an engine or bring the spares holding at that base up to (nearer) its reserve limit.

A typical life for such a fleet would be 30-50 years. It might take from 5-10 years to build up the fleet to its full compliment of aircraft and possibly 5 years to run it down or decommission it. Let us assume that we have already decided which components should or should not be repaired and whether they should be given a soft or hard life and what each of these should be. We then only have to look at how many spares we need of each component and where we should hold these. As there are some 3000 different types of components in an engine, we will make life simple and only consider the "rotables" or repairables, i.e. the engine and its modules. Let us also assume that there will be no modifications to any of these rotables so that we do not have to consider obsolescence.

The task then is simply to decide how many spare engines and how many of each type of module should be held at each base and the depot each year the fleet is in operation.

SOLUTION:

Suppose an engine is made up of 20 modules and the fleet is to be operated (including build-up and run-down) for 30 years.

No. of components $= 1 + 20 = 21$
No. of sites $= 10 + 1 = 11$

No. of re-order points $= 30$ (1 each year).

No. of variables to be optimised $= 21 \times 11 \times 30 = 6930$

Let $N_{i,j,k} =$ the no. of spares of type i, at site j at the start of year k.

If we assume that the only costs are the purchase prices of the spares ($= C_i$) then we can very easily determine the cost of any combination of spares as

$$C = \sum_{i=1,21} \left\{ C_i * \max_{k=1,30} \left\{ \sum_{j=1,11} N_{i,j,k} \right\} \right.$$

Unfortunately, in order to determine the "availability" of the given scenario we will have to run the simulation.

Suppose the availability for the given scenario is $A_{1,S1}$ where S1 represents the (21x11x30) matrix of values.

We can now plot one point on the Cost-Availability graph at point $(C_{1,S1}, A_{1,S1})$ where $C_{1,S1}$ is the cost of the spares for the given scenario.

If we were to run the same scenario again, but using different random numbers we would get the point $(C_{2,S1}, A_{2,S1})$ noting that $C_{1,S1} = C_{2,S1}$ but, $A_{1,S1}$ is certainly not necessarily equal to $A_{2,S1}$. If we run the simulation a large number of times (each time with different random numbers but the same input parameters) then we could plot a histogram of values of $A_{i,S1}$ and maybe fit a distribution to them. If we now vary one, or more, of the parameters, we could repeat this exercise and then determine, with a certain level of confidence, whether "S1" is "better" than "S2". In theory, we could continue doing this until we are "confident" we have found the set of parameters Si which minimise the cost C_{Si} such that the expected availability A_{Si} is within acceptable limits of the required availability.

The above example raises a number of points. The first is the size of the problem. Here we had a sample space of nearly 7000 dimensions in which to search for the solution. If we had considered individual parts instead of modules the number of dimensions would increase to around 700,000. Adding in the fact that many of these parts will undergo several modifications could easily put the number over one million.

The second problem, and one, which is common to all stochastic simulations, is that we cannot determine exact values of the function. In this case, whilst we know the exact cost, we need the simulation to determine the availability of the system. If we only run the simulation once with a given set of parameters then we cannot know whether the estimate of the availability is too high, too low or close to the expected value (for the given scenario). This means that if we use these estimates in steepest descent/hill climbing type optimisation methods, we could easily head in totally the wrong direction because it is quite possible for $A_{1,S1} < A_{1,S2}$ when $A_{S1} > A_{S2}$.

One possible approach is to use evolutionary techniques such as Genetic Algorithms. This method is known to work reasonably well with large numbers of variables and has the advantage that it does not use the actual function values to determine the size and direction of the next step. It uses these values to decide which members of a particular generation to allow to be parents of the next generation. Because of the variance between passes, it is possible to "kill" the parents which might "breed" the optimal solution but, it is also possible that such parents may be born again in a later generation.

A particular difficulty with this particular example is that of deciding which members of the current generation should be allowed to "breed". In general, we general we are trying to minimise the cost for a given availability. The question that arises, however, is whether a solution with a high cost and acceptable availability will make a better parent than one with a low cost but unacceptable availability. For example if the desired availability is 90% and scenario 1 has an (estimated) availability of 95% but a cost of 100 whereas scenario 2's availability is 85% but its cost is only 50. Which of these two is the "better" solution?

We would expect the availability-cost curve to be convex. One method might be to assume a certain shape for this curve and use two or three points (estimated from several passes, say) to estimate the parameters of the equation of this curve. From this we could determine the distance of each point from this curve and rank each according to this distance recognising that a point above the curve is clearly better than one below it. No doubt many further refinements could be made.

In practice, the biggest problem is one of time. If the simulation takes minutes to perform one pass then using genetic algorithms is going to be a very long process. If it also takes the user several minutes to make the changes to the input data set between each new scenario then the whole process will almost certainly be too slow to be practical.

One method, that has worked well in practice, is to use the simulation to determine a reasonable starting position. By running a simulation of the operation, maintenance and support activities with effectively an infinite number of spares, the model can give a good indication of the numbers that will be needed. This is done by only using those spares which it needs by operating a last-in-first-out (LIFO) regime.

The same model is run again but now being supplied with only the number of spares it has recommended for the required level of availability. This will show, in fact, that this number is almost certainly not sufficient as each component is now being used for more of the time and hence the number of (age-related) arisings will have increased. If this process is repeated, eventually it should converge to a recommendation that meets the availability requirement.

This recommendation is unlikely to be optimal but it should be reasonably stable in so far as running the model again with different random numbers should still give an availability reasonably close to the required level. Note, in the above, each "run" will normally consist of at least ten passes.

If time permits, the next stage is to try running the model with fewer spare engines recognising that, in general, the engine will be considerably more expensive than any of the modules. If this results in a failure to meet the availability then increasing the numbers of those modules which suffer the most rejections may restore it to an acceptable level. This sort of trial and error approach may be improved by using marginal estimated back orders although this is likely to require changes to the model to produce the outputs required.

In Chapter 10 we will return to the question of availability and look at some of the problems that can arise.

Chapter 9

Integrated Logistic Support

Engineering judgement means, they are going to make up some numbers

Richard Feynman

It is generally recognised that a very large proportion (often put as high as 90-95%) of the life-cycle costs (LCC) is determined during the concept and design stages, before the system has been manufactured and often long before entry into service (EIS). Having said that, however, there is still considerable opportunity for minimising the life-cycle cost during the operational life of the system but this optimisation is constrained by the inherent reliability, maintainability and supportability of the system and its components. The role of integrated logistic support (ILS) is very much concerned with both of these areas: of ensuring the system is conceived, designed and manufactured to be operationally effective and; to provide through life support of the system to ensure that it remains so, even unto its grave (or disposal).

ILS is the management and technical process through which supportability and logistics support considerations of systems/equipment are integrated from the early phases of and throughout the life cycle of the product (Hillman, 1997). ILS is sometimes referred to as a "cradle to grave" activity but, to have maximum effectiveness, it should really be "lust to dust". By the time the system has been "born", it is far too late to have the influence needed to make certain that it will meet the operators' needs. And, increasingly, the costs of ecologically acceptable disposal will come as a nasty shock to the owners if they have not been recognised and considered during the earlier stages of the life of the system.

There are very few components of any system, let alone the systems themselves, whether they are military or commercial, public or private, hardware or software, mechanical, chemical, electrical or electronic that will never fail, never need maintenance or never need support. Indeed, if such a system does exist, then it has almost certainly been "over-engineered" or is so trivial as to be of no interest or relevance.

The end of the Cold War brought also the end of "arms at any cost". The role of the armed forces is much more likely to be one of policing trouble spots than acting as a deterrent through "superior" firepower and technology. Defence ministries can no longer use the latest Mig, Stealth Bomber, ICBM or laser gun to justify research and development budgets that exceed the GDP of small and sometimes not so small countries. In the civil/commercial world, this has rarely, if ever, been an issue – few system operators have or are ever likely to have the capital or desire to buy systems which are likely to be uneconomical although it is extremely unlikely you could find a single operator who would complain that the systems he/she is operating is too reliable, too maintainable, too supportable, too available or too cheap to run. Even with the most modern gas turbine engines which have been known to stay on the wing for over 40,000 hours, it takes at least 4 economy class passengers on every 10-hour flight to pay for the cost of maintenance of these (two) engines and that does not cover the cost of the engines, the fuel and oil or any so called non-basic failures (i.e. ones not directly attributable to the engine such as bird strikes, stone or ice ingestion, etc.).

9.1   HISTORY OF ILS

Traditionally, military projects have been completed late and over budget. When they did arrive, the systems were quite likely to fail to meet the users' requirements. They might suffer from poor quality, be unreliable, unmaintainable and unsupportable. A common criticism was that they spent more time in a state of failure than in a state of functioning.

It was generally thought that the primary cause of this was that the designers failed to give due consideration to the post-design stages: manufacture, operation and disposal. Failures, or poor reliability, would be blamed on the quality of the material used, the manufacturing processes, inadequate pass-off inspection, improper use, poor maintenance, in fact on anything except the design itself. A classic quote by the operators of British Rail (before it was privatised) was, "Our trains would run on time if they didn't have to stop to pick up passengers." This epitomised the prevalent attitude.

As with most things in life, regulations were brought in that caused the pendulum to swing to the opposite extreme. MIL-STD 1388 was the US Department of Defence's answer. This laid out, in minute detail, exactly what tasks and when in the life-cycle they had to be performed. It is probably fair to say that if all of the books written on MIL-STD 1388 were piled on top of one another, they would almost equal the amount of paperwork required to be produced for a single project!

Despite all the plans, analyses and reports many projects were still coming in over budget and late. The systems were often still suffering from many of the problems that ILS was supposed to correct.

Here is not the place to discuss the merits and failings of MIL-STD 1388, in particular, or ILS in general. However, before we look at the activities that are generally listed under the heading of ILS, it is perhaps worth making a few observations.

1. Most of the activities are described as separate, effectively standalone, tasks that are performed once at a specific point in the sequence of activities which make up the total ILS process.
2. Once a task has been "done", there is a tendency to "put a tick in the box" and forget it.
3. Different tasks are often the responsibility of different departments with little, or no, communication between them with the result that analyses may be made using different assumptions and different methods resulting in duplication of effort and often producing contradictory results.
4. Reliability data, or more specifically, "constant failure rates", are often "determined" by an allocation process starting with a "target" failure rate for the system or sub-system. These allocated rates may bare little relationship to the actual failure times.
5. In the case of electronic units, the failure rate (for the unit) is likely to be "calculated" using published tables (e.g. MIL-HDBK-217) of "failure rates" for individual components (e.g. resistors, capacitors and transistors) often ignoring the connections, conditions and supplier data.
6. The tasks using this inadequate and inaccurate reliability data will almost certainly produce invalid results and recommendations, although it may be many years before these errors are discovered.
7. There is little, or no, incentive to collect and analyse in-service data in order to feed this back into the process to improve the data for the next project.

Table 9.1 lists the various tasks to be accomplished during various stages of the life cycle (Hillman, 1997).

Table 9.1 ILS considerations during different stages of ILS

Life cycle phase	ILS Activity to be Accomplished
Pre-feasibility phase	Identify support resource constraints dictated by maintenance concepts, level of skill available, capabilities and capacities.
Feasibility Phase	Incorporate logistics experts, take into account potential logistic support, manpower, and training requirements and constraints. Consider the logistic support required identify ownership and related support matters such as facilities, personnel etc.
Project Definition Phase	Establish a consistent set of measurable objectives for availability, reliability, maintainability and other logistics support parameters.
Design and Development Phase	Verify attainment of the objective availability, reliability, maintainability and other logistic support parameters. Ensure ILS considerations are given appropriate weight.
Production Phase	Validate and deliver ILS elements to meet the requirements. Correct supportability deficiencies and validate corrective actions.
In-Service Phase	Establish and maintain the ILS management system. Analyse field data related to logistic support, identify and develop availability improvements. Identify deficiencies in the system and evaluate trade-offs prior to modifications,

9.1.1   The concept of ILS

ILS is many things to many people but, essentially, it is about achieving an acceptable balance between whole life cost, performance and operational availability. Put simply, we are looking for a design such that the system will be able to perform all the tasks required of it, at minimum cost (over its entire life) and that it will always be available when required.

For most commercial enterprises, it is generally possible to convert all of these desired attributes into monetary terms. This makes it, at least conceivable, to find an overall optimal solution, albeit, that the problem is likely to be beyond the capability of even the fastest computers available today.

9.1.2   The Case Study of Airline Costs

If we consider a commercial airliner then its performance will be measured in terms of range, speed and payload. Possible measures would be "passenger-miles per hour" or "passenger-miles per flight". An airliner, which can carry 500 passengers, at an average speed of 500 mph over a range of 5000 miles, would therefore be able to achieve 250,000 passenger-miles per hour or 2,500,000 passenger-miles per flight. Using a price per mile of 10p, say, we can see that this aircraft would be capable of earning £25,000 per hour or £250,000 per flight. Of course, the actual revenue per flight will inevitably be less than this because, it is extremely unlikely that every flight will be full. For most aircraft, certainly of this size, there will be a wide range of ticket prices and routes of differing distances, prices and popularity.

The life-cycle cost is even more involved. Here we have to consider the acquisition cost (say £200,000,000) plus the fuel costs (at say, £1 per mile) plus maintenance and support costs (at maybe £150 per hour) plus landing fees, booking costs, in-flight entertainment (including meals and drinks, etc.), advertising, crew training, salaries and wages and numerous sundry costs. Some of these costs will be fixed, some variable, based on miles, hours or passengers carried and others will be a combination of both fixed and variable costs.

Finally, there is the cost of availability, or rather, of unavailability. Although it may theoretically be possible to keep an aircraft in the air for 100 or even 200 hours by employing flight-refuelling, there is the small matter of getting the passengers and their luggage on and off at the desired locations. Typically, turnaround times are of the order of one hour. That is from the time the aircraft docks, the passengers alight, their luggage is unloaded, the cabins are cleaned, all post and pre-flight maintenance checks are performed, the aircraft is re-fuelled, the galleys are re-stocked, the next set of passengers are boarded and their luggage loaded is around 60 minutes.

Before an aircraft is allowed to take-off, it must first negotiate a "landing slot" with the airport/air traffic controllers at its next destination. Although this is a nominal time when it is expected that the aircraft will have reached its destination, it generally does not allow very much room for error. If the aircraft arrives too early, it may be stacked in a holding pattern until a slot becomes available. If, on the other hand, the aircraft's take-off is delayed then, depending on how late it is likely to be reaching its destination, it may have to re-negotiate another landing slot. Such a slot may be several hours after the first, indeed, if the aircraft is due to land shortly before the evening curfew (due to noise restrictions), the aircraft may have to wait until the next morning before it can land. Similarly, if it was due to take-off shortly before the curfew, it may be barred from taking-off until the next day. This means the passengers will have to be found over-night accommodation, fed, watered and possibly compensated. They may also decide to change to another airline.

Delays may be due to built-in test equipment warnings, (critical) equipment malfunction, weather conditions (anywhere along the route) or even a passenger not turning up at the departure gate. With the constant threat of sabotage, few airlines will now risk taking off with unaccompanied baggage. If a passenger has booked in his or her luggage but has then failed to show, the baggage will have to be unloaded, the passengers will also have to leave the aircraft and be asked to identify their luggage. Any unclaimed item will then be checked against the passenger list and removed to a safe area where, if it is not reclaimed within a given time, it will be treated as a potential bomb.

The cost of a delay is not an easy figure to estimate, as it can be several weeks before the given aircraft can get back onto its proper timetable. The turnaround times include virtually no margin so if the aircraft lands late at the first destination, it will almost inevitably be late taking off for the next destination. Because it cannot take-off on time, it is likely to be subjected to further delays, as its normal landing slot may no longer be available. In the height of the charter season, typically between Easter and October, aircraft may be on a schedule in which the only prolonged or over-night stopover is once in 21 days. This means that potentially several thousand passengers will be delayed and inconvenienced. If they lose part of their holiday or miss connecting flights or are late for important meetings, they may decide not to use that particular airline again. There are very few routes where there is no alternative airline ready to step in and offer passengers incentives to switch. Given that an airline knows how much they have to pay out in incentives to win customers from other airlines then they can make an estimate of the costs of losing customers.

It is also becoming increasingly common for airlines to negotiate package deals with their prime suppliers such that the supplier may be forced to pay substantial compensation claims if their products or services fail to achieve their targets. Operators are also making increasingly greater use of leasing companies, particularly during peak demand periods or when they have aircraft on the ground for prolonged periods.

With fewer new aircraft being produced and every new sale has to be hard-won against fierce competition, manufacturers are increasingly taking over much of the maintenance and support (after-market) activities often charging the operator a fixed price per flying hour. Rolls-Royce plc, for example, has negotiated what it calls a Power-by-the-Hour™ agreement on the support and maintenance of their engines with a number of airlines. These deals remove much of the risk and uncertainty and the need to carry large stocks of spares from the operators and, of course, can make their budgeting task that much easier. By linking such deals with "guaranteed" "availability", e.g. no more than a given number of delays and cancellations (D&C) per 100 landings, airlines are also able to mitigate against unreliability or poor quality service. This, of course, does not come free as the service provider will wish to remain in business and hence will include provision for these factors in the rate levied.

Disposal costs, for most of the larger, more successful airlines, will generally be negative. In most cases, airlines will dispose of their aircraft by selling them to another operator in much the same way as car hire companies sell their cars when they believe they have become uneconomical to keep them.

9.1.3 Case Study on Air Force's Cost

The problem of balancing performance with life-cycle cost with availability becomes very much more difficult when we consider the military environment. Although we will consider the specific case of a combat aircraft, most of the comments and observations are equally applicable to almost any military vehicle.

The performance of a bomber aircraft could be measured as the weight of high explosives it can deliver to the target. We should, however, take into account the probability it will reach the target and then return safely so it can perform the next mission. Whilst we could, maybe, put a cost on the failure to complete the mission (in terms of the loss of, or damage to, the aircraft and crew), it is another matter putting a price on the successful completion on the mission or, indeed, the cost of not landing the explosives on the target.

The cost of unavailability is equally difficult to measure in monetary terms. If the aircraft is not ready to fly at a given time it could mean the mission's success will be jeopardised, possibly to the point where the mission has to be postponed or even cancelled. We have all seen films of aircraft being blown up on the runway because they could not get airborne quickly enough. There is also a probability that an aircraft that has set off on a mission may not reach the target due to faults and failures on the way. Other faults may not stop it reaching the target but might stop it from deploying its weapons either through some sub-system failure or because the pilot/crew were unable to find or identify the target.

In both cases, we could consider deploying more aircraft, so that the probability of the required number taking off, reaching their target(s) and deploying their weapons is sufficiently high to be acceptable, although, deciding what that level has to be is, in the end, down to the generals and politicians. In effect, we are using system level redundancy to ensure mission success.

From an optimisation point of view, there is a major difficulty. If we cannot determine a cost of failure, i.e. what the cost of not carrying out the mission successfully would be, or how much "better" a probability of one percentage point higher would be then it becomes an intractable problem, mathematically. In the end, all we can do, is try to find the cheapest options for any given probability of mission success or level of availability.

Determining the life-cycle cost is also not as easy as in the case of the commercial operator. Very often, one, or a consortium of governments, will fund a major military system development. Their primary aim is to have a weapons platform, or whatever, that is "better" than the "enemy's" that meets the various requirements of each of the armed forces that may have to deploy it and that will, ultimately be affordable and, of course, good value for money.

Typically, high capital, military projects are likely to have a life of at least 30 years and may be expected to remain operational and efficient for 50 or even more years. Of course, during that time, the system will not remain unchanged. The electronics will probably be changed every 2-5 years, the types of weapons and their guidance systems, possibly a little less frequently and, most of the other systems may undergo one major change at the "mid-life update". Minor components will generally only be changed if they have poor reliability or if the manufacturers have improved their design and/or manufacture, maybe by using newer materials or techniques.

Whilst many of these changes will have to be cost-justified, some will be as a result of components becoming obsolete and no longer available. This is a particular problem with commercial-off-the-shelf (COTS) items where the military customer is likely to have an almost insignificant influence. A classic example of this occurred when it was decided to fit engine monitoring computers in a subset of the aircraft in operation. There was a need to produce five units with special computer circuits. The manufacturer approached one of the major "chip" manufacturers and asked, "Can you produce this design, please?"

"Certainly, sir. How many would you like?"
"Five please", came the response.
"What, five million? No problem!"
"No, just 5".
The response was along the lines of "you cannot be serious".

Governments also tend to have this rather strange notion of using net present value based on the assumption that a pound invested today will attract 6% compound interest per annum until it is needed in say 20 (or 30) years time. This means that it will, according to their calculations, be worth £3:20 (or £5:74) when it is needed. Or, more particularly, if a part costs £1 today, then they only have to invest 31p (or 17p) now to cover its cost in 20 (or 30) years time when it is needed.

There are a number of problems with this:

1. In practice, no government (to our best knowledge), invests money today to pay for something in even 2 – 3 years time, let alone 20 – 30 years.
2. Unless the component is required frequently throughout the period, it is unlikely that the drawings, let alone the jigs and tools, will still be available and, there will almost certainly be no one around who can remember how it was manufactured. This means that it will effectively have to be re-designed and hand made at a cost of possibly several orders of magnitude higher than the original.
3. It also tends to favour cheap, unreliable designs. The cost of the original equipment has to be paid in today's money so there is no discount factor applied to it. A replacement part in 20 years time, as we have seen, only costs 31p today so, using this argument, the older the system gets the cheaper it becomes to maintain and support it, which, strangely tends to be contrary to most people's opinions and, indeed, contradicts the ministry's figures.

Unlike the "power-by-the-hour" rate, which is likely to be a fixed price contract over up to ten years, the maintenance and support costs that the air force will have to pay is a lot less definite. The reliability of combat aircraft tends to be several orders of magnitude less than that of commercial airliners. In addition, military commanders are naturally reluctant to allow aircraft to operate with known faults and failures, even if these do not compromise the current mission. Although combat aircraft tend to spend most of their time in peacetime operation, it is a requirement that they are maintained in battle-ready condition at all times: asking the enemy to wait a couple days or weeks while our aircraft can be made ready is not considered a viable option.

9.2.   RELIABILITY PLANS

As we have seen in Chapter 5, "maintenance is the management of failure and the assurance of availability" (Hessburg, 1999). The main driver for all maintenance and support during the operational phase is unreliability. However, unlike performance, weight, size, payload, range, speed, turning circle, acceleration, rate of climb, and many more factors, reliability is not something that can be measured, at least not until it is too late.

We have talked about probabilities of failure and probabilities of being operational after a given time and said that these are measures of (un)reliability. This is true, but they depend on the assumption that the times to failure for every component (of a given type) are independent and identically distributed. They make the assumption that each of these components will experience similar conditions throughout its life and that what has happened, in the past, is an accurate predictor of what will happen in the future.

In reality, no two components are truly identical. There will be small variations between their molecular structure due to impurities and variations due to the smelting, casting, forging or rolling processes. There may be slight variations due to the machining, even though they may all be within the required tolerance. Further variations may be introduced in the way they are handled or stored, an invisible scratch or nick can act as a *stress raiser* – a point of weakness from which a crack may later start to propagate. Then, once the components enter service, their lives could be very different even though they may be in identical systems.

Example of variations in stress

Some years ago, it was decided that the type of flying the RAF *Red Arrows* aerobatics team was doing could be particularly hard on some of the components in the Turbomecca *Adour* engines. With only one per aircraft, it was felt this pose an unnecessary safety hazard so, at the first opportunity, this squadron was fitted with, what was then, the latest *engine usage monitoring system (EUMS)*.

Essentially, this consisted of a specially adapted portable cassette tape-recorder, which received digitised signals from a number of probes. These recorded the temperatures and pressures at various points in the engine and the spool speeds (angular velocities of the HP and LP turbine discs in rpm). Measurements of these parameters were taken at so many times per second and recorded onto the cassette, which was inserted in the deck by the pilot before each mission (and removed at the end). These tapes were then played back at a special ground station where they were checked for consistent signals and then analysed using complex algorithms, which used the measurements to determine the amount of stress each of the safety-critical components (e.g. discs) experienced during the mission.

The result of the analysis showed that there was a very large, but essentially consistent, variation between the stresses on certain components depending on where in the formation the aircraft was positioned. The difference between the leader (No. 1) and the No. 9 aircraft was a factor of around 25 to 30. That is the components in the No. 9 aircraft were using up their stress cyclic life at 25 to 30 times as fast as the lead aircraft. Although reasons for this could be found, the magnitude of the difference came as a surprise to everyone concerned. Primarily, the lead flies at a steady speed, constant throttle setting but, the tail man has to keep adjusting his position and hence his speed and throttle setting to stay in the correct position in the formation and it is this throttle movement that puts a lot of stress on the engine (or at least its rotating components).

Converting the *predicted safe cyclic life (PSCL)* from cycles into hours, as had always been the practice before EUMS, meant that, in this case, the components in the lead aircraft would be life-expired far too early or, more seriously, those in the tail aircraft could be left in past their safe limit if the same *cyclic-exchange rate* was used for all aircraft in the squadron. (In practice, even before EUMS, aircraft (and hence the engines) were quite regularly rotated so that each received relatively similar amounts of flying in each position.

So, with variations in materials, manufacture, handling and usage, it is not surprising that "identical" components do not last exactly the same length of time. If, on top of all the other factors, we add environmental conditions then, these variations can be quite significant. It is no coincidence that much of the equipment used in the "Gulf War" experienced very poor reliability. During the "Cold War" era, the enemy was always presumed to be USSR and that most of the battles would be fought on the northern plains of Europe so, neither land nor air vehicles were designed to cope with the sands and heat of Asia Minor.

9.2.1   Reliability Demonstrations

In the early concept and design stages, the only reliability data available is the target MTBF for the system which may be expressed as x failures per thousand operational hours or, more recently, possibly as a probability of surviving a given period without the need for any corrective or preventative maintenance. There may be some feedback from similar systems currently in operation or recently retired, although, in most cases the prospective buyers and operators will be looking for a significant improvement on past (reliability) performance so this data may be of limited applicability.

Part of the failure modes effects and criticality analyses will consider how the system requirement should be apportioned between the sub-systems and on down to the lowest level components. It will also identify those components whose single point failures can cause catastrophic consequences along with other components whose failures will have the most impact on the system's availability.

Much work will be expended on safety-critical components to make the probability of failure as low as possible. In many cases, the failures of these components, especially those in gas turbine engines, will be primarily due to an accumulation of stress or cyclic fatigue. Most metals remain elastic if subjected to small positive and negative bending forces (see Figure 9.1). However, if these forces are repeatedly applied over a long period then cracks are likely to propagate along the line of bending, usually starting from either end. If the sample is also subjected to heating and cooling at the same time, cracking is likely to start very much earlier.

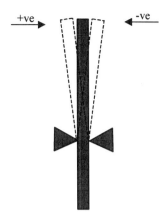

Figure 9.1 Cyclic Fatigue

9.2.2   Case Study of a Turbine Disc

Compressor and turbine discs, in gas turbine engines are subjected to high centrifugal forces and large temperature variations across the disc. The traditional design of turbine disc will have a large number of blades spaced equally around the circumference. These are held in place by a dovetail type joint – in some cases, this has the appearance of a Christmas tree or fir tree and, is know for that reason as a "fir-tree-root"(see Figure 9.2).

The blades that fit into these fir-trees can exert forces of several tonnes. They also have to have a small gap between the bottom of the blade and the disc so that cooling air can pass through the small holes in the root of the blade and out through holes along the leading edge and across the back of the blade. Without this cooling air the blades would melt the first time the engine was brought up to operating temperature as the gases impinging onto the blades are considerably hotter than the melting point of the material from which the blades are manufactured. Around a disc could be as many as 100 plus blades so the distances between them may only be a few millimetres. The blades may only weigh a few grams, being mainly titanium, but the disc is likely to be rotating at angular velocities of 15,000 rpm and may be 300 to 600 mm in diameter with forces acting on the blades measured in mega-newtons. The temperature at the centre of the disc could be as low as −60 °C whereas the temperature on the circumference could be nearer to 2000 °C. The speed of the disc will vary considerably as the pilot accelerates and decelerates. The disc will also be subjected to G-forces ranging from −4½ to +9.

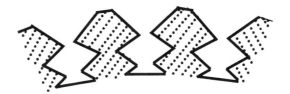

Figure 9.2 Fir-Tree-Root

The most common form of disc failure, which is actually extremely rare, is for two blades to be released (as the small piece of metal between them breaks). The most serious failure, however, is a disc burst as cracks propagate from the centre radially. If this happens, the energy in the pieces will almost certainly cause them to pass through the engine casing, the Kevlar shielding and the aircraft skin – indeed, in tests, they have been known to penetrate 3 metres of reinforced concrete. Uncontained failures such as these have a high probability of being catastrophic so every effort will be made to minimise the likelihood.

To minimise this probability, these discs will be given a life limit (known as *hard life*). This sets the maximum age, in stress cycles, which the disc is allowed to achieve. To determine this limit, a number of discs, made to the same specifications as those that will be used on the final product, will be tested to destruction. Typically, they will be put in a "spin rig" and spun. Heating and cooling may be applied to try to simulate the conditions inside the engine. The speeds will also be varied according to set sequences simulating the variations likely to be experienced in practice, although, in most cases, the "profile" will be compressed – see Figure 9.3. The left-hand graph shows a typical mission profile, in terms of the spool speeds. The right-hand graph shows how this may be compressed. Both have the same amount of acceleration and deceleration and at the same rates but the second one takes approximately half the time. In practice, accelerated testing can achieved compressions of around a factor of 4 for military aircraft and very much higher for civil.

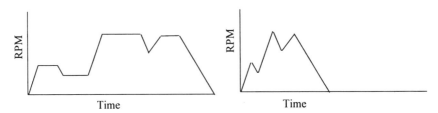

Figure 9.3 Mission Profile

By counting the stress-cycles received by the disc, it is possible to get an estimate of the age at the time of failure, in stress-cycles. To determine the number of stress cycles requires the application of some rather complex algorithms, which take spool speed, temperatures and pressures into account.

Having obtained these times-to-failure for a sample of discs, it is possible to determine what is called the *predicted safe cyclic life (PSCL)*. This is based on a statistical analysis of the times and uses *a priori* estimates of the type of distribution and its variance.

Using accelerated testing obviously reduces the time the disc will spend on the rig before it fails. Unfortunately, the stresses received during an accelerated test and those in reality may not be exactly correlated. Acceleration and deceleration of the spool are, alas, not the only forces acting – temperature, G-forces, engine vibrations are just some of the others. To overcome these deficiencies, the *release life* for the first few discs is usually set at a value close to half the PSCL. When a disc reaches this release life in service, it will be removed from the engine and put on test to check that the times to failure are compatible with the original estimates. If this proves satisfactory, the release life will be increased. After several iterations, the release life should equate to the PSCL. If the tests show a significant difference, it may be deemed necessary to modify the discs to improve their reliability and hence their (hard) lives.

To achieve a PSCL of the equivalent of 4000 hours, say, will require a mean time to failure of around 16,000 hours, which could be reduced to approximately 4,000 hours using accelerated testing. This represents 6 months of continuous testing per trial. If the times to failure are too low, then testing would have to start all over again on any modification. With several discs in an engine, there will clearly be a need for many hours of testing time.

9.2.3   Reliability testing

In the above case-study, we considered discs, which are tested because their failure is likely to have catastrophic consequences. The loss of just one aircraft and its crew would almost certainly be placed at a higher value than the 4,000 hours of testing per disc. Where the consequence of failure is less than catastrophic, the cost of testing to establish an estimate of the reliability of a component must be seriously considered.

In this case, we are not trying to fix an age limit to minimise the risk of a catastrophic accident but, merely establish that the component is as reliable as the manufacturer has claimed or as the operator is demanding.

Example 9.1

Suppose we wish to be 90% confident that the mean time to failure of a component is (at least) 1000 hours. We know from past experience that these components have a time to failure that can be described by a Weibull distribution with a shape (β) = 3. We would like to test 10 simultaneously, how long should we run the test?

SOLUTION:

For a MTTF = 1000, η = 1000/0.892 = 1121 when β = 3

We therefore need to establish that η is at least 1121 with a 90% confidence.

If we only test one unit then if it lasted until at least the time at which we would have expected 90% to have failed then we would be 90% confident the mean time to failure was η.

Now the time at which 90% would be expected to have failed is:

$$t = \eta(-\ln(1-p))^{\frac{1}{\beta}}$$
$$= 1121(-\ln(1-0.9))^{\frac{1}{3}}$$
$$= 1480$$

Now if the probability a unit fails after time t is p, the probability that 10 of them fail after that time is p^{10}. So to have a probability of 10% that 10 (out of 10) units will have failed after a given time we need a value of p such that $p^{10} = 0.1$, i.e. p = 0.7943.

Now

$$t = \eta(-\ln((1-p)^{\frac{1}{n}}))^{\frac{1}{\beta}}$$

$$= 1121(-\ln((1-0.9)^{\frac{1}{10}}))^{\frac{1}{3}}$$

$$= 1121(-\ln(0.7943))^{\frac{1}{3}}$$

$$= 1121*0.613$$

$$= 687$$

So, if we put 10 "identical" units on test and they all survive 1856 hours we can be 90% confident that their characteristic life is 1121 and hence their MTTF is 1000, given they are Weibull distributed with a shape of 3.

Example 9.2

Suppose we are required to establish with a 90% confidence that a complex system will have a 98% probability of surviving an MFOP of 150 hours using 5 systems.

Solution:

Given no information about the times to failure, we will have to make an assumption about their distribution. Let us therefore assume they follow a Weibull distribution $W[\beta, \eta]$.

For the system to survive 150 hours with a probability of 98% we have:

$$p = \exp(-(\frac{t}{\eta})^{\beta})$$

That gives,

$$\eta = \frac{t}{(-\ln(p))^{1/\beta}} = \frac{150}{(-\ln(0.98))^{1/\beta}}$$

Let us consider the cases when $\beta = 1$ and $\beta = 3$

For $\beta = 1$

$$\eta = -\frac{150}{\ln(0.98)} = 7425$$

If we have 5 systems then to establish that the MTTF is at least 7425 with a confidence of 90%

$$t = -\eta(\ln((1-p)^{\frac{1}{n}}))$$

$$= -7425\ln((1-0.9)^{\frac{1}{5}})$$

$$= 7425 * 0.4605$$

$$= 3420$$

This means that there will need to a total of 17,100 hours testing. In fact, if the times to failure are exponential ($\beta = 1$) then the total testing time will always need to be 17,100 no matter how many units are tested.

For $\beta = 3$

$$\eta = \frac{150}{(-\ln(0.98))^{\frac{1}{3}}} = 551$$

If we have 5 systems then to establish that the characteristic life is at least 551 with a confidence of 90%

$$t = \eta(-\ln((1-p)^{\frac{1}{5}}))^{\frac{1}{\beta}}$$

$$= 551(-\ln(.631))^{\frac{1}{3}}$$

$$= 425$$

In this case, the total testing time is 2,125 hours but this only applies if each unit is tested for 425 hours.

Note: if one of the 5 fails before reaching the required testing time, then we can no longer be 90% confident of the mean time to failure or that the system will have a 98% probability of surviving 150 hours. In fact the confidence now becomes:

$$CL = 1 - 5p^4 q$$
$$= 1 - 5 * 0.631^4 * 0.369$$
$$= 1 - 0.29$$
$$= 71\%$$

This result is independent of the distribution and is calculated using the binomial distribution.

9.3. WHOLE LIFE COST

Whole life cost, life-cycle cost, cost of ownership, through life cost are all variations on the same theme. They are an attempt to calculate how much the operator will have to pay out. Since different texts give these different definitions, it is essential that before embarking on trying to determine the value you make sure that both you and the organisation asking for the value understand what costs should or should not be included. As we have used the term life cycle cost (LCC) elsewhere, we keep to this.

Clearly, the LCC will depend on the reliability, maintainability and supportability of the system. It will also depend on how the system is used, maintained and supported and, it will depend on what level of availability is required out of the system.

During the pre-EIS (entry-into-service) phase, most of these values will be unknown. There will however be target values, at least in the case of military customers for some of these factors, e.g. reliability, maintainability, supportability and usage. Towards the end of this phase, we may have some confidence in the reliability and maintainability, depending on the level of testing/demonstration. Before we can start calculating the LCC, however, we will need to consider such questions as which components of the system will be repaired and where this should be done and which discarded, when will their replacements be bought and where will they be stored.

Since all of these factors affect the LCC and the customer requires the supplier to minimise the LCC, it clearly makes sense to consider the effects of different policies on the LCC. The whole process needs to be an iterative one in which the LCC is one element of the cost function, which needs to be optimised. The other elements are the performance and availability.

There tends to be a very strong temptation to allow each section to optimise their own little bit and then simply add up all of these parts. Unfortunately, even when the cost function only has one stationary point (maximum or minimum) there is no guarantee that a piecewise approach will find it, in fact, the probability of doing so is infinitesimal. In the case of the

LCC, there is very good evidence that it will have many stationary points – i.e. it will have many local maxima and minima. If we also recognise that we will need to decide for every component in the system what maintenance and support policies should be employed at every stage throughout its life then we will soon realise that the problem of optimising the LCC is likely to be one of searching a hyperspace of a very great many dimensions, certainly exceeding 10^6. If we were to only consider 2 values for each variable then and we would still have to perform $2^{1,000,000}$ calculations. Assuming we could evaluate the cost function in say one micro-second then it would take just $2^{50,000}$ (approx $10^{15,000}$) seconds which is about 10^{2000} years (or 10^{666} millennia).

Even if we were lucky and actually found the optimum solution, it would only be valid for the given inputs. As soon as any of the variables changed, such as if a new conflict arose requiring additional operations or, a component was found to be more (or less) reliable than had been assumed, or the operators decided to close a base, or the suppliers offer a new level of support, or the level of inflation or interest changes then the solution will no longer be optimal.

The major point with LCC is that the calculated value will be wrong. The best we can hope for is that it will give a reasonable estimate of the order of magnitude (i.e. whether it is $1 billion, $10 billion or $100 million). Even if it is being used for comparison (or more likely selection) it is important to recognise that the margin of error will almost certainly be greater than the difference between alternative products.

9.4.   MAINTENANCE PLANNING

The task of maintenance planning is to decide what maintenance will need to be done and where it would best be located. Having decided that, the next part is to determine what facilities, equipment and resources will be needed to enable this maintenance to be done in the most cost-effective manner.

9.4.1   Repair or Replace

At the system level, the decision as to whether to repair or replace is generally relatively straight forward, at least while the system is still young. If we believe that all failures are "random" (i.e. independent of time/age) then we must accept that all repairs will restore the system to an "as-good-as-new" condition. Based on these assumptions, then we would never choose to replace the system unless the cost of repairing it exceeds the price of replacing it. In practice, however, system reliability tends to deteriorate with age because many parts in any system (except possibly software systems) will corrode or erode. Of course, there are other factors that need to be considered as well, including capability, obsolescence or simply status. Many people who buy new cars replace them long before they have worn out or become expensive to maintain.

At the component level, the decision to repair or replace will need to take into account the expected frequency of the task, the amount of work needed, the length of time the system and component will be out-of-service, the type of person required (e.g. skill level), the facilities (whether it can be done on the wing, by the side of the road or does it need a "clean room"), what equipment (e.g. a screwdriver and sledgehammer or a sophisticated master computer) and what instructions (e.g. technical publications including detailed diagrams). The decision should also consider the effectiveness of the repair – will it restore that part of the system to "as-good-as-new", "same-as-old", somewhere in between or possibly, better than new.

Repair effectiveness is an important issue. In many cases, a component can only be repaired so many times. Traditionally compressor and turbine rotors were constructed from separate blades and discs. The blades were jointed into the discs. A recent development has made it possible to construct a "blisk" from a single piece of metal so there are no separate blades, as such. If one of the "blades" is damaged, possibly due to foreign objects, then it can be cut off and a new blade friction welded in its place, however, this can only be done once (for any given blade position). If the new blade becomes damaged or worn then the whole blisk would have to be replaced. Friction welding is a delicate and very costly procedure requiring specialist equipment and highly skilled operators. It is not yet known whether a replaced blade has the same reliability as the original. Also, if the original was rejected due to wear, the expected time to the next rejection (due to wear) of one of the other original blades will be very much less than the time to the first rejection, therefore it may not make good economical sense to carry out the repair, rather the blisk should be replaced.

With large gas turbine engines on commercial airliners, the time between engine removals can be several years. However, once they have been removed, it then becomes necessary to decide how much maintenance should be done. In many cases, components will not have failed but, the amount of wear and stress they have experienced may mean that they are likely to need to be replaced in a relatively short time. In some cases, there will be physical/detectable signs that can be used to assist this decision process, in others, the only "evidence" will be based on estimated time-to-failure probability distribution parameters.

Often operators will expect restored items to be given a warranty – i.e. the restorer will repair or replace the item if it fails within the warranty period (given certain caveats). Airlines may also impose penalties for "poor availability" (delays and cancellations). Rail service providers may be required or expected to pay out compensation to customers who have had to wait more than a certain time. Customers may choose to change suppliers ("churn") if they do not get the level of service they would like.

During the decommissioning phase, particularly with military systems, it may not make economical sense to replace a component with a new one, if it is known that the parent system will be scrapped in the near future. With aircraft, however, repairs can only be carried out if an airworthiness authority has approved the repair and its accompanying procedure. Such approval may be both costly and time-consuming. At the same time, there may not be any spare parts available and, if their usage has been low, the capability to make new ones may no longer be available either. Under these circumstances, it may be necessary to resort to cannibalisation – decommissioning one unit and "break" it up for spares.

9.5.   TECHNICAL PUBLICATIONS

At it simplest, technical publications can be considered as user manuals. They tell the operator how to operate the system, the maintainer what preventative maintenance will be needed and when and what corrective maintenance can be done and how and, they tell the supporter what parts are available and from whom.

In addition, they will cover configuration control, when required. In many cases, as the system develops over time, parts will be redesigned and modified. Often, there will restrictions on which modification standard of part can be fitted to a particular modification standard of the parent. Anyone who has been to a motor factors for a spare will, if lucky, have been asked for details about the car (year of manufacture, model, engine capacity, etc.) from which they will have ascertained which part number or modification is likely to be needed. If they were unlucky, they will have got home, tried to fit the part and found that the boltholes do not align, the connectors are of a different design or it simply does not fit.

9.6.   SOFTWARE CONFIGURATION CONTROL

With software, it can become a major problem if different users have different versions. I am sure we have all experienced the situation where we have received a copy of a document (on disc or through the e-mail) that, when we have tried to read it, we cannot and that all we get is screen after screen of "gobbledegook" or "hieroglyphics". This is usually because the version of the software used to create the document is incompatible with the version being used to read it.

Most software suppliers will only support the latest version and, maybe, the previous version but usually only for a limited time. This means that if you have a problem with an old version, maybe because you have changed your computer or the operating system, then you cannot expect to receive any help from the supplier and, indeed, it is very likely that the old version will not even run under a new (version of the) operating system.

To ensure everyone was using the latest version of a program used to be an almost impossible task, particularly if the author/supplier was not directly involved in the installation and upgrading. It is also a major problem when the users are operating on different equipment or even when they are using different directory naming conventions. With the move towards Web-based software, where the one and only authorised copy is held centrally on a network file server, many of these problems have been resolved. In this situation, none of the users have their own copy of the software, instead everyone users the same source. As and when a change is made, if affects the way the program is accessed, executed, the inputs or the outputs then, the user will be obliged to download the latest changes before being able to submit a run.

9.7.   PACKAGING, HANDLING, STORAGE AND
TRANSPORTATION (PHS&T)

Although the components in many systems are subjected to temperature extremes, vibration, contaminated air, sand blasting and salt water, this does not mean that they do not need to be protected when being moved around the world either for maintenance or as serviceable spares. Gas turbine engines, for example, are actually quite vulnerable to damage when not installed in an aircraft, ship or power station. The pipe work on the outside of the engine has, in general, not been designed to be used as handles, lifting points or as foot holds. If the distances between two pipes has been reduced below a safe margin, vibrations during operation can cause them to come into contact, possibly leading to chafing and ultimately serious damage.

Missiles, in particular, may be held in storage for many years before they are either used or disposed of (safely). Their propellants are often quite corrosive but it is not always practical to store them separately. These items pose a difficult problem because the only way it is possible to tell whether they are still operational is to fire them, which is a bit like striking a match to see if it works. Assuming the bits can be found, it is generally not a viable proposition to rebuild them. To gain some assurance that they are still operational, it is necessary to select a sample for firing at certain intervals and record what proportion were successful. Using the binomial distribution, it is then possible to decide how confident one can be that the remainder will still be functional when required. The deterioration of the propellant may not cause the missile to misfire but could affect its range and maybe its accuracy so it is also necessary to replace the warheads with telemetry equipment, which can record (or transmit) data about the flight.

Similar considerations will need to be taken with uninhabited [combat] air vehicles (U[C]AV). The difference with these is that they can be tested without having to destroy them (hopefully). The intention with these "aircraft" is that they should have a fairly limited life, maybe 250 hours partly because the probability of such a vehicle surviving more than a certain number of missions is very small (due to enemy action) and partly because there is considerably less need to fly them on training/familiarisation missions. The majority of flying done by military aircraft is to keep the pilots in a constant state of readiness. With UCAV it should be possible to do almost all of the training using simulators and simulation.

The problem of PHS&T then becomes one of how to store these craft so that the rate of corrosion is kept to a minimum but, at the same time, that the time to "unpack" them is also minimised. It is not a lot of use having a UCAV stored, fully inhibited if it takes weeks to restore them to full operational capability. If they are going to be required then generally, they will only be truly effective if they can be made operational within a few hours, at most. One would also need to consider whether they would be transported to the forward position from where they will be launched or whether they should fly there under their own power. In the former case, the "transporter" may also be the "launch pad" as in the case of an aircraft carrier or it may be a cargo/heavy lift aircraft, which will mean the UCAV has to be capable of being packed into a small volume.

The means of transporting spares can have a significant effect on the cost/availability. Although using ships, particularly container ships, may be the cheapest method of transport (dollar per kilogram), it will not always be the most cost-effective. David Pearcy (1999) showed that when all of the factors, including out-of-service times, shipping costs, special packaging, potential loss at sea, etc. then airfreight could prove significantly cheaper than using ships.

9.8. SUPPORT AND TEST EQUIPMENT

Many systems require special equipment to support their operation. Aircraft, for example, need fuel bowsers, tugs, jetways or mobile stairways/lifts for getting the passengers on and off. They may also need special ground stations to interrogate the on-board test, prognostic and diagnostic equipment to enable fault detection, prevention and isolation. The mechanics in the pits at Formula 1 races need special jacks, air-powered spanners (to remove and replace the wheel nuts), fuel pipes that can be quickly and safely connected and disconnected heating jackets to warm the tyres to near operating temperatures, computers that can receive and analyse data from the cars (as they are racing round) and many other items which have been carefully designed to minimise the time spent in the pits (and hence, hopefully, maximise the probability of winning the race).

To recover an aircraft's gas turbine engine, it is normally required to fit it into a turnover stand. This allows the mechanic(s) to remove the modules from either the front or the back quickly with the minimum handling of the engine. Depending on the depth of strip (how many modules have to be removed), the type of engine and the availability of spare modules, it may stay on the stand for the total time it is being recovered or the stand may be freed between the end of the stripping process and the start of the rebuild. Since these stands are quite expensive and bulky, the maintenance facilities would prefer not have more than they need. However, since their unavailability will increase the logistics delay time for the engines and hence mean that the engine availability reduces or more spare engines may have to be acquired, the cost of not having sufficient may be significantly greater. To determine the optimum number requires a process similar to that of determining how many spare LRU's (engines) are required.

To determine the number of items of a particular type of support equipment at a particular location will require some knowledge of the maintenance that will be done there. The estimate will need to take into account the frequency of the maintenance tasks, the length of time the equipment will be needed (for each task), how many people can perform the task at any given time and whether they will all need their own piece of equipment, the cost of the item and the costs of not having the item when required.

In many cases, there is a need to test the system, sub-system or component after it has been recovered to be sure that (a) it has cured the original problem/fault, (b) that it is working, i.e. it has been properly restored and (c) that it has been fitted correctly. Sometimes this can be done safely by actually operating the system. Ships after they have had a major re-fit or overhaul are usually sent to sea on "shakedown trials" to make sure that everything is working satisfactorily and so that the crew can become familiar with any new equipment that has been installed. With aircraft, the engines and many other components cannot be tested safely in this way. Instead, it is common practice to run them on a test bed or through a simulator so that if they do not perform as required there will be no danger to aircraft, its crew or any other third party.

With electronic components and again occasionally with aircraft engines and other complex sub-systems, it may be necessary to run them through a test facility to isolate the cause of a failure. The on-board test equipment will normally be able to determine which LRU (line replaceable unit) is responsible for the loss of the system's functionability but, it may not have the necessary sophistication to determine which component(s) within the LRU are not performing correctly. To do this may require the LRU to be connected to some form of simulator so that parameters can be varied in a controlled manner and hence by studying the effects isolate the faulty component(s). Such equipment can be extremely expensive, for example, an engine test facility is likely to cost several million US dollars. If the LRU is to be recovered at second line then either every such site will need its own test facility or, it will need to able to transport the LRU's quickly and cheaply between itself and a centralised test facility.

As electronics becomes more sophisticated, more compact and, to a certain extent, cheaper, so the possibility of installing built-in-test equipment (BITE) becomes an increasingly more practical and attractive proposition. In many cases, fault isolation can now be done down to part level or, at least, to a level at which the component is considered to beyond economic repair. Typically, components (e.g. resistors, capacitors and transistors) on a printed circuit board (PCB) would not normally be replaced, as it is usually more economical to replace the PCB.

9.9.   SOFTWARE TESTING

Computer programs generally have their own problems and their own solutions. A common practice with complex programs is to create user interfaces (sometimes optimistically referred to as "user friendly front ends") which run a number of checks and cross-checks on the data being submitted to the program to minimise the chances of incorrect data getting through. There is however a limit to how much this can do, for example, if the program is expecting a probability then it can check to see if the value is between 0 and 1 but it cannot check to see if you meant 0.67 rather than 0.76, say.

If the (main) program fails during execution, this will be due one of two possible causes: there is one, or more, errors in the data or; there is one, or more, errors in the program. Given that the program has been well tested and has been in use for some considerable time then it is very much more likely that the error is in the data but, the alternative cannot be ruled out completely.

With many (computer programming) languages, the source code can be compiled in a number of different ways. For fast operation (minimum run times) it is normal to use the (time) optimisation option. Unfortunately, this generally gives very little diagnostic information following a failure. The best you can hope for is that it will tell you in which subroutine it failed. Sometimes this can be sufficient, particularly if the subroutines are short but, often it would be a lot more helpful if you could find out on which line it failed and what values the variables had at that time. To do this, it may be necessary to use a compiler option that gives you full checking, however, as we have said, this will tend to cause the program to take considerably longer to execute.

One possibility is to recompile the program and run the data through it again. A technique that we found particularly useful in a fairly large, complex, Monte Carlo simulation was to include a large number of strategically placed write statements which gave the location (in the program), the (simulation) clock time and the values of the "key" variables. These write statements are by-passed during normal execution but can be activated at certain (simulation clock) times by the user as part of the standard data input.

Another option that is available with Simscript II.5™ is to compile the program with the diagnostics facility enabled. This allows you to selectively "step through" parts of the program as it is executing giving you the path it has taken and the values of the variables as they are changed within the program. This can be an extremely time-consuming process particularly if the failure occurred quite late in the simulated time but, it is extremely powerful if all else fails and is particularly useful during the development stages of the model.

9.10.   TRAINING AND TRAINING EQUIPMENT

Operational Training

For many years pilots have done much of their training and familiarisation on flight simulators. Although these highly sophisticated pieces of machinery are by no means cheap, they are still a lot less expensive when they fail than the real aircraft. It is also possible to set them up to repeatedly simulate conditions that would be potentially lethal if the pilots tried to do it on a real aircraft.

Shortly after the Boeing 767 went into service, an aircraft flying across Canada ran out of fuel approximately half way some 200 miles from Winnipeg (due to confusion in converting between imperial and metric measures). Without either of the engines, there is no electrical power to light up the digital display screens in the cockpit or activate the servo motors that help the pilots move the control surfaces. This meant that they could not tell their altitude, air speed, pitch or even in which direction they were heading. To mitigate such a situation, the '767, as with most other airliners, is fitted with a RAT – ram air turbine – which drops down into the air-stream and provides enough power to illuminate the display but not sufficient for the servos. The pilots were unable to reach Winnipeg but managed to land the aircraft at a disused war-time airstrip at a place called Gimli. It happened that the pilot was an experienced glider pilot so was able to use his skills to keep the aircraft in the air long enough to get to this strip and to slow it down sufficiently upon arrival to avoid crashing into the ground. It also happened that the co-pilot had spent some time in the Canadian Air Force and had remember this airstrip at Gimli.

As an exercise, it was decided by the airline to feed in the scenario into one of their flight simulators and invite pilots to try their hand at repeating what their colleagues had done. Out of the eleven that took up the challenge, not one landed the aircraft safely.

Although pilots still have to spend some time on real aircraft to maintain their licences and to gain their certificates to allow them to fly aircraft types which they have not previously flown, the time spent on a simulator can very significantly reduce this in the air time. This is also true for pilots of military aircraft but to a lesser extent. It is still necessary to put them in real situations as it is still not possible to simulate the affects of high positive and negative G-forces and close combat with the enemy.

Flight simulators can also play a vital role in accident investigations. If it is suspected that the accident was caused by the failure or malfunction of one (or possibly more) parts, it is often possible to use the simulator to see what the actual effects might be and compare these to the information extracted from the flight "black box" recorders and any eyewitnesses. Unfortunately, with some 4,000,000 parts in an aircraft, it is not practical to use the simulator to determine which part, or combination of parts, may have caused the accident.

What can be done for aircraft, is just as feasible for any vehicle and, indeed, for most systems. Many people have learnt the basics of driving a car in a simulator, although, as yet, it is still not possible to pass your driving test this way.

Maintenance Training

Most expensive, complex systems are now designed using computer-aided design and computer-aided manufacture (CAD/CAM). This has reached a level of sophistication where it is possible to "walk round" the "finished" product long before any metal has been cut. Images in 3-D can be displayed on screens or through virtual reality equipment.

This allows the designers to ensure parts, that need to be accessed can be accessed, that fasteners can be easily fastened and unfastened and that the parts can be removed and refitted without interfering with others. This has two advantages from the maintainers point of view: it ensures that the system is designed to be maintainable, assuming the maintainability engineers are involved and; it can provide them with training simulators and interactive maintenance manuals.

As systems become more reliable, the number of maintenance actions will decrease. In many cases, major sub-systems may remain maintenance-free for several years. This means that, even for operators with relatively large fleets, many mechanics will never even see some of maintenance tasks. For those that do have to perform them, their training may have several years earlier so they are unlikely to be familiar with the processes involved. However, with a laptop computer and access to the Web, the maintainer can quickly access the latest on-line maintenance manuals which can describe the process verbally accompanied by animated pictures to actually show her what she will have to do, what tools she will need and in what order.

If we now provide our mechanic with a (small) camera, she can send back pictures of the damage and get immediate help from an expert should the situation not correspond to that displayed on the on-line help. Using a barcode reader, she will be able to record, instantly, which parts she has removed and, at the same time, order any replacements that are needed. The workshops can be alerted to get their mechanics ready if any of the removed items will need to be recovered.

With all of this accurate data, it will be possible to generate much more accurate estimates of time-to-failure distribution parameters and demand patterns. This can then be used to help the designers of the next system to make them even more reliable, maintainable and supportable. It will also help make the logistician's task easier in terms of predicting which parts will be needed and when, the likely pattern of (maintenance facility) shop loading and, the likely costs.

Having a better understanding of the time-to-failure distributions will also mean that maintenance can be better managed. For parts that have age-related failures, opportunistic, preventative maintenance can be prescribed. Parts can be replaced when their parents are being recovered (for some other reason) to reduce the number of shop visits which can help achieve higher levels of availability and, in some cases, reduce the likelihood of expensive repairs (because the part has not failed and caused damage to others as a result).

Training Logistician and Supportability Specialists

As we have seen in several parts of this book, deciding how many parts will be needed, where and when and whether parts should be replaced before they fail at a given age or based on their condition is not a trivial exercise. In addition it is necessary to keep track of the configuration, the age, location, and status of every part in every operational unit, maintenance facility, store and in-transit.

In the same way that the operators and maintainers are benefiting from simulators, there are also considerable potential benefits for the supporters/logisticians to benefit from some form of interactive model. In this case, it is more likely to be a simulation rather than a simulator. There are already arisings simulation models (in fact these have been around since 1969, in the case of aero-engines). These can help determine the numbers of spares required, the likely shop loading figures, the costs and benefits of modifications, the effects of different maintenance and support policies and help estimate the numbers of ground support equipment, special tools and facilities.

Simulation languages, such as Simscript II.5™, can be programmed to provide animated graphics of the movement of items, the sizes of queues, distributions of failure, maintenance and support times, in fact all of the information the logistician needs to help her make the right decisions. In addition, these models can also be written so that the user can interact with them. This allows the user to make a decision and see what the consequences are likely to be over the next few days, months or years, all in a matter of minutes. They can even go back to the decision point and change the decision and follow that through. This way they can learn what the effects are likely to be of the various options and from this improve their skills.

9.11.   LOGISTIC SUPPORT ANALYSIS (LSA)

There are numerous cases where failure to co-ordinate information during the design phase of a system resulted in technical not matching equipment, spare parts not being interchangeable, training courses which failed to address actual equipment design and the provision of useless, expensive and unnecessary support equipment. As a result Logistic Support Analysis was developed with to meet the following objectives (Knotts, 1996).

1. To influence the equipment design from the point of view of supportability.
2. To identify support problems and related cost drivers early enough in the design process to enable the design to rectify deficiencies or eliminate the associated support problems.
3. To identify support resource requirements.
4. To develop and provide a single database for all analyses.

In summary, LSA to be applied in the early phase of the project to define economical support, influence equipment design and to establish a RMS database.   LSA also provides a baseline information for logistics documentation and technical publications.

Chapter 10

Availability

There is nothing in this world constant, but inconsistency

Jonathan Swift

Availability is used to measure the combined effect of reliability, maintenance and logistic support on the operational effectiveness of the system. A system, which is in a state of failure, is not beneficial to its owner; in fact, it is probably costing the owner money. If an aircraft breaks down, it cannot be used until it has been declared airworthy. This is likely to cause inconvenience to the customers who may then decide to switch to an alternative airline in future. It may disrupt the timetables and cause problems for several days.

As mentioned in Chapter 9, most large airliners have a very high utilisation rate with the only down time being to do a transit check, unload, clean the cabin, refuel, restock with the next flight's foods and other items, and reload with the next set of passengers and baggage. The whole operation generally takes about an hour. Any delay may cause it to miss its take off slot and more significantly its landing slot, since an aircraft cannot take-off until it has been cleared to land, even though this may be 12 hours later. Many airports close during the night to avoid unacceptable levels of noise pollution. If the particular flight was due to land just before the airport closes, missing its slot could mean a delay of several hours.

An operator of a system would like to make sure that the system will be in a state of functioning (*SoFu*) when it is required. Designers and manufacturers know that they are unlikely to remain in business for very long if their systems do not satisfy the customers' requirements in terms of operational effectiveness. Many forms of availability are used to measure the effectiveness of the system. Inherent availability, operational availability and achieved availability are some of the measures used to quantify whether an item is in an operable state when required. Availability is defined as:

The probability that an item is in state of functioning at a given point in time (point availability) or over a stated period of time (interval availability) when operated, maintained and supported in a prescribed manner.

It is clear from the above definition that availability is a function of reliability, maintainability and supportability factors (Figure 10.1).

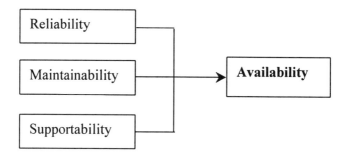

Figure 10.1 Availability as a function of reliability, maintainability and supportability

In this chapter, we look at few important availability measures such as point availability, interval availability, steady state inherent availability, operational availability and achieved availability.

10.1.   POINT AVAILABILITY

Point availability is defined as the probability that the system is in the state of functioning (*SoFu*) at the given instant of time t. We use the notation $A(t)$ to represent the point availability. Availability expressions for systems can be obtained by using stochastic processes. Depending on the time to failure and time to repair distributions, one can use Markov chain, renewal process, regenerative process, semi-Markov process and semi-regenerative process models to derive the expression for point availability. For example, consider an item with constant failure rate λ and constant repair rate μ. At any instant of time, the item can be in either the state of functioning (say, state 1) or in the state of failure (say, state 2). As both failure and repair rates are constant (and thus follow exponential distribution), we can use a Markov chain to model the system to derive the availability expression.

Let $p_{ij}(h)$ denote the transition probability from state i to state j during the interval 'h'($i,j = 1, 2$). Define, $P_i(t+h)$, as the probability that the system would be in state i at time $t+h$, for $i = 1, 2$. The expression for $P_1(t+h)$ can be derived using the following logic:

1. The system was in state 1 at time t and continues to remain in state 1 throughout the interval h.
2. The system was in state 2 at time t and it transits to state 1 during the interval h.

The corresponding expression can be written as:

$$P_1(t + h) = P_1(t) \times p_{11}(h) + P_2(t) \times p_{21}(t) \tag{10.1}$$

Using similar logic, the expression for $P_2(t+h)$ can be written as:

$$P_2(t + h) = P_1(t) \times p_{12}(h) + P_2(t) \times p_{22}(h) \tag{10.2}$$

$p_{11}(h)$ is the probability of remaining in state 1 during the interval h. The probability $p_{11}(h)$ is given by

$$p_{11}(h) = \exp(-\lambda h) \approx 1 - \lambda h \text{ for } \lambda h \ll 1$$

$p_{21}(h)$ is the probability of entering state 1 from state 2 during the interval h. The corresponding expression is given by

$$p_{21}(h) = 1 - \exp(-\mu h) \approx \mu h \text{ for } h\mu \ll 1$$

$p_{12}(h)$ is the probability of entering state 2 from state 1 during the interval h. The probability $p_{12}(h)$ is given by

$$p_{12}(h) = 1 - \exp(-\lambda h) \approx \lambda h \text{ for } h\lambda \ll 1$$

$p_{22}(h)$ is the probability of remaining in state 2 during the interval h. The probability $p_{22}(h)$ is given by:

$$p_{22}(h) = \exp(-\mu h) \approx 1 - \mu h \text{ for } h\mu \ll 1$$

Substituting the values of $p_{ij}(h)$ in equation (10.1) and (10.2), we get

$$P_1(t + h) = P_1(t) \times (1 - \lambda h) + P_2(t) \times \mu h$$

$$P_2(t+h) = P_1(t) \times (\lambda h) + P_2(t) \times (1 - \mu h)$$

By rearranging the terms and setting $h \to 0$, we have

$$\underset{h \to 0}{Lt} \frac{P_1(t+h) - P_1(t)}{h} = \frac{dP_1(t)}{dt} = -\lambda P_1(t) + \mu P_2(t)$$

$$\underset{h \to 0}{Lt} \frac{P_2(t+h) - P_2(t)}{h} = \frac{dP_2(t)}{dt} = \lambda P_1(t) - \mu P_2(t)$$

On solving the above two differential equations, we get

$$P_1(t) = \frac{\mu}{\lambda + \mu} + \frac{\lambda}{\lambda + \mu} \times \exp(-(\lambda + \mu)t)$$

$P_1(t)$ is nothing but the availability of the item at time t, that is the probability that the item will be in state of functioning at time t. Thus, the point availability $A(t)$ is given by:

$$A(t) = \frac{\mu}{\lambda + \mu} + \frac{\lambda}{\lambda + \mu} \times \exp(-(\lambda + \mu)t) \qquad (10.3)$$

Substituting $\lambda = 1/MTTF$ and $\mu = 1/MTTR$ in the above equation, we get

$$A(t) = \frac{MTTF}{MTTF + MTTR} + \frac{MTTR}{MTTF + MTTR} \times \exp(-(\frac{1}{MTTF} + \frac{1}{MTTR})t) \,(10.4)$$

When the time to failure and time to repair are not exponential, we can use a *regenerative process* to derive the availability expression. If $f(t)$ and $g(t)$ represent the time-to-failure and time-to-repair distributions respectively, then the point availability $A(t)$ can be written as (Birolini, 1997):

$$A(t) = 1 - F(t) + \int_0^t \sum_{n=1}^{\infty} [f(x) * g(x)]^n [1 - F(t - x)]dx$$

where $[f(x)*g(x)]^n$ is the n-fold convolution of $f(x)*g(x)$. The summation $\sum_{n=1}^{\infty} [f(x)*g(x)]^n$ gives the renewal points $f(x)*g(x)$, $f(x)*g(x)*f(x)*g(x)$, ... lies in $[x, x+dx]$, and $1 - F(t-x)$ is the probability that no failures occur in the remaining interval $[x, t]$.

10.1.1 Average Availability

Interval availability, $AA(t)$, is defined as the expected fractional duration of an interval $(0, t]$ that the system is in state of functioning. Thus,

$$AA(t) = \frac{1}{t}\int_0^t A(x)dx \qquad (10.5)$$

where $A(x)$ is the point availability of the item as defined in equation (10.3) and (10.4). For an item with constant failure rate λ and constant repair rate μ, the average availability is given by:

$$AA(t) = \frac{\mu}{\lambda + \mu} + \frac{\lambda}{(\lambda + \mu)^2 t}[1 - \exp(-(\lambda + \mu)t)] \qquad (10.6)$$

10.1.2 Inherent Availability

Inherent availability (or steady-state availability), A_i, is defined as the steady state probability (that is, $t \to \infty$) that an item will be in a state of functioning, assuming that this probability depends only on the time-to-failure and time to repair distributions. It is assumed that any support resources that are required are available without any restriction. Thus, the inherent availability is given by:

$$A_i = \underset{t \to \infty}{Lt} \, A(t) = \frac{MTTF}{MTTF + MTTR} \qquad (10.7)$$

The above result is valid for any time to failure function $F(t)$ and any time to repair distribution $G(t)$ (*Birolini*, 1997). Also, in the case of constant failure rate λ and constant repair rate μ, the following inequality is true.

$$|A(t) - A_i| \leq \exp(-t / MTTR) \tag{10.8}$$

Example 10.1

Time to failure distribution of a digital engine control unit (DECU) follows an exponential distribution with mean time between failures 1200 hours and the repair time also follows an exponential distribution with mean time to repair 400 hours.

1. Plot the point availability of the DECU.
2. Find the average availability of the DECU during first 5000 hours.
3. Find the inherent availability.

SOLUTION:

1. The point availability of the DECU is calculated using the equation (10.4). Figure 10.2 depicts the point availability of the system.

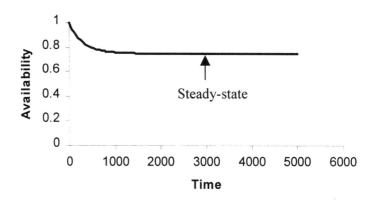

Figure 10.2 Point availability of DECU

2. The average availability of the system during 5000 hours of operation is given by:

$$AA(t) = \frac{\mu}{\lambda + \mu} + \frac{\lambda}{(\lambda + \mu)^2 t}[1 - \exp(-(\lambda + \mu)t)]$$

Substituting the values of λ (= 1/1200) and μ (=1/400), we get the value of the average availability during 5000 hours as 0.7649.

3. The inherent availability is given by

$$A_i = \frac{MTTF}{MTTF + MTTR} = \frac{1200}{1200 + 400} = 0.75$$

Thus, the steady state availability of the system is 0.75 or 75%.

10.1.3 System Availability of different reliability block diagrams

Availability of a system with series reliability block diagram with n items is given by

$$A_s(t) = \prod_{k=1}^{n} A_i(t) \tag{10.9}$$

where $A_i(t)$ is the point availability of ith item. The inherent availability of the system is given by

$$A_{i,s} = \prod_{k=1}^{n} \frac{MTTF_i}{MTTF_i + MTTR_i} \tag{10.10}$$

For a series system with all the elements having constant failure and repair rates, the system inherent availability

$$A_{i,s} = \frac{MTTF_s}{MTTF_s + MTTR_s} \tag{10.11}$$

$MTTF_s$ and $MTTR_s$ are system mean time to failure and system mean time to repair respectively. Let λ_i and μ_i represent the failure rate and repair rate of item i respectively. $MTTF_s$ and $MTTR_s$ are given by

$$MTTF_s = \frac{1}{\sum\limits_{i=1}^{n} \lambda_i}$$

$$MTTR_s = \sum_{i=1}^{n} \frac{\lambda_i MTTR_i}{\lambda_s}, \text{ where } \lambda_s = \sum_{i=1}^{n} \lambda_i$$

Availability of a parallel system with n items is given by

$$A_s(t) = 1 - \prod_{i=1}^{n}[1 - A_i(t)] \qquad\qquad (10.12)$$

Example 10.2

A series system consists of four items. The time to failure and the time to repair distributions of the different items are given as given in Tables 10.1 and 10.2. Find the inherent availability of the system.

Table 10.1. Time to failure distribution for different items.

Item Number	Distribution	Parameters
Item 1	Weibull	$\eta = 2200$ hours $\beta = 3.7$
Item 2	Exponential	$\lambda = 0.0008$ per hour
Item 3	Weibull	$\eta = 1800$ hours $\beta = 2.7$
Item 4	Normal	$\mu = 800$ hours $\sigma = 180$ hours

Table 10.2. Time to repair distribution for different items

Item number	Distribution	Parameters
Item 1	Lognormal	$\mu_l = 3.25$ and $\sigma_l = 1.25$
Item 2	Normal	$\mu = 48$ hours $\sigma = 12$ hours
Item 3	Lognormal	$\mu_l = 3.5$ and $\sigma_l = 0.75$
Item 4	Normal	$\mu = 72$ hours $\sigma = 24$ hours

SOLUTION:

First we calculate $MTTF_i$ and $MTTR_i$ for different items:

$$MTTF_1 = \eta \times \Gamma(1 + \frac{1}{\beta}) = 2200 \times \Gamma(1 + \frac{1}{3.7}) = 2200 \times 0.902 = 1984.4$$

$$MTTF_2 = 1/\lambda = 1/0.0008 = 1250, \quad MTTF_3 = 1600.2, \quad MTTF_4 = 800$$

$$MTTR_1 = \exp(\mu_l + \sigma_l^2 / 2) = 56.33 \text{ hours}, \qquad MTTR_2 \quad = 48 \qquad \text{hours}$$
$$MTTR_3 = \exp(\mu_l + \sigma_l^2 / 2) = 43.87 \text{ hours}, \ MTTR_4 = 72 \text{ hours}$$

Inherent availability, A_i, for item i can be calculated using the equation (10.11). Substituting the values of $MTTF_i$ and $MTTR_i$ in equation (10.11), we have

$$A_1 = 0.9723, \quad A_2 = 0.9630, \quad A_3 = 0.9733, \quad A_4 = 0.9174$$

The system availability is given by

$$A_s = \prod_{i=1}^{4} A_i = 0.8362$$

10.2.   ACHIEVED AVAILABILITY

Achieved availability is the probability that an item will be in a state of functioning (*SoFu*) when used as specified taking into account the scheduled and unscheduled maintenance; any support resources needed are available instantaneously. Achieved availability, A_a, is given by

$$A_a = \frac{MTBM}{MTBM + AMT} \tag{10.13}$$

MTBM is the mean time between maintenance and AMT is active maintenance time. The mean time between maintenance during the total operational life, T, is given by:

$$MTBM = \frac{T}{M(T) + T / T_{sm}} \tag{10.14}$$

M(T) is the renewal function, that is the expected number of failures during the total life T. T_{sm} is the scheduled maintenance interval (time between scheduled maintenance). The above expression is valid when after each scheduled maintenance, the item is 'as-bad-as-old' and after each corrective maintenance the item is 'as-good-as-new'. The active maintenance time, AMT, is given by:

$$AMT = \frac{M(T) \times MTTR + (T/T_{sm})MSMT}{M(T) + T/T_{sm}} \qquad (10.15)$$

MTTR stands for the mean time to repair and MSMT is the mean scheduled maintenance time.

Example 10.3

Time to failure distribution of an engine monitoring system follows a normal distribution with mean 4200 hours and standard deviation 420 hours. The engine monitoring system is expected to last 20,000 hours (subject to corrective and preventive maintenance). A scheduled maintenance is carried out after every 2000 hours and takes about 72 hours to complete the task. The time to repair the item follows a lognormal distribution with mean time to repair 120 hours. Find the achieved availability for this system.

SOLUTION:

Mean time between maintenance, MTBM, is given by

$$MTBM = \frac{T}{M(T) + T/T_{sm}} = \frac{20000}{M(20000) + 20000/2000}$$

M(20000) for normal distribution with mean 4200 hours and standard deviation 420 hours is given by

$$M(20000) = \sum_{n=1}^{\infty} \Phi(\frac{20000 - n \times 4200}{\sqrt{n} \times 420}) = 4.1434$$

$$MTBM = \frac{20000}{4.1434 + 10} \approx 1414 \text{ hours}$$

The active maintenance time is given by:

$$AMT = \frac{M(T) \times MTTR + (T/T_{sm})MSMT}{M(T) + T/T_{sm}}$$

$$= \frac{4.1434 \times 120 + 10 \times 72}{4.1434 + 10} \approx 86.06$$

The achieved availability of the system is given by:

$$A_a = \frac{MTBM}{MTBM + AMT} = \frac{1414}{1414 + 86.06} = 0.9426$$

10.3. OPERATIONAL AVAILABILITY

Operational availability is the probability that the system will be in the state of functioning (*SoFu*) when used as specified taking into account maintenance and logistic delay times. Operational availability, A_o, is given by

$$A_o = \frac{MTBM}{MTBM + DT} \tag{10.16}$$

where, MTBM is the mean time between maintenance (including both scheduled and unscheduled maintenance) and DT is the Down time. The mean time between maintenance during the total operational life, T, is given by:

$$MTBM = \frac{T}{M(T) + T/T_{sm}} \tag{10.17}$$

M(T) is the renewal function, that is the expected number of failures during the total life T. T_{sm} is the scheduled maintenance interval (time between scheduled maintenance). The system down time DT is given by:

$$DT = \frac{M(T) \times MTTRS + (T/T_{sm})MSMT}{M(T) + T/T_{sm}} \tag{10.18}$$

MTTRS stands for the mean time to restore the system and MSMT is the mean scheduled maintenance time. MTTRS is given by

$$MTTRS = MTTR + MLDT$$

where MLDT is the mean logistic delay time for supply resources. In the absence of any scheduled maintenance the operational availability can be calculated using the following simple formula

$$A_O = \frac{MTBF}{MTBF + MTTR + MLDT} \tag{10.19}$$

Example 10.4

In the previous example, assume that whenever a system fails it takes about 48 hours before all the necessary support resources are available. Find the operational availability.

SOLUTION

MTBM is same as in the previous example and is equal to 1414 hours. The mean time to restore the system is given by

$MTTRS = MTTR + MLDT = 120 + 48 = 168$ hours

The system down time is given by

$$DT = \frac{M(T) \times MTTRS + (T/T_{sm})MSMT}{M(T) + T/T_{sm}}$$

$$= \frac{4.1434 \times 168 + 10 \times 72}{14.1434} = 100.12 \text{ hours}$$

The operational availability of the system is given by

$$A_O = \frac{MTBM}{MTBM + DT} = \frac{1414}{1414 + 100.12} = 0.9338$$

Chapter 11

Design for Reliability, Maintainability and Supportability

A few observations and much reasoning lead to error; many observations and a little reasoning to truth

Alexis Carrel

One of the important tasks of the design process is to translate the overall functional requirements for a new system into its physical requirements in relation to performance, power consumption, cost, reliability, maintainability, supportability etc. Reliability, maintainability and supportability should be designed into the product. The Design Phase is particularly important for any product, as the decisions made during this stage will have a major influence on the operational effectiveness of the system throughout its life. In this chapter, we would like to some of the tools and techniques that can be used at the design stage to improve the RMS characteristics.

11.1. RELIABILITY ALLOCATION

Reliability allocation is a process by which the system's reliability requirements are divided into sub-system and component reliability requirements. The main advantage of an allocation process are listed below:

1. Permits system-level requirements to be disseminated to a lower indenture level.
2. Enables designers and subcontractors to work and achieve appropriate targets for their subsystem.
3. Permits the initial evaluation of the feasibility of achieving system requirements.
4. Reduces inappropriate design efforts.

Reliability allocation must be considered at the very early stage of the system design. Baseline system reliability requirement should be used as one of guidelines when various design alternatives are considered. The main objective of reliability allocation is to make sure that the following inequality is valid:

$$f(R_1(t), R_2(t),..., R_n(t)) \geq R^*(t) \tag{11.1}$$

$R^*(t)$ is the reliability requirement at the system level at time t, and f is the function that relates the component reliability to system reliability [Ebeling, 1997]. The function 'f' depends on the reliability block diagram of that particular system. Alternatively, if the system reliability requirement is $MTTF^*$, then the following inequality must hold:

$$\int_0^t f(R_1(t), R_2(t),..., R_n(t))dt \geq MTTF^* \tag{11.2}$$

Figure 11.1 illustrates the process of reliability allocation [Anon, 1984]. Allocation process translates the system reliability requirements into a lower level (sub-system, component) reliability requirement.

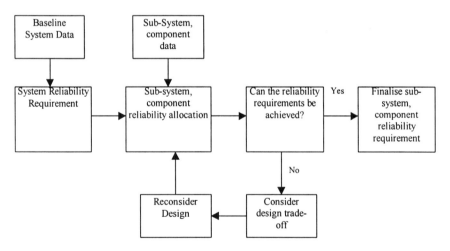

Figure 11.1 Reliability Allocation Process

The allocation process starts by examining the baseline data from historic experience to quantify the system reliability requirements. Allocations are developed using the functional equipment breakdown, for example, functional breakdown for a car is illustrated in Figure 11.2.

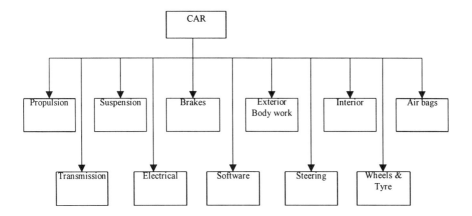

Figure 11.2 Functional breakdown for reliability allocation

11.1.1   AGREE Method for Reliability Allocation

Reliability allocation method developed by AGREE (Advisory Group on Reliability of Electronic Equipment) is one of the simplest ways allocating reliabilities to components and subsystem. AGREE method assumes that the reliability block diagram of the system has a series structure with n independent sub-systems, where each subsystem i has n_i modules.

Assume that t denote the system operating time and t_i represents the operating time of sub-system i. Also, it is assumed that not all failure would result in system failure, thus we introduce a weighing factor, w_i, that represents the probability that the system will fail given that the sub-system i has failed.

Assuming constant failure rate, the failure probability of i-th subsystem is given by:

$$1 - \exp(-t_i / MTTF_i) \tag{11.3}$$

where, $MTTF_i$ is the mean time to failure of the subsystem i. The probability that the system survives and i-th subsystem fails is given by:

$$1 - w_i[1 - \exp(-t_i / MTTF_i)] \tag{11.4}$$

The AGREE methods allocate an equal share of system reliability to each module in the system. Suppose the sub-system i has n_i modules, then the i-th sub-system's contribution to system reliability is given by:

$$[R^*(t)]^{n_i/N} \tag{11.5}$$

where, $R^*(t)$ is the target reliability and,

$$N = \sum_{i=1}^{n} n_i$$

Thus, for the i-th subsystem the following relation is valid:

$$1 - w_i[1 - \exp(-t_i / MTTF_i)] = [R^*(t)]^{n_i/N} \tag{11.6}$$

Solving for $MTTF_i$ in equation (11.6), we get:

$$MTTF_i = -t_i \times \left[\ln\left[1 - \frac{1 - [R^*(t)]^{n_i/N}}{w_i}\right] \right]^{-1} , i = 1,2,...,n \tag{11.7}$$

Equation (11.7) gives the allocated MTTF for subsystem i.

Example 11.1

Mobility of an armoured vehicle is provided by five subsystems: Engine, Gearbox, Suspension Unit, Road Wheels, and Batteries. The reliability requirement of the mobility function is 0.95 for 500 hours of operation. The weighing factor, operating time of individual subsystem and the number of modules are given in Table 11.1. Perform the reliability allocation using AGREE method.

Table 11.1 Data for example 11.1

Subsystem	Weighing Factor, w_i	Subsystem Operating Time, t_i	Number of modules, n_i
Engine	1.0	500	120
Gearbox	1.0	500	80
Suspension Unit	0.9	400	40
Road Wheels	0.8	300	45
Battery	0.7	100	40

SOLUTION:

$N = 120 + 80 + 40 + 45 + 40 = 325$

$$[R^*(t)]^{n_i/N} = [0.95]^{n_i/325}$$

The MTTF$_i$ for various subsystems are given by the expression:

$$MTTF_i = -t_i \times \left[\ln\left(1 - \frac{1-[0.95]^{n_i/325}}{w_i}\right) \right]^{-1}, i = 1,2,...,5$$

Table 11.2 shows the reliability allocation for various sub-systems

Table 11.2 Allocation of MTTF using AGREE method

Subsystem	$MTTF_i$
Engine	26,400.46
Gearbox	39600.69
Suspension Unit	57004.95
Road Wheels	33762.51
Battery	22146.31

11.1.2 Equal Apportionment Technique

The equal apportionment technique assigns equal reliabilities to all the items in order to achieve a specified level of reliability for the whole system. The system is assumed to consist of *n* items in series. The main drawback of this method is that the item reliability goals are not assigned in accordance with the degrees of difficulty associated with meeting them.

Let $R_s(t)$ be the required system reliability and R_i be the reliability for the item i, then

$$R_s(t) = \prod_{i=1}^{n} R_i(t)$$

or

$$R_i(t) = \left(R_s(t)\right)^{1/n}, \qquad\qquad i = 1,2,...,n \qquad\qquad (11.8)$$

Example 11.2

Consider a system consisting of four items, each of which must function if the system is to function. What reliability requirement should be assigned to each item in order to meet a system requirement of 0.895?

SOLUTION:

Using equation (11.8) we have

$$R_i(t) = \left(R_s(t)\right)^{1/n} = (0.895)^{1/4} = 0.97$$

Thus a reliability requirement of 0.97 should be assigned to each item of the system.

11.1.3   The ARINC Apportionment Technique

The ARINC method assumes that the items are connected in series with constant failure rates, that any item failure causes a system failure, and that the item mission time equal the system mission time. The apportionment technique requires the expression of the required reliability in terms of the failure rates. Thus

$$\sum_{i=1}^{n} \lambda_i \leq \lambda_s$$

where λ_i is the failure rate allocated to item i, $i = 1,2,\ldots,n$, and λ_s is the required system failure rate.

The following steps summarise this technique:

Determine the item failure rate, λ_i, form the past data

1. Assign a weighting factor, ω_i, to each item according to failure rates determined in step 1, where ω_i is given by

$$\omega_i = \frac{\lambda_i}{\displaystyle\sum_{i=1}^{n} \lambda_i} \tag{11.9}$$

2. Compute the item failure rate requirements using

$$\lambda_i = \omega_i \lambda_s$$

It is clear that this method allocates the new failure rates based on relative weighting factors that are functions of the past failure rates of the items.

Example 11.3

Consider a system composed of 3 items with estimated rates of $\lambda_1 = 0.002$, $\lambda_2 = 0.003$, and $\lambda_3 = 0.005$ failure per hour. The system has mission time of 100 hours. A system reliability of 0.95 is required. Find the reliability requirements for the items.

SOLUTION:

Using equation (11.9), we compute the weighting factors:

$$\omega_1 = \frac{0.002}{0.002 + 0.003 + 0.005} = 0.2$$

$$\omega_2 = \frac{0.003}{0.002 + 0.003 + 0.005} = 0.3$$

$$\omega_3 = \frac{0.005}{0.002 + 0.003 + 0.005} = 0.5$$

We know that system failure rate can be determined using the following expression

$$R_s(100) = \exp[-\lambda_s(100)] = 0.95$$

$$\lambda_s = 0.00513 \text{ failure per hour}$$

Therefore the failure rates for the items are

$$\lambda_1 = \omega_1 \lambda_s = 0.2 \times 0.00513 = 0.0001026$$
$$\lambda_2 = \omega_2 \lambda_s = 0.3 \times 0.00513 = 0.0001539$$
$$\lambda_3 = \omega_3 \lambda_s = 0.5 \times 0.00513 = 0.0002565$$

The corresponding apportioned reliability for each item is:

$$R_1(100) = \exp[-0.0001026(100)] = 0.9999989$$
$$R_2(100) = \exp[-0.0001539(100)] = 0.9999984$$
$$R_3(100) = \exp[-0.0002565(100)] = 0.9999974$$

11.2.   MAINTAINABILITY ALLOCATION

Maintainability allocation is a process by which the system's maintainability requirements are divided into sub-system and component maintainability requirements. Maintainability requirement is usually stated using the mean time to repair, *MTTR*. For example, system maintainability requirement might be that the system mean time to repair, $MTTR_s$, should be less than or equal to, say, 50 minutes. Consequently, when the first ideas about a new product emerge, it is necessary to decompose the system requirement into individual requirements for consisting subsystems, modules and components. The maintainability measure become the target for the designers and maintainability engineers which should be achieved by their design solutions. Clearly, the maintainability allocation can only be performed in conjunction with allocation tasks regarding other system requirements, like, reliability and availability.

Over the years, several methods have been developed for maintainability allocation. However, it is necessary to stress that as most of the basic reliability allocation models are based on the assumption that the components failures are independent with constant failure rate, the maintainability allocation models are also limited by these factors. One of the simple models for maintainability allocation is based on the inherent availability requirement at system level. Assume that the system has n independent subsystem connected in series. Suppose the system level availability requirement is A_s, by assigning equal availability to subsystem, the subsystem level availability requirement is given by:

$$A_i = [A_s]^{1/n} \tag{11.10}$$

We know that the expression for inherent availability is given by:

$$A_i = \frac{MTTF_i}{MTTF_i + MTTR_i}$$

Rearranging the above expression, we get:

$$MTTR_i = \frac{1 - A_i}{A_i} \times MTTF_i$$

That is, the allocated MTTR for subsystem i satisfies the following inequality:

$$MTTR_i \leq \frac{1 - A_i}{A_i} \times MTTF_i \tag{11.11}$$

Example 11.4

In example 11.1, if the require inherent system availability is 0.99, find the upper bounds for MTTR of various subsystems.

SOLUTION:

By assigning equal availability to all subsystems, the subsystem availability requirement is give by:

$$A_i = [0.99]^{1/5} = 0.997992$$

Using equation (11.11), the upper bounds for *MTTR* for different components can be evaluated. Table 11.3 gives the upper bound for *MTTR$_i$* for different subsystems.

Table 11.3. *MTTR* allocations

Subsystem	*MTTF$_i$*	*MTTR$_i$*
Engine	26,400.46	53.12
Gearbox	39,600.69	79.68
Suspension Unit	57004.95	114.69
Road Wheels	33762.51	67.93
Battery	22146.31	44.56

11.3. SUPPORTABILITY ALLOCATION

Supportability allocation is a process by which the system's supportability requirements are divided into sub-system and component supportability requirements. Supportability allocation for an item can be carried as follows.

The operational availability of an item can be written as:

$$A_O = \frac{MTTF}{MTTF + MTTR + MTTS} \tag{11.12}$$

Where MTTS denotes the mean time to support the item. Rearranging equation (11.12), we get:

$$MTTF = \frac{1 - A_O}{A_O} MTTF - MTTR$$

That is, the allocated MTTR for subsystem i satisfies the following inequality:

$$MTTS \leq [\frac{1-A_O}{A_O} \times MTTF] - MTTR \qquad (11.13)$$

11.4. FAILURE MODES, EFFECTS AND CRITICALITY ANALYSIS

The FMEA, which we discuss in Chapter 6, can be extended to include an evaluation of the failure criticality- an assessment of the severity of the failure effect and its probability of occurrence. Thus, this procedure is the result of two steps which, when combined, provide the FMECA.

- Failure Modes and Effects Analysis (FMEA)
- Criticality Analysis (CA)

The FMEA can be accomplished without a CA, but CA requires that the FMEA has previously identified critical failure modes for items in the system design. The FMECA is potentially one of the most beneficial and productive tasks in a well-structured reliability program.

The failure modes, effects and criticality analysis (FMECA) is a systematic method for examining all modes through which a failure can occur, potential effects of these failures on the system performance and their relative severity in terms of safety, extent of damage, and impact on mission success. FMECA is primarily performed to identify potential safety and reliability problems but, as a result of this, can also identify maintainability and supportability issues. If used properly, It is an excellent methodology for identifying and investigating potential product weaknesses.

FMECA establishes a detailed study of the product design, manufacturing operation or distribution to determine which features are critical to various modes of failure. The FMECA concept was developed by US defence industries in the 1950s, to improve the reliability of military equipment. Since then, FMECA has become an important tools applied by almost all industries around the world to improve the reliability, maintainability and supportability of their product. It is claimed that a more rigorous FMECA analysis would have avoided the disastrous explosion of the Challenger launch on 28th January 1986.

The three principal study areas in FMECA analysis are the failure mode, failure effect and failure criticality. The first two areas were briefly discussed in Chapter 5.

1. Failure mode analysis lists all possible ways in which (system) failure can occur taking into account the condition, the components involved, location etc. For example, some of the modes of failure for a turbine blade, in a gas turbine engine are melting (if its cooling holes become clogged), breaking off at its root (due to either resonance/vibration or impact from a foreign object), breaking at some other point (possible due to creep leading to contact with the casing).
2. The failure effect analysis includes the study of the likely impact of failure on the performance of the whole product and/or the process. For example, if our turbine blade breaks off at its root, it will cause a small deterioration in the performance of the engine. It is very likely to cause vibration in the disc, which if left, could cause the ultimate disintegration of the engine. It can also cause significant damage to other parts of the engine, depending on its exit path.
3. The criticality analysis examines how critical a failure would be for the operation and safe use of the product. The criticality might range from minor failure, through degradation of performance, loss of mission capability to catastrophic failure. This analysis is best utilised during the early design and development phase of new systems, and in the evaluation of existing system [D Verma, 1993].

The actual FMECA performed could be both quantitative and qualitative based on the information available to the analyst. Input requirements for FMECA include reliability data, their modes of failure, and the estimated criticality of the failures. Additionally, the probabilities of detection for the various failure modes are also required. A prerequisite for the successful completion of FMECA is good knowledge of, and familiarity with the product/process being analysed and its design and functionality [D Verma, 1993].

11.4.1   Procedural Steps in the FMECA analysis

The procedural steps in FMECA analysis depend to a certain extent on what product or process is being examined. The sequence of steps followed to accomplish the failure modes, effect and criticality analysis is depicted in Figure 11.3. The following are the key steps involved in the FMECA analysis:

1. Identification of the system requirements, by defining the basic requirements for the system in terms of input criteria for design.

During the system requirement definition, the following tasks should be addressed (Refer to Blanchard and Fabrycky 1999 for detailed discussion).

- What is expected from the system in terms of operation and performance.
- What is the customer requirements with respect to reliability, maintainability and supportability
- How the system is used in terms of hours of operation/number of cycles per day etc.
- What are the requirements for disposal after the system is withdrawn from service.

2. Accomplish functional analysis (Functional analysis is a systematic approach to system design and development, which employs a functional approach as a basis for identification of design requirements for each hierarchical level of the system. Functional analysis is accomplished through a functional flow diagram that portrays the system design requirements illustrating series and parallel relationships and functional interfaces).

3. Accomplish requirements allocation, that is for a specified requirement at system level, what should be specified at unit and assembly level. System effectiveness factors such as reliability, maintainability and supportability specified at system level are allocated to unit and assembly level.

4. Identification of all possible failure modes for the system as well as the subsystem, modules and components.

5. Determine causes of failures, which could be due to design and manufacturing deficiency, ageing and wear-out, accidental damage, transportation and handling, maintenance induced failures.

6. Identify the effects of failure. Effect of failure might range from catastrophic failure to minor performance degradation.

7. Assess the probability of failure. This can be achieved by analysing the failure data and identifying the time-to-failure distribution.

8. Identify the criticality of failure. Failure criticality can be classified in any one of four categories, depending upon the failure effects as follows

a) *Minor failure* – Any failure that doesn't have any noticeable affect on the performance of the system.

b) *Major failure* – Any failure that will degrade the system performance beyond an acceptable limit.

c) *Critical failure* – Any failure that would affect safety and degrade the system beyond an acceptable limit.

d) *Catastrophic failure* – Any failure that could result in significant system damage and may cause damage to property, serious injury or death.

9. Compute the *Risk Priority Number* (RPN) by multiplying the probability of failure, the severity of the effects and the likelihood of detecting a failure mode.

10. Initiate corrective action that will minimise the probability of failure or effect of failure that show high RPN.

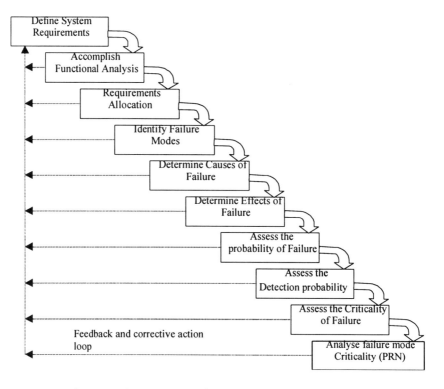

Figure 11.3 Sequence of steps involved in FMECA

11.4.2   Risk Priority Number

Risk Priority Numbers play a crucial role in selecting the most significant item that will minimise the failure or effect of failure. As mentioned earlier,

RPN is calculated by multiplying the probability of failure, the severity of the effects of failure and likelihood of failure detection. That is:

$$RPN = FP \times FS \times FD \qquad (11.14)$$

Where, FP is the Failure probability, FS is the failure severity and FD denotes the failure detection probability. Tables 11.4 – 11.6 gives possible ratings for probability of failure, severity of failure and failure detection. Note that, the ratings given in the Tables 11.4-11.6 are only suggested ratings.

Table 11.4. Rating scales for occurrence of failure

Description	Rating
Remote probability of occurrence	1
Low probability of occurrence	2 - 3
Moderate probability of occurrence	4 - 6
High probability of occurrence	7 - 8
Very High probability of occurrence	9 - 10

Table 11.5 Rating scales for severity of failure

Description	Rating
Minor failure	1 – 2
Major Failure	3 – 5
Critical Failure	6 – 9
Catastrophic Failure	10

Table 11.6. Rating scales for detection of failure

Description	Rating	Probability of Detection
Very high probability of detection	1	0.76 – 1.00
High probability of detection	2 - 3	0.36 – 0.75
Moderate probability of detection	4 - 5	0.16 – 0.35
Low probability of detection	6 - 8	0.06 – 0.15
Remote probability of detection	9 - 10	0.00 – 0.05

Assume that a failure mode has following ratings for probability of failure, failure severity and failure detection:

Failure probability = 7

Failure severity = 4

Failure detection = 5

Then the risk priority number for this particular failure mode is given by
7 × 4 × 5 = 140. Risk priority number for all the failure modes are
calculated and priority is given to the ones with the highest RPN for
eliminating the failure. This is usually achieved using Pareto analysis with a
focus on failure mode, failure cause and failure criticality. Outputs from a
properly conducted FMECA can be used in developing a cost effective
maintenance analysis, system safety hazard analysis, and logistic support
analysis.

11.5.   FAULT TREE ANALYSIS

Fault tree analysis is a deductive, top-down approach involving graphical
enumeration and analysis of the different ways in which a particular system
failure can occur, and the probability of its occurrence. It starts with a top-
level event (failure) and works backwards to identify all the possible causes
and therefore the origins of that failure. During the very early stages of the
system design process, and in the absence of information required to
complete a FMECA, fault tree analysis (FTA) is often conducted to gain
insight into critical aspects of selected design concepts.

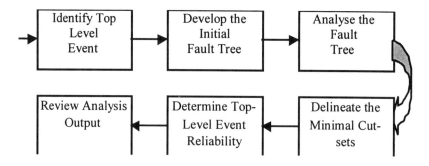

Figure 11.4 Steps involved in a fault tree analysis

Usually, a separate fault tree is developed for every critical failure mode or undesired Top-Level event. Attention is focused on this top-level event and the first-tier causes associated with it. Each first-tier cause is next investigated for its causes, and this process is continued. This '*Top-Down*' causal hierarchy and the associated probabilities, is called a Fault Tree.

One of the outputs from a fault tree analysis is the probability of occurrence of the top-level event or failure. If this probability is unacceptable, fault tree analysis provides the designers with an insight into aspects of the system to which redesign can be directed or compensatory provisions be provided such as redundancy. The FTA can have most impact if initiated during the conceptual and preliminary design phase when design and configuration changes can be most easily and cost effectively implemented.

The logic used in developing and analysing a fault tree has its foundations in Boolean Algebra. The following steps are used to carry out FTA (Figure 11.4).

1. *Identify the top-level event* – The most important step is to identify and define the top-level event. It is necessary to be specific in defining the top-level event, a generic and non-specific definition is likely to result in a broad based fault tree which might be lacking in focus.
2. *Develop the initial fault tree* – Once the top-level event has been satisfactorily identified, the next step is to construct the initial causal hierarchy in the form of a fault tree. Techniques such as *Ishikawa's cause and effect diagram* can prove beneficial. While developing the fault tree all hidden failures must be considered and incorporated. For the sake of consistency, a standard symbol is used to develop fault trees. Table 11.4 depicts the symbols used to represent the causal hierarchy and interconnects associated with a particular top-level event. While constructing a fault tree it is important to break every branch down to a reasonable and consistent level of detail.
3. *Analyse the Fault Tree* – The third step in FTA is to analyse the initial fault tree developed. The important steps in completing the analysis of a fault tree are 1. Delineate the minimum cut-sets, 2. Determine the reliability of the top-level event and 3. Review analysis output.

Table 11.7. Fault tree construction symbols

Symbol	Description
	The Ellipse represents the top-level event (thus always appears at the very top of the fault tree).
	The rectangle represents an intermediate fault event. A rectangle can appear anywhere in a fault tree except at the lowest level in the hierarchy.
	A circle represents the lowest level failure event, also called a basic event.
	The diamond represents an undeveloped event, which can be further broken. Very often, undeveloped events have a substantial amount of complexity below and can be analysed through a separate fault tree.
	This symbol represents the AND logic gate. In this case, the output is realised only after all the associated inputs have been received.
	This symbol represents the OR logic gate. In this case, any one or more of the inputs need to be received for the output to be realised.

11.6. FAULT TREE ANALYSIS CASE STUDY – PASSENGER ELEVATOR

In this section we discuss a case study on fault tree analysis of a passenger elevator (Main source, D Verma, 1993). Consider a passenger elevator depicted in Figure 11.5. We consider two major assemblies for FTA 1. Control assembly and 2. Drive/suspension assembly. All drive assembly failures are generalised as '*motor failures*' and '*other failures*' while control unit failures are generalised as '*hardware failures*' and '*software failures*' for the sake of simplicity.

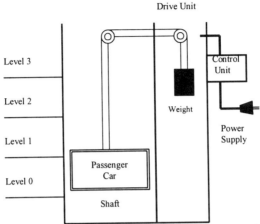

Figure 11.5 Schematic diagram of a passenger elevator

The control assembly consists of a microprocessor, which awaits an operator signal request to move the car to a certain level. The control unit activates the drive unit that moves the car to that level and opens the elevator door once the car comes to a stop. Switches exist at each level and inside the car allowing the controller to know where the car is at any time. Drive/suspension assembly holds the car suspended within the shaft and moves it to the correct level as indicated by the control unit. The Drive unit moves or stops the car only when prompted to do so by the control unit. The brake unit is designed to hod the car stationary when power is removed and to allow the motor shaft to turn when power is applied.

We define the top-level event in this case is '*passenger injury occurs*'. The following are the possible system operating conditions:

A. Elevator operating properly.
B. Car stops between levels.
C. Car falls freely.
D. Car entry door opens in the absence of car.
    In this case, operating conditions 'C' and 'D' are of concern. The initial fault tree is shown in Figure 11.6.

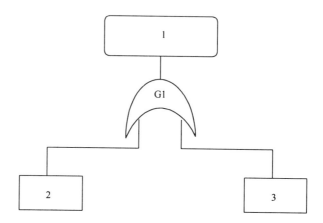

Figure 11.6 Initial fault Tree

In Figure 11.4, G1 represents the OR logic gate and the events 1, 2 and 3 are as defined below:

Event 1 – Passenger injury occurs
Event 2 – Car free falls
Event 3 – Door opens without car present.

Thus, the top-level event (passenger injury occurs) can be either due to car free fall or door opens without the car present. The probability of occurrence for the top event is given by the probability of the event $E_2 \cup E_3$. Where E_1 and E_2 denote the events 2 and 3 respectively. That is,

$$P \text{ [Passenger injury Occurs]} = P [E_2 \cup E_3] = P[E_2] + P[E_3] - P[E_2 \cap E_3]$$

Now the event, car free fall, can be further analysed by treating it as a top-level event, resulting in a fault tree depicted in Figure 11.7. In Figure 11.7, G2 is again a OR gate and the events 4, 5 and 6 are defined below:

Event 4 – Cable slips off pulley
Event 5 – Holding brake failure
Event 6 – Broken cable

Event 4 and 6 are *undeveloped events*, which can be further broken down, using separate fault trees. Event 5 is an *intermediate event*.

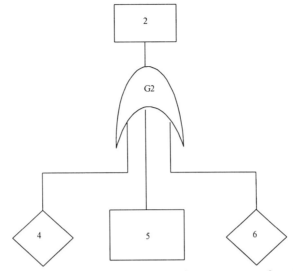

Figure 11.7 Further FTA analysis of the event car free fall

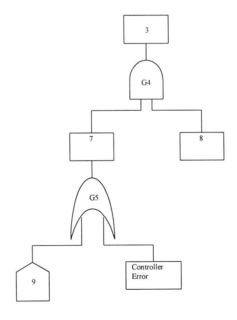

Figure 11.8 Fault tree for the event, the door opens erroneously

The event 3 can be further analysed to find the causes, Figure 11.8 depicts FTA for the event 3, door opens without the car present. This can be caused by the following events:

Event 7 – Door close failure

Event 8 – Car not at level
Event 9 – Latch failure

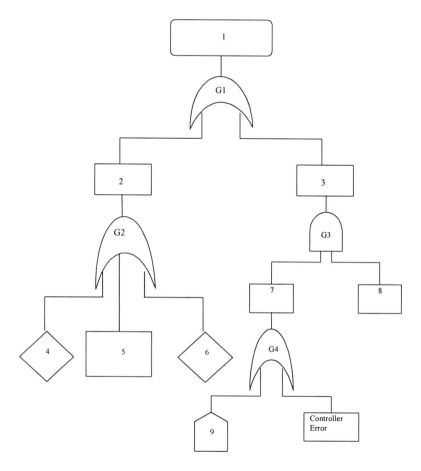

Figure 11.9 Fault tree for the event, passenger injury

For the event, *door opens erroneously*, to occur, events 7 and 8 must happen, thus we have an AND gate G3. The door close failure can be caused either due to the latch failure or due to controller error (denoted by OR gate, G4). Combining fault trees depicted in Figures 11.6-11.8, we can construct an almost complete FTA for the event, passenger injury, as shown in Figure 11.9. Note that events 4, 5, 6, 8 and 9 can be further expanded to find the causes using fault tree analysis. The probability for the occurrence of the top-level event can be calculated once the time-to-failure and probability of occurrence of all the events are known. If the derived top-level probability is unacceptable, necessary redesign or compensation efforts should be identified and initiated.

11.7.   FAULT TOLERANT SOFTWARE

For safety related software, it is very important to design the system in such a way that software failures do not lead to total system failure. That is, the software should be able to survive failures. Fault tolerant software are used in avionics of space shuttle and fighter aircraft where high system reliability is desired. Fault tolerance is a technique buy, which enables the system to complete the function even when there is a failure within the software. Two of the most widely discussed fault tolerant schemes in the literature are *N-version programming* also known as multi-version and *recovery blocks*. Both N-version programming and recovery block scheme achieve fault tolerance by increasing redundancies within the software.

11.7.1   Recovery Blocks

A recovery block scheme requires n versions of a program and a testing segment.   Whenever a version fails, the testing segment activates the succeeding version. The function of the testing segment is to ensure that the operation performed by a version is correct. If the output of the version is incorrect then the testing segment recovers the initial state and activates the next version. (Here initial state refers to the initial values of the input variables just before entering the recovery block. The flow of a recovery block is illustrated in Figure 11.10.

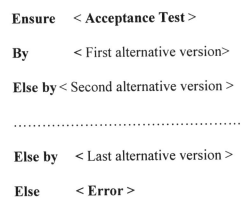

Ensure   < **Acceptance Test** >

By   < First alternative version>

Else by < Second alternative version >

....

Else by   < Last alternative version >

Else   < **Error** >

Figure 11.10 Flow of recovery blocks

The reliability of a recovery block can be derived as follows. Assume that:

n Number of versions in the recovery block.

$P(.)$ Probability of the event (.).

p_i Failure probability of version i.

t_1 Probability that the testing segment cannot perform successful recovery of the input state.

t_2 Probability that the testing segment rejects correct result.

t_3 Probability that the testing segment accepts incorrect result.

R_n Reliability of a recovery block with n versions.

Upon invocation, a recovery block executes version 1 first. If the testing segment detects an error in the output then the state of the system before entering version 1 is recovered and version 2 is activated. This procedure is repeated until acceptance or the testing segment rejects the output of the last version. The function of the testing segment is:

1. To ensure that the output from the block is correct.
2. To recover the initial state and switch to the next version if the output from the present version is not correct.

Here it is assumed that the versions are independent that is the additional versions are not merely a copy of the main version (version 1), so that it can cope with the circumstances that caused the failure of the main version. Three different types of errors that can result in system failure of a recovery block are given by [Dinesh Kumar, 1999]:

1. A version produces correct result, but the testing segment labels it as incorrect.
2. A version produces incorrect result but the testing segment labels it as correct.
3. The testing segment cannot perform successful recovery upon failure of a version.

For the following discussion it is convenient to express reliability of a recovery block scheme as follows. Let,

Y_i be the event that version i produces a correct result and the testing segment accepts the correct result.

X_i be the event that either version i produces an incorrect result and the testing segment rejects it or the version produces a correct result and the testing segment rejects it; in either case the testing segment performs a successful recovery.

The corresponding probabilities for the above two events are given by:

$$P(Y_i) = (1 - p_i)(1 - t_2)$$
$$P(X_i) = (1 - t_1)[p_i(1 - t_3) + (1 - p_i)t_2]$$

Then the reliability of a recovery block scheme with a single version called version 1, R_1, is given by:

$$R_1 = P(Y_1)$$

In general, reliability of a recovery block with n versions is given by

$$R_n = P(Y_1) + \sum_{i=2}^{n}\left[\prod_{k=2}^{i-1} P(X_k) \right] P(Y_i), \quad n \geq 2 \qquad (11.15)$$

Recursively, for $n \geq 2$,

$$R_n = R_{n-1} + [\prod_{k=1}^{n-1} P(X_k)]P(Y_n) \qquad (11.16)$$

11.7.2 N-Version Programming

In an *N*-version programming scheme a number $(N \geq 2)$ of independently coded versions for a given program are run concurrently and the results are compared. A voting scheme recognises the majority as "correct output". N-version programming is used in space shuttle avionics and fighter aircraft like new *Eurofighter*. A specific limitation of *N*-version programming is the requirement of N computers that are hardware independent and able to communicate very efficiently to compare the outputs. Recovery block scheme however can be utilised on a single computer. The reliability of a recovery block can be derived as follows [Dinesh Kumar, 1999]:

Let

$P(G_n) = P$ (2 or more outputs agree)

$P(G_c) = P$ (recurring output is correct)

$P(D_n) = P$ (all the outputs are different).

Reliability of N-version program is given by:

$$RC_n = P(G_n)P(G_c)$$

$$= (1 - P(G_n))P(G_c) \tag{11.17}$$

Note that $P(D_n)$ is the probability that at least $(n - 1)$ versions of the n versions fail. Therefore,

$$P(D_n) = \sum_{i=1}^{n} \frac{1}{p_i} [\prod_{k=1}^{n} p_k](1 - p_i) + \prod_{i=1}^{n} p_i \tag{11.18}$$

$$P(D_n) = P(D_{n-1})p_n + (\prod_{i=1}^{n-1} p_i)(1 - p_n) \tag{11.19}$$

The success of the fault tolerant software depend on the ability to develop statistically independent versions of software, in some cases, this is extremely difficult.

11.8. LIFE CYCLE COSTING

Several factors such as competition, increasing operating and maintenance costs, increasing cost effectiveness awareness among users, introduction of different type of procurements such as the Private Finance Initiative (PFI), prime contracting, and client demand for known costs of ownership could increase the need to apply life cycle costing. Life cycle costing could be defined as a technique for examining and determining all the costs- in money terms - direct and indirect, of design and development, constructing and/or manufacturing and utilising a system throughout its entire service life. Life cycle cost analysis is an economic evaluation tool for choosing among alternative system investments and operating maintenance and support strategies by comparing all of the significant differential costs of ownership over a given time period in equivalent economic terms. The primary aim of life cycle costing is to evaluate and optimise the life cycle costs of the system while satisfying the user and the system requirements. The objectives of life cycle costing are therefore to;

1. Provide a useful input to the decision making process throughout all phases of a system's life cycle;
2. Determine the cost drivers and evaluate all the costs arising during the life span of a system;
3. Provide an equitable comparison on a quantitative basis amongst competing designing options within the same decision-making process, in order to determine and selection of the most appropriate and cost-effective system design option
4. Obtain one all-embracing figure that represents the investment position of the client.
5. Forecast future running expenditure.
6. Reduction of risks and increasing of certainty of estimating whole life cycle costing
7. Increase value for money

11.8.1 Components of Life Cycle Costing

The life cycle of a construction project is the period of time over which a series of phases such as design and development, construction/ manufacturing, commissioning, operation, maintenance and support, and ultimate project disposal or demolition, constitute the total existence and entire scenario of a system. Life cycle costing is therefore all costs associated with the system as applied to the defined life cycle. The total cost of a system, could be broken down into four categories [Blanchard et al, 1991]:

- design and development cost
- production / manufacturing cost
- utilisation cost
- retirement and disposal cost

Each of these categories could be broken down into several cost elements.

Design cost

Design cost includes research and development, engineering design, design documentation, and associated management functions.

Production / manufacturing cost

Production / manufacturing cost includes construction costs, commissioning costs, management costs, and overhead costs.

Utilisation costs

Utilisation costs are the cost incurred throughout the life of the system. They fall into three categories.

Operating cost

Operating cost includes all ongoing costs required to successfully operate a system. For example the significant cost elements of operating cost of a building includes; utility cost such as energy cost and water cost, external and internal cleaning cost, insurance, rates, security, management and administration cost, and overheads.

Maintenance cost

Maintenance cost includes all money spent on keeping the system up to the acceptable standard. Maintenance cost is the cost associated with failure-based, time-based, condition-based maintenance tasks. Maintenance cost is related to direct cost of maintenance resources such as labour, materials, equipment, etc., and indirect costs such as management and administration needed for the successful completion of the task. The breakdown of Maintenance costs were discussed in Chapter 5.

Support cost

Support cost includes all money spent on supporting the system throughout its entire service life. Support cost is related to the direct and indirect cost of providing support elements such as, supply support, test and support equipment, transportation and handling, personnel and training, facilities, data and computer resources. These elements are discussed in Chapter 7.

Retirement and disposal cost

Retirement and disposal cost is the anticipated value at the end of the expected life of a system. The disposal cost includes; demolish cost, recycle elements cost, reused elements cost, and salvage value.

Thus, life cycle cost for a system can be calculated by

$$LCC = D_c + P_c + \sum_{i=1}^{n}(\sum_{j=1}^{m} Oc_j) + \sum_{i=1}^{n}(\sum_{j=1}^{m} Mc_j)$$

$$+ \sum_{i=1}^{n}(\sum_{j=1}^{m} Sc_j) + Di_c \qquad (11.18)$$

where, $Dc =$ Design and development cost, $Pc =$ production / manufacturing cost, $Oc =$ Operating cost, $Mc =$ Maintenance cost, $Sc =$ Spport cost, $Di_c =$ Disposal cost, $n =$ number of years (expected life of system), and $m =$ number of cost elements or tasks

11.8.2 Life Cycle Costing Procedures

According to BS 5760: Part 23, LCC analysis is conducted based on the following steps:

a) Create or adopt a cost breakdown structure that identifies all relevant cost categories in all appropriates life cycle phases which will generates costs.

b) Evaluation of the impact of alternative courses of action (such as design, operating, maintenance, support alternatives) on the LCC of an asset/system

c) Identify cost elements, which will have a significant impact on the overall LCC of the system.

d) Select method(s) for estimating the cost associated with each cost element.

e) Determine the data required developing these estimates, and identifying possible sources for the data. Validate the LCC model with available historical data, if possible

f) Identify any uncertainties, which are likely to be associated with the estimation of each cost elements. In this step sensitivity analyses to examine the assumptions and cost elements uncertainties on LCC should be carried out.

g) Integrate the individual cost elements into a unified overall LCC model will provide the LCC outputs required to meet the analysis objectives

h) Review LCC outputs against the objectives defined in the analysis plan to ensure that all goals have been fulfilled, and that sufficient information has been provided to support the required decision.

11.8.3   Factors affecting LCC appraisal

While the principles of life cycle costing, LCC, have been demonstrated in theory, there are difficulties in using the techniques in practice. These difficulties are related to many factors such as predicting the life of project elements, the lives of the various components used, together with their repair intervals and costs, the discount rates, the rates of interest, the rate of inflation, taxation, and the influences of future government economic policies. Other factors may also affect LCC such as unforeseen use of the project, and change of ownership. These factors should be carefully examined as they will fundamentally effect decisions regarding capital cost and future costs. The main factors affect the life cycle cost appraisal are discussed below.

The Time Horizon

The time horizon relates to the period over which the LCC appraisal is to be carried out. There are two ways to define a project life expectancy:

expected physical life - the period over which a project and its elements can be maintained in an acceptable physical condition;

economic life - the period over which the life cycle cost will be considered.

The determination of a suitable time horizon will depend on the client's expectations and the nature of the system. The difficulty of forecasting the physical and economical life of a project stems from many factors such as deterioration rate, mean time between failures, economic, functional, technological, social and legal regimes, location, fashion, and environmental obsolescence. Consideration of these variables can be a source of many complexities in undertaking a LCC appraisal. The anticipated time horizon of the system and its constituent parts should be carefully examined, as it will fundamentally affect decisions regarding future costs. In many cases, economic life is different from physical life. As far as management and investors are concerned, it is economic life, which is significant. Slater states that a life cycle cost calculation is an economic appraisal based on the economic life rather than on the physical life of a project. However, the physical life of a project has a significant role in the LCC analysis, which quite often depends on the estimation of anticipated lives of the entire project and its constituent components.

Interest/discount rate

The discount rate is one of the critical variables in LCC analysis; the decision to proceed with a project will be crucially affected by which discount rate is chosen. The discount rate selected for an LCC analysis has therefore a large effect on the final results. The discount rate has two functions [Blanchard et al, 1991]:

1. it enables future costs over a time horizon to be equated with their present value;
2. by converting future costs which occur at both regular and irregular intervals to today's equivalent it is possible to directly compare different alternative design options.

The choice of interest rate for discounting depends on whether or not inflation is ignored, on financial circumstances such as whether the client is financing the project through borrowed money or from capital assets, and on the objectives of the client. The higher the discount rate selected, the lower the present value of future costs [Bull, 1993]. Using a high rate emphasises initial costs over future costs: it does not pay to introduce designs with lower operation and maintenance costs when the interest rate is high. These conditions impact heavily on the decision to invest in more durable materials and equipment. Using a discount rate that is too low, does just the opposite: the future costs will be exaggerated.

Inflation

Inflation can have a profound effect on the financial performance of alternative design options. Given that the purpose of financial appraisal is to minimise future expenditure, whilst taking into account the time value of money, it is obvious that inflation must be considered seriously. When dealing with periods as long as the life of an asset such as a construction project, which is usually more than 50 years, it is clear that inflation can have a significant effect. Currently estimates of future inflation rates are based on trends from the past, predicted economic conditions, and judgement. However, it is dangerous to apply a single rate to the whole project because some of cost elements such as of labour, materials, fuel for example may inflate at different rates. Since inflation rate estimates may change considerably with general economic conditions, cost estimates should be reviewed and adjusted as required [Blanchard et al, 1991]. Inflation factors should be estimated on a year-to-year basis if at all possible.

Taxation

Tax affects almost all of the cost elements of the LCC and can change significantly the outcome of the LCC analysis. LCC should take into account tax charges that they bear and also any tax reliefs (capital allowances in construction projects) which are attached to the system or asset. Tax regulations change from year to year, not merely in respect of rates of tax that are applied, but also in terms of the type of project that may be eligible for particular tax allowances.

Uncertainty

Life cycle cost teams are not only attempting to estimate capital cost options which is a problem in itself, but they are also predicting future costs of maintenance, operating, and other related costs as well as revenues. On top of this, they must forecast the economic factors to be used, interest rates, inflation rates, system life and the lives of the various components used, together with their repair intervals and replacement cycles, unexpected use of the project, unusual events such as change of ownership, change in technology and fashion, changes in use, and the influences of future fiscal policies, and environmental and social changes. Generally speaking, these factors can be grouped into the following categories; economic, technical, social and political and others. The major difficulties in applying life cycle costing in practice are related to the prediction of the future behaviour of the above factors. These factors are never constant, but vary from year-to-year. Some of them can, at least, be considered, analysed and evaluated, others can not even be imagined today. Life cycle cost decisions therefore involve a considerable amount of uncertainty, which makes it very difficult to carry out economic evaluations with a high degree of reliability. Examples of the techniques used to deal with uncertainty are conservative benefits and cost estimating, sensitivity analysis, risk-adjusted discount rate, mean-variance criterion and coefficient of variance, decision analysis, and simulation [Flanagan et al, 1983]. No single technique can be labelled as best for treating uncertainty and risk. What is best will depend on many things such as availability of data, availability of resources and computational aids, user understanding, level of risk exposure of the project, and the size of the investment [Marshall, 1991]. If the reliability of LCC analysis is to be improved, the sources of uncertainties, which mentioned above must be dealt with as an integral part of the whole life cycle process.

It is important that The LLC team should be familiar with all phases of system life cycle, typical cost elements, sources of cost data, financial principles, clear understanding of the factors affecting the application of LCC and the methods of assessing the uncertainties associated with cost estimation.

In many instances, the LCC is used as an aid to choose between competing bids. In is particularly important in these cases that the costs to be included, the discount factors (interest and inflation) to be applied, the methods to be used and the assumptions that may be used are all clearly stated by the buyer and agreed with the prospective suppliers. In practice, it is unlikely that this will be done as most major suppliers have developed their own LCC models and will be unwilling (or even unable within the time scales) to change these models to suit every prospective customer. A possible alternative is to use an independent group that can perform the LCC but, this also has practical difficulties as most suppliers are reluctant to divulge sensitive (company confidential) information to a third party, particularly if there is any chance it could find its way into the hands of the competitors.

Life cycle costs and leasing arrangements

As suppliers chase more and more of the potential revenue from a system, such as the after-market (maintenance, management and support during the operational phase) so new ways are being introduced to spread the risk. An increasingly popular approach is that of leasing in which the support service costs are recovered by levying a charge on the use of the system. One such case is "power-by-the-hour" in which the cost of maintaining and supporting the engines (on aircraft) is recovered usually over a fixed period and at a fixed price per hour of (engine) operation (or running). Since all aircraft operators are legally obliged to keep track of the engine hours, this unit of measure is easily obtained. It also has the advantage that, because the majority of the support costs are reasonably linearly related to the use of the engines, provided this usage is reasonably consistent across all flights, these charges will be relatively unaffected by variations in usage. It is, of course, tempting to use the LCC models to determine the rates (per hour) that will be offered to the customer.

The danger with this is that whilst it may be (fairly) reasonable to use MTBF (or even MTTF), MTTR and MTTS over the life of the system, there can be very significant variations in the actual times over shorter periods. For example, it is to be hoped that there will be very few failures in the first few years of operation (i.e. within the normal warranty period) so the MTTF is likely to be lower for this period than the "steady-state" value. On the other hand, the mechanics may be on something of a learner curve in the first few years so they may not be able to achieve the expected maintenance times. The lead time for spares may be considerably lower in the early years while new operational units (engines) are still be manufactured (to meet aircraft delivery programs) but, once the production line has been closed down, it may take a lot longer to produce the spares required.

The main drivers of engine removals tend to have very strongly age-related failure rates (hazard functions) which means that there should be a long period when there will be relatively little maintenance required. A problem with this is that at the end of this period, there could a very high peak in the number of engine removals. This means there will be a lot more spare engines needed at this time and, all of the maintenance facilities, equipment and resources will be stretched to their limit. Also, of course, the outlay, at this time will be significantly higher, although, in theory, this should not be a problem if the income has been properly invested in the preceding years.

Basically, such fleet hour arrangements require models that are generally significantly more sophisticated than those that may have been developed to meet the LCC requirements. There is also a need to be able to use these models to optimise such variables as maintenance and support policies. For example, when an engine comes in for (corrective) maintenance, is it more cost-effective to thoroughly overhaul it and return it into service in a state which is close to as good as new or, do the minimum amount of repair and expect the time to the next engine removal to be that much shorter. This decision may depend on the length of time the fixed price contract has to run and, hence, the time until a new rate can be renegotiated.

Chapter 12

Analysis of Reliability, Maintenance and Supportability Data

Often statistics are used as a drunken man uses lamp posts... for support rather than illumination.

To predict various reliability characteristics of an item, as well as its maintainability and supportability function, it is essential that we have sufficient information on the time to failure, time to repair (maintain) and time to support characteristics of that item. In most cases these characteristics are expressed using theoretical probability distributions. Thus, the problem which every logistician face is the selection of the appropriate distribution function to describe the empirical data (obtained from data capturing sources) using theoretical probability distributions. Once the distribution is identified, then one can extract information about the type of the hazard function and other reliability characteristics such as mean time between failures and failure rate etc. In the case of maintenance and supportability data, we would identify the maintainability and supportability function as in the case of reliability data and then compute MTTR and MTTS.

To start with we look at ways of fitting probability distributions to in-service data, that is the data relating to the age of the components at the time they failed while they were in operation (in maintenance and logistic support we analyse the data corresponding to the maintenance and support task completion times). We look at three popular tools; 1. Probability papers, 2. Linear regression, and 3. Maximum likelihood estimates to identify the best distribution using which the data can be expressed and to estimate the corresponding parameters of the distribution. In the section on "censored data" we recognise that very often we do not have a complete set of failure data. We may wish to determine whether a new version of a component is more reliable than a previous version to decide whether we have cured the problem (of premature failures, say). Often, components will be replaced before they have actually failed, possibly because they have started to crack, they have been damaged or they are showing signs of excessive wear. We may have a number of systems undergoing testing to determine whether the product is likely to meet the various requirements but we need to go into production before they have all failed. There is useful data to be gleaned from the ones that have not failed as well as from the ones that have failed. If a component is being used in a number of different systems, it may be reasonable to assume that the failure mechanism in each of these instances will be similar. Even though the way the different systems operate may be different, it is still likely that the shape of the failure distribution will be same and that only the scale will be different.

Even relatively simple systems can fail in a number of different ways and for a number of different reasons. Suppose we wish to fasten two pieces of metal together using a nut and bolt. If we over-tighten the nut, we might strip the thread or we might shear the bolt. If we do not put the nut on squarely, we could cross the threads and hence weaken the joint. If the two pieces of metal are being forced apart then the stress on the nut and bolt may cause the thread to strip either inside the nut or on the outside of the bolt or it may cause the bolt to exceed its elastic and plastic limits until it eventually breaks. If the joint is subject to excessive heat this could accelerate the process. Equally, if it is in very low temperatures then the bolt is likely to become more brittle and break under less stress than at normal temperatures. If the diameter of the bolt is towards the lower limit of its tolerance and the internal diameter of the nut is towards the upper limit then the amount of metal in contact may not be sufficient to take the strains imposed. As the two components age, corrosion may cause the amount of metal in contact to be even further reduced. It may also change the tensile strength of the metals and cause premature failure.

Components may therefore fail due to a number of failure modes. Each of these modes may be more or less related to the age. One would not expect corrosion to be the cause of failure during the early stages of the component's life, unless it was subjected to exceptionally corrosive chemicals. On the other hand, if the components have been badly made then one might expect to see them fail very soon after the unit has been assembled.

Very often, a possibly small, number of components may fail unexpectedly early. On further investigation it may be found that they were all made at the same time, from the same ingot of metal or by a particular supplier. Such a phenomenon is commonly referred to as a *batching* problem. Unfortunately, in practice, although it may be possible to recognise its presence, it may not always be possible to trace its origin or, more poignantly, the other members of the same batch or, indeed, how many there may be.

In deciding whether a new version of a component is more reliable than the old one, we need to determine how confident we are that the two distributions are different. If they both have the same (or nearly the same) shapes then it is a relatively straightforward task to determine if their scales are different. In some cases, the primary cause of failure of the origin version may have been eliminated or, at least, significantly improved but, another, hitherto rarely seen cause, may have become elevated in significance. This new primary cause may have a distinctly different shape than the first one that often makes it very difficult to decide between the two.

In this chapter, we first look at the empirical approaches for finding estimates for MTTF, MTTR and MTTS as well as failure function, maintainability and supportability functions. Rest of the chapter describes some of the well-known methods for selection of the most relevant theoretical distribution functions for the random variables under consideration.

12.1. RELIABILITY, MAINTENANCE AND SUPPORTABILITY DATA

A very common problem in reliability engineering is the availability of failure data. In many cases getting sufficient data for extracting reliable information is the most difficult task. This may be due the fact that there is no good procedure employed by the operator (or supplier) to collect the data or the item may be highly reliable and the failure is very rare. However, even without any data, one should be able to predict the time-to-failure distribution if not the parameters. For example, if the failure mechanism is corrosion, then it cannot be an exponential distribution. Similarly if the failure cause is 'foreign object damage' then the only distribution that can be used is exponential. The main problem with insufficient failure data is getting an accurate estimate for the shape parameter. Fortunately, we don't have such problems with maintenance and supportability data. These are easily available from the people who maintain and support the item. The reliability data can be obtained from the following sources:

1. Field data and the in-service data from the operator using standard data capturing techniques. There are standard *failure reporting* forms for the purpose of capturing desired information regarding the reliability of the item under consideration. Unfortunately, all these forms are flawed, as they record only MTBF (or MTTR and MTTS in case of maintenance and support). Just the value of

MTBF alone may not be enough for many analyses concerning reliability (similarly, in the case of maintenance (support), information on MTTR (MTTS) is not enough for complete analyses).

2.   From *life testing* that involves testing a representative sample of the item under controlled conditions in a laboratory to record the required data. Sometimes, this might involve '*accelerated life testing*' (ALT) and '*highly accelerated life testing*' (HALT) depending on the information required.

As mentioned earlier, in some cases it is not possible to get a complete failure data from a sample. This is because some of the items may not fail during the life testing (also in the in-service data). These types of data are called '*censored data*'. If the life testing experiment is stopped before all the items have failed, in which cases only the lower bound is known for the items that have not failed. Such type of data is known as '*right censored data*'. In few cases only the upper bound of the failure time may be known, such type of data is called '*left censored data*'.

12.2.   ESTIMATION OF PARAMETERS - EMPIRICAL APPROACH

The objective of empirical method is to estimate failure function, reliability function, hazard function, MTTF (or MTTR and MTTS) from the failure times (or repair and support times). Empirical approach is often referred as n*on-parametric approach* or *distribution free approach*. In the following sections we discuss methods for estimating various performance measures used in reliability, maintenance and support from different types of data.

12.2.1   Estimation of Performance Measures - Complete Ungrouped Data

Complete ungrouped data refers to a raw data (failure, repair or support) without any censored data. That is, the failure times of the whole sample under consideration are available. For example, let t_1, t_2, ..., t_n, represents n ordered failure times such that $t_i \leq t_{i+1}$. Then the possible estimate for failure function (cumulative failure distribution at time t_i) is given by:

$$\hat{F}(t) = \frac{i}{n} \qquad (12.1)$$

A total of i units fail by time t out of the total n in the sample. This will make $F(t_n) = n/n = 1$. That is, there is a zero probability for any item to survive beyond time t_n. This is very unlikely, as the times are drawn from a sample and it is extremely unlikely that any sample would include the longest survival time. Thus the equation (12.1) underestimates the component survival function. A number of mathematicians have tried to find a suitable alternative method of estimating the cumulative failure probability. These range from using n+1 in the denominator to using -0.5 in the numerator and +0.5 in the denominator. The one that gives the best approximation is based on *median rank*. Bernard's approximation to the median rank approach for cumulative failure probability is given by

$$\hat{F}(t_i) = \frac{i - 0.3}{n + 0.4} \qquad (12.2)$$

Throughout this chapter we use the above approximation to estimate the cumulative failure distribution or failure function. From equation (12.2), the estimate for reliability function can be obtained as

$$\hat{R}(t_i) = 1 - \hat{F}(t_i) = 1 - \frac{i - 0.3}{n + 0.4} = \frac{n - i + 0.7}{n + 0.4} \qquad (12.3)$$

The estimate for the failure density function $f(t)$ can be obtained using

$$\hat{f}(t) = \frac{\hat{F}(t_i) - \hat{F}(t_{i+1})}{t_i - t_{i+1}}, \qquad t_i \le t \le t_{i+1} \qquad (12.4)$$

Estimate for the hazard function can be obtained by using the relation between the reliability function $R(t)$ and the failure density function $f(t)$. Therefore,

$$\hat{h}(t) = \hat{f}(t) \Big/ \hat{R}(t) \quad \text{for} \quad t_i < t < t_{i+1} \qquad (12.5)$$

An estimate for the mean time to failure (or mean time to repair or mean time to support) can be directly obtained from the sample mean. That is,

$$\hat{MTTF} = \sum_{i=1}^{n} \frac{t_i}{n} \tag{12.6}$$

Estimate for the variance of the failure distribution can be obtained from the sample variance, that is

$$s^2 = \sum_{i=1}^{n} \frac{(t_i - \hat{MTTF})^2}{n-1} \tag{12.7}$$

Estimate for MTTR (MTTS) and Variance of time to repair distribution (time to support distribution) can be obtained by replacing failure times by repair times (support times) in equation (12.6) and (12.7) respectively.

12.2.2   Confidence Interval

It is always of the interest to know the range in which the measures such as MTTF, MTTR and MTTS might lie with certain confidence. The resulting interval is called a *confidence interval* and the probability that it contains the estimated parameter is called its *confidence level* or *confidence coefficient*. For example, if a confidence interval has a confidence coefficient equal to 0.95, we call it a 95% confidence interval.

To derive a (1-α) 100% confidence interval for a *large sample* we use the following expression:

$$\hat{MTTF} \pm z_{\alpha/2} \left(\frac{\sigma}{\sqrt{n}} \right) \tag{12.8}$$

Where $z_{\alpha/2}$ is the z value (standard normal statistic) that locates an area of α/2 to its right and can be found from the normal table. σ is the standard deviation of the population from which the population was selected and n is the sample size. The above formula is valid whenever the sample size *n* is greater than or equal to 30. The 90%, 95% and 99% confidence interval for MTTF with sample size n ≥ 30 are given below:

90% confidence $\hat{MTTF} \pm 1.645 \times \left(\dfrac{\sigma}{\sqrt{n}} \right)$ $\tag{12.9}$

$$95\% \text{ confidence } \hat{MTTF} \pm 1.96 \times \left(\frac{\sigma}{\sqrt{n}} \right) \qquad (12.10)$$

$$99\% \text{ confidence } \hat{MTTF} \pm 2..58 \times \left(\frac{\sigma}{\sqrt{n}} \right) \qquad (12.11)$$

When the number of data is small (that is when n is less than 30), the confidence interval is based on t distribution. We use the following expression to calculate $(1-\alpha)100\%$ confidence interval.

$$\hat{MTTF} \pm t_{\alpha/2} \left(\frac{s}{\sqrt{n}} \right) \qquad (12.12)$$

where $t_{\alpha/2}$ is based on $(n-1)$ degrees of freedom and can be obtained from *t* distribution table (refer appendix).

Example 12.1

Time to failure data for 20 car gearboxes of the model *M2000* is listed in Table 12.1. Find:

1. Estimate of failure function and reliability function.
2. Plot failure function and the reliability function.
3. Estimate of MTTF and 95% confidence interval.

Table 12.1. Failure data of gearboxes in miles

1022	1617	2513	3265	8445
9007	10505	11490	13086	14162
14363	15456	16736	16936	18012
19030	19365	19596	19822	20079

SOLUTION:

The failure function and reliability function can be estimated using equations 12.2 and 12.3. Table 12.2 shows the estimated values of failure function and reliability function.

Table 12.2. Estimate for failure and reliability function.

Failure data	$\hat{F}(t_i)$	$\hat{R}(t_i)$
1022	0.0343	0.9657
1617	0.0833	0.9167
2513	0.1324	0.8676
3265	0.1814	0.8186
8445	0.2304	0.7696
9007	0.2794	0.7206
10505	0.3284	0.6716
11490	0.3774	0.6225
13086	0.4264	0.5736
14162	0.4754	0.5246
14363	0.5245	0.4755
15456	0.5735	0.4265
16736	0.6225	0.3775
16936	0.6716	0.3284
18012	0.7206	0.2794
19030	0.7696	0.2304
19365	0.8186	0.1814
19596	0.8676	0.1324
19822	0.9167	0.0833
20079	0.9657	0.0343

The failure function and the reliability function graph are shown in Figure 12.1 and 12.2 respectively.

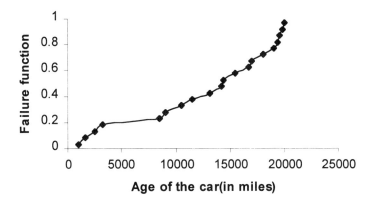

Figure 12.1 Estimate of failure function for the data shown in Table 12.1

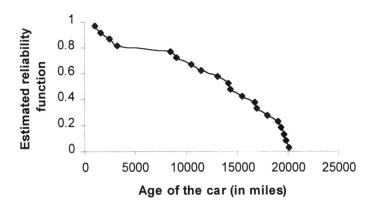

Figure 12.2 Estimated reliability function for the data given in Table 12.2

The estimate for mean time to failure is given by:

$$\hat{MTTF} = \sum_{i=1}^{20} \frac{t_i}{20} = 12725.5 \ \text{miles.}$$

Estimate for the standard deviation is given by

$$s = \sqrt{\sum_{i=1}^{n} \frac{(t_i - \overset{\wedge}{MTTF})^2}{n-1}} = 14827.16 \text{ miles}$$

As the sample data is less than 30, we use equation (12.12) to find the 95% confidence level. From t-table the value of $t_{0.025}$ for $(n-1) = 19$ is given by 2.093. The 95% confidence level for MTTF is given by:

$$\overset{\wedge}{MTTF} \pm t_{\alpha/2}\left(\frac{s}{\sqrt{n}}\right) = 12725.5 \pm 2.093(14827.16/\sqrt{19})$$

That is, the 95% confidence interval for MTTF is (5605.98, 19845.01).

Example 12.2

Time taken to complete repair tasks for an item is given in Table 12.3. Find the cumulative time to repair distribution and mean time to repair. Find 95% confidence level for MTTR.

Table 12.3. Time to repair data

28	53	71	90
30	56	72	92
31	58	74	94
33	59	75	95
35	61	79	97
40	65	81	99
41	67	82	100
44	68	84	103
49	69	85	108
51	70	89	110

Maintainability function can be estimated using following expression:

$$\hat{M}(t_i) = \frac{i - 0.3}{n + 0.4} = \frac{i - 0.3}{40.4}$$

Figure 12.3 shows the estimated maintainability function.

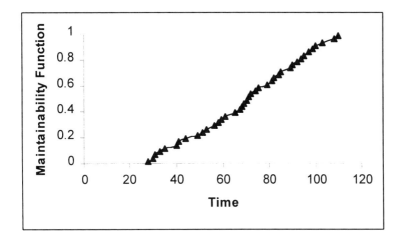

Figure 12.3.Maintainability function for the data given in Table 12.3.

Mean Time to Repair is given by:

$$\hat{MTTR} = \sum_{i=1}^{40} \frac{t_i}{40} = 69.7 \text{ hours}$$

Standard deviation for repair time is given by

$$s^2 = \sqrt{\sum_{i=1}^{n} \frac{(t_i - \hat{MTTR})^2}{n-1}} = 23.43 \text{ hours}$$

Since $n > 30$, we use equation (12.10) to calculate 95% confidence interval for MTTR. 95% confidence level for MTTR is given by

$$\hat{MTTR} \pm 1.96\left(\frac{s}{\sqrt{n}}\right) = 69.7 \pm (1.96)\left(\frac{23.43}{\sqrt{40}}\right) = (62.43, 76.96)$$

12.2.2   Analysis of Grouped Data

Often failure data is placed into time intervals when the sample size is large. The failure data are classified into several intervals. The number of intervals, NI, depends on the total number of data n. Following equation can be used as guidance for determining the suitable number of intervals:

$$\lfloor NI \rfloor = 1 + 3.3 \times \log_{10}(n) \tag{12.13}$$

$\lfloor NI \rfloor$ denotes that the value is rounded down to the nearest integer.

The length of each interval, LI, is calculated using:

$$LI = \frac{(x_{max} - x_{min})}{\lfloor NI \rfloor} \tag{12.14}$$

where x_{max} is the maximum recorded failure time and x_{min} is the minimum recorded failure time. The lower and upper bound of each interval is calculated as follows:

$$X_{min,i} = x_{min} + (i-1) \times LI$$
$$X_{max,i} = x_{min} + i \times LI$$

$X_{min,i}$ is the lower bound of the ith interval and $X_{max,i}$ is the upper bound value of the ith interval. Let $n_1, n_2, \ldots n_n$ be the number of items that fail in the interval i. Then the estimate for cumulative failure distribution is given by

$$\hat{F}(X_{max,i}) = \frac{\sum_{k=1}^{i} n_k - 0.3}{n + 0.4} \tag{12.15}$$

Estimate for the reliability function R(t) is given by:

$$\hat{R}(X_{max,i}) = 1 - \hat{F}(X_{max,i}) = \frac{\sum_{k=i+1}^{n} n_i + 0.7}{n + 0.4} \tag{12.16}$$

Estimate for the failure density is given by:

For $X_{max,i+1} < t < X_{max,i}$

$$\hat{f}(t) = \frac{\hat{F}(X_{max,i+1}) - F(X_{max,i})}{X_{max,i+1} - X_{max,i}} = \frac{n_{i+1}}{(n+0.4) \times (X_{max,i+1} - X_{max,i})}$$

The MTTF is estimated using the expression:

$$\hat{MTTF} = \sum_{i=1}^{NI} \frac{X_{med,i} \times n_i}{n} \tag{12.17}$$

where $X_{med,i}$ is the midpoint in the ith interval and n_k is the number of observed failures in that interval. Estimate for sample variance is given by

$$s^2 = \sum_{i=1}^{NI} (X_{med,i} - \hat{MTTF})^2 \times \frac{n_i}{n} \tag{12.18}$$

Example 12.3

Results of 55 observed values of the duration of support tasks in hours are given in Table 12.4. Calculate the Mean Time to Support (MTTS).

Table 12.4. Time to support data

3	56	9	24	56	66	67	87	89	99	4
26	76	79	89	45	45	78	88	89	90	92
99	2	3	37	39	39	77	93	21	24	29
32	44	46	5	46	46	99	47	77	79	89
31	78	34	67	86	86	75	33	55	22	44

SOLUTION:

First we need to find the number of groups using equation (12.13). The number of intervals is given by:

$$\lfloor NI \rfloor = 1 + 3.3 \times \log_{10}(55) = \lfloor 6.74 \rfloor = 6$$

The length (range) if each interval (group) is given by:

$$LI = \frac{x_{max} - x_{min}}{\lfloor NI \rfloor} = \frac{99 - 2}{6} = 16.17$$

Table 12.5 shows the various calculations associated in computing the mean time to support.

Table 12.5. Analysis of grouped data given in example 12.3

i	$LI\ (x_{min,I} - x_{max,i})$	n_i	$x_{med,i}$	$X_{med,i} \times n_i$
1	2 - 18.17	6	10.08	60.51
2	18.17 – 34.34	10	26.25	262.55
3	34.34 - 50.51	11	42.42	466.67
4	50.51 - 66.68	5	58.59	292.97
5	66.68 - 82.85	9	74.76	672.88
6	82.85 - 99	14	90.92	1272.95

MTTS is given by:

$$\hat{MTTS} = \sum_{i=1}^{NI} \frac{X_{med,i} \times n_i}{n} = \sum_{i=1}^{6} \frac{X_{med,i} \times n_i}{55} = 55.06$$

12.3.   ANALYSIS OF CENSORED DATA

In many cases, the complete data may not be available due to the reasons such as all the items may not have failed or the manufacturer may wish to get interim estimates of the reliability etc. The mechanism for censoring may be based on a fixed age, on a fixed number of failures or at some arbitrary point in time. In practice, provided the times at the time of failure or, at the time of suspension (censor) are known, the reason for terminating the test is not important. We will assume that the times of failure are known precisely. We will look at cases in which we do not know the exact time, only that the failure occurred sometime between the last inspection and the current age later. In this section we derive estimates for failure function, reliability function when the data is multiple censored. We denote t_i to represent a complete data and t_i* to denote a censored time.

The only difference between the estimation of parameters in complete data and the censored data is the calculation of median ranks. Now we will need to adjust the ranks in order to take account of the components that have not failed. The rank adjustment is done in the following two steps:

1. Sort all the times (failures and suspensions) in ascending order and allocate a sequence number i starting with 1 for the first (lowest) time and ending with n (the sample size for the highest recorded time). Now we discard the suspended times as it is only the (adjusted rank) of the failures with which we are concerned.

2. For each failure calculate the adjusted rank as follows:

$$R_i = R_{i-1} + \frac{n+1-R_{i-1}}{n+2-S_i} \tag{12.19}$$

where, R_i is the adjusted rank of the ith failure, R_{i-1} is the adjusted rank of the $(i-1)$th failure, that is the previous failure. R_0 is zero and S_i is the sequence number of the ith failure.

As a quick check, the adjusted rank of the ith failure will always be less than or equal to the sequence number and at least 1 greater than the previous adjusted rank. If there is no suspensions, the adjusted rank will be equal to the sequence number as before. These adjusted ranks are then substituted into the Benard's approximation formula to give the median rank and the estimate for cumulative probability is given by:

$$\hat{F}(t_i) = \frac{R_i - 0.3}{n + 0.4}$$

Example 12.4

The following data were observed during the data capturing exercise on 12 compressors that are being used by different operators. Estimate the reliability and failure function (* indicates that the data is a censored data)

2041, 2173, 2248*, 2271, 2567*, 2665*, 3008, 3091, 3404*, 3424, 3490*, 3716

SOLUTION:

We need to calculate the adjusted rank of the failure times using equation (12.19), once this is done, then the failure and reliability function can be estimated using equations (12.2) and (12.3) respectively. The estimated failure and reliability functions are shown in Table 12.6.

Table 12.6 Estimated failure and reliability function

S_i	t_i	j	$R_j = R_{j-1} + [(n+1- R_{j-1}) / (n+2 - S_i)]$	$F(t_i)$	$R(t_i)$
1	2041	1	1	0.0565	0.9435
2	2173	2	2	0.1370	0.8630
3	2248*				
4	2271	3	3.1	0.2258	0.7742
5	2567*				
6	2665*				
7	3008	4	4.51	0.3395	0.6605
8	3091	5	5.92	0.4532	0.5468
9	3404*				
10	3424	6	7.69	0.5960	0.4040
11	3490*				
12	3716	7	10.34	0.8097	0.1903

12.4.   FITTING PROBABILITY DISTRIBUTIONS GRAPHICALLY

The traditional approach for measuring reliability, maintenance and supportability characteristics is using a theoretical probability distribution. It should however, be borne in mind that failures do not occur in accordance with a given distribution. These are merely convenient tools that can allow us to make inferences and comparisons in not just an easier way but also with known levels of confidence. In this section we will look at a graphical method that can be used to not only to fit distributions to given data but also help us determine how good the fit is. To illustrate the graphical approach we use the following failure data observed on 50 tyres.

To draw a graph we obviously need a set 'x' and 'y' co-ordinates. Sorting the times-to-failure in ascending order will give us the 'x' values so all we need is to associate a cumulative probability to each value. This is done using the median rank approach discussed earlier, that is 'y' axis values are given by the cumulative failure probabilities calculated using the equation (12.2). Now, we can plot the values $[t_i, F(t_i)]$. In Figure 12.4 we can see the result of this for the 50 tyre time-to-failure.

Table 12.7. Failure data for 50 tyres

1022	14363	20208	26530	31507
1617	15456	20516	28060	33326
2513	16736	20978	28240	33457
3265	16936	21497	28757	35356
8445	18012	24199	28852	35747
9007	19030	24582	29092	36250
10505	19365	25512	29236	36359
11490	19596	25743	29333	36743
13086	19822	26102	30620	36959
14162	20079	26163	30924	38958

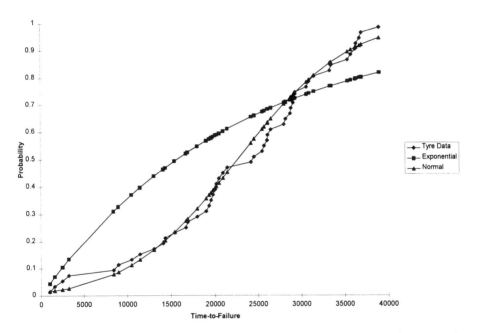

Figure 12.4 Tyre Data compared to Exponential and Normal Distributions

The two additional lines on this graph have been plotted to show what an exponential distribution (with the same mean as the sample) would look like and similarly for a normal distribution with the sample mean and standard deviation. This indicates that the exponential distribution is not a very good fit whereas the normal is certainly better. What it does not tell us, however, is how much better or, indeed, whether another distribution gives an even better fit.

A measure of how good the curve fits the data would be the *correlation coefficient* but, this only applies to straight line fits. Similarly we could use the *Kolmogorov-Smirnov* test but this really only tells us whether the there is a significant difference between the data and that which would be expected if the data were exponentially or normally distributed.

There are, in fact, two standard approaches to fit the data to a probability distribution graphically: to use "probability paper" or to transform either the "x" or "y" (or both) data so that the resulting graph would be a straight line if the data were from the given distribution. Actually both methods are essentially the same because to create probability paper the axes have been so constructed as to produce straight lines plot if the data is from the given distribution. If we can determine the necessary transforms then we can easily construct the probability paper.

12.4.1   Fitting an exponential distribution to data graphically

The cumulative probability density function for the exponential distribution is given by

$$F(t) = \begin{cases} 0, & t < 0 \\ 1 - \exp(-\lambda t), & t \geq 0 \end{cases}$$

Since we are only considering positive failure times, we can, without loss of generality, omit the expression for $t < 0$. If we replace $F(t)$ with p then we get

$$p = 1 - \exp(-\lambda t)$$

Rearranging and taking natural logarithm we get

$$\ln[\frac{1}{1-p}] = \lambda t \tag{12.20}$$

This is a linear function in t such that the slope of the line is the reciprocal of the MTTF. Figure 12.5 is an example of "Exponential Graph Paper" (for the failure date from Table 12.7). The y-scale is given as percentages rather than probabilities. The x-scale is linear.

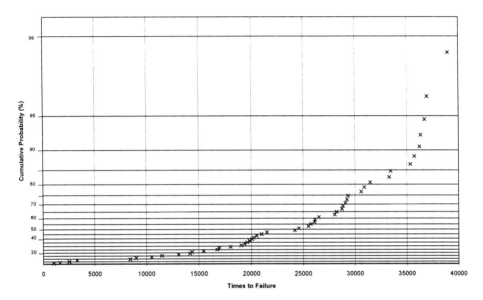

Figure 12.5 Data Plotted on Exponential Graph Paper

If the data forms a straight line in the exponential probability paper, then we can find the value of MTTF by using the relation F(MTTF) = 0.632. That is, we find the time to failure from the paper for which the percentage failures is 63.2.

12.4.2   Fitting a Normal Distribution Graphically

We will now see how good a fit the normal distribution gives. Again we can plot the times-to-failure on special normal (probability) paper. Such paper is becoming increasingly more difficult to obtain commercially. It can, however, be created using a proprietary spreadsheet package. Figure 12.6 shows how the tyre example failure times (and their respective median ranks) would appear on "normal paper".

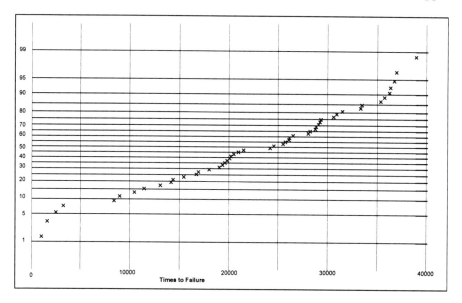

Figure 12.6 Times-to-Failure plotted on Normal Paper

The cumulative density function for the normal distribution is not as simple to transform to a linear form as the exponential.

$$F(t) = p = \int_{-\infty}^{t} \frac{1}{\sqrt{2\pi}\sigma} e^{-\frac{1}{2}\left(\frac{x-\mu}{\sigma}\right)^2} dx$$

However, we can obtain the standardised normal variable $z = (\frac{x-\mu}{\sigma})$, for any given value of p ($F(t)$) either from tables or, using the *NORMSINV* function in MicroSoft™ Excel®, for example. Now we can plot this value(as the y co-ordinate) against the corresponding time-to-failure (as the x co-ordinate). The value of μ and σ can be found by using the relation, $F(\mu) = 0.5$ and $F(\mu+\sigma) = 0.84$.

12.4.3   Fitting a Log-Normal Distribution Graphically

Essentially the log-normal distribution is the same as a normal distribution excepting that the (natural) logarithm of the x-values are used in place of the actual values. Figures 12.7 and 12.8 show log-normal plot for the data given in Table 12.7.

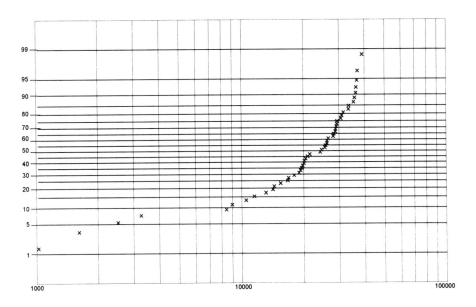

Figure 12.7 Times-to-Failure plotted on Log-Normal Paper

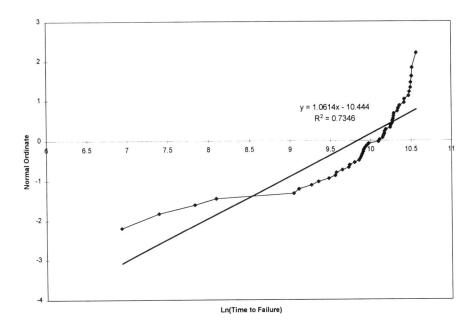

Figure 12.8 Fitting a Log-Normal Distribution Graphically

Here the plotted points form a concave curve to which the straight line is not a particularly good fit although it is still better than the exponential fit. The mean in this case is 18,776 which is considerably lower than the mean from the previous graphs but, this is because it is the geometric mean (the nth root of the product of the TTFs) and not the arithmetic mean with which we are more familiar.

12.4.4   Fitting a Weibull Distribution Graphically

The cumulative density function of the Weibull distribution is similar to that of the exponential, indeed the latter is the (mathematically) degenerative form of the former.

$$F(t) = p = \begin{cases} 0 & \text{for } t < 0 \\ 1 - e^{-\left(t/\eta\right)^{\beta}} & \text{for } t \geq 0 \end{cases}$$

By re-arranging and taking natural logarithms

$$-\ln(1-p) = \left(\frac{t}{\eta}\right)^{\beta}$$

which is still not in a linear form so we have to take logs again to give:

$$\ln(-\ln(1-p)) = \beta \ln(t) - \beta \ln(\eta)$$

So if we plot $ln(-ln(1-p))$ against $ln(t)$ an estimate of the shape parameter (β) of the Weibull will be given by the slope of the straight line drawn through the plotted points. To get an estimate of the scale parameter (η) we need to carryout a transform on the intercept:

$$\eta = e^{-c/\beta}$$

where c is the intercept of the regression line with the x-axis. Figures 12.9 and 12.10 shows Weibull plot for the data given in Table 12.7.

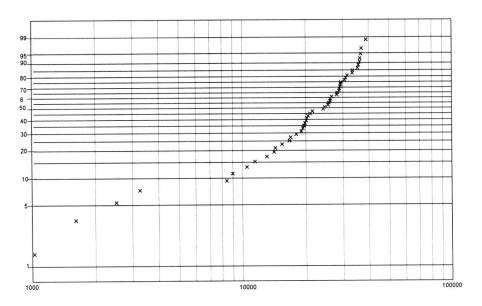

Figure 12.9 Times-to-Failure Fitted on Weibull Paper

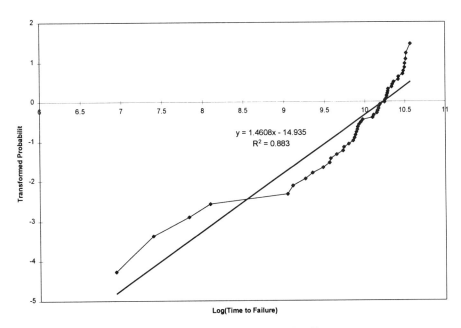

Figure 12.10 Fitting a Weibull distribution graphically

Again the Weibull distribution does not give as good a fit as the normal (distribution) but it is better than either the exponential or the log-normal.

The slope (1.48) indicates that there could be a certain amount of age-relatedness to the failures.

12.5.   REGRESSION

The models used to relate a dependent variable *y* to the independent variables x are called regression models. The simplest regression model is the one that relates the variable *y* to a single independent variable *x* (*linear regression model*). Linear regression provides predicted values for the dependent variables (y) as a linear function of independent variable (*x*). That is, linear regression finds the best-fit straight line for the set of points (*x y*). The objectives of linear regression are:

1. To check whether there is a linear relationship between the dependent variable and the independent variable.
2. To find the best fit straight line for a given set of data points.
3. To estimate the constants 'a' and 'b' of the best fit y = a + bx.

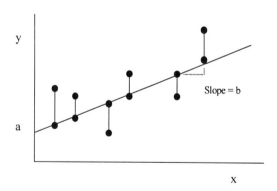

Figure 12.11 Least square regression.

The standard method for linear regression analysis (fitting a straight line to a single independent variable) is using the *method of least squares*. Least square regression is a procedure for estimating the coefficients 'a' and 'b' from a set of X, Y points that have been measured. In reliability analysis, the set X is the set of time to failures (or function of TTF) and set Y is their corresponding cumulative probability values (or function of cumulative distribution). Figure 12.11 illustrates the least square regression. The measure of how well this line fits the data is given by the correlation coefficient. If we construct a line such that it passes through the point (x, y) where x is the mean of the x values and y is the mean of the y values then the sum of the distances between each point and the point on the line vertically above (-ve) or below (+ve) will always be zero (provided the line is not parallel to the y-axis). The same holds for the horizontal distances provided that the line is not parallel to the x-axis. This means that any line passing through the means (in the way described) will be an unbiased estimator of the true line.

If we now assume that there is a linear relationship between the x's (x ∈ X) and y's (y ∈ Y), that the x's are known exactly and that the "errors" in the y values are normally distributed with mean 0 then it can be shown that the values of a and b which minimises the expression:

$$\sum_{i=1}^{n}(y_i - a - bx_i)^2 \tag{12.21}$$

Will give the best fit. The expression $(y_i - a - bx_i)$ gives the vertical distance between the point and the line. Cutting out lot of algebra, one can show that the values of a and b can be found by solving the following equations:

$$na + b\sum_{i=1}^{n} x_i = \sum_{i=1}^{n} y_i \tag{12.21}$$

$$a\sum_{i=1}^{n} x_i + b\sum_{i=1}^{n} x_i^2 = \sum_{i=1}^{n} x_i y_i \tag{12.22}$$

'a' is the estimate of the intercept (of the line with the y-axis) and 'b' is the estimate of the slope – i.e. y = a + bx is the equation of the line giving:

$$b = \frac{n\sum_{i=1}^{n} x_i y_i - \sum_{i=1}^{n} x_i \sum_{i=1}^{n} y_i}{n\sum_{i=1}^{n} x_i^2 - (\sum_{i=1}^{n} x_i)^2} \tag{12.23}$$

$$a = \sum_{i=1}^{n} \frac{y_i}{n} - b\sum_{i=1}^{n} \frac{x_i}{n} \tag{12.24}$$

Note also that these expressions are not symmetrical in x and y. The formula quoted here gives what is called *"y on x" regression* and it assumes the errors are in the y-values.

By replacing each x with a y and each y with an x we can perform "x on y" regression (which assumes the errors are in the x-values). If c is the estimate of the intercept so obtained and d is the estimate of the slope then to get estimates of *a* and *b* (the intercept and slope of the original graph):

$$b = \frac{1}{d} \text{ and } a = -\frac{c}{d}$$

Note: unless the points are collinear, the "x on y" estimates will not be the same as the "y on x" estimates. In the special case where you want to force the line through the origin (i.e. the intercept is zero), the least squares formula for the slope becomes:

$$b = \frac{\sum_{i=1}^{n} x_i y_i}{\sum_{i=1}^{n} x_i^2} \tag{12.25}$$

Note this line does not pass through the means (unless it is a perfect fit).

12.5.1   Correlation Co-efficient

A measure of the dependence between two variables is given by the correlation coefficient. The correlation coefficient, r is given by:

$$r = \frac{n\sum_{i=1}^{n} x_i y_i - \sum_{i=1}^{n} x_i \sum_{i=1}^{n} y_i}{\sqrt{n\sum_{i=1}^{n} x_i^2 - (\sum x_i)^2} \times \sqrt{n\sum_{i=1}^{n} y_i^2 - (\sum y_i)^2}} \tag{12.26}$$

The correlation coefficient always lies between -1 and $+1$. A value of $+1$ or -1 means that x and y are exactly linearly related. In the former case y increases as x increases but for $r = -1$, y decreases as x increases. Note that if x and y are independent then $r = 0$, but $r = 0$ does not mean that x and y are independent. The best fit distribution is the one with maximum r value (close to one). To find the best fit, regression analysis is carried out on the popular distribution such as exponential, Weibull, normal and log-normal. The one with highest correlation coefficient is selected as the best. The coordinates (x, y) and the corresponding parameters for different distributions are listed given in the following sections.

12.5.2   Linear Regression for Exponential Distribution

To fit a data to an exponential distribution, we transform the co-ordinates $(t_i, F(t_i))$ such a way that, when plotted, it gives a straight line. Here t_i is the observed failure times and $F(t_i)$ is the estimated cumulative distribution function. The cumulative distribution of exponential distribution is given by:

$$F(t) = 1 - \exp(-\lambda t)$$

that is,

$$\ln\left[\frac{1}{1 - F(t)}\right] = \lambda t \tag{12.27}$$

Equation (12.27) is a linear function. Thus, for an exponential distribution, the plot of $(t, \ln\left[\frac{1}{1 - F(t)}\right])$ provides a straight line. Thus, if $t_1, t_2, \ldots, t_n$ are the observed failure times, then to fit this data into an exponential distribution, we set:

$$x_i = t_i \tag{12.28}$$

$$y_i = \ln\left[\frac{1}{1 - F(t_i)}\right] \tag{12.29}$$

Substituting (x_i, y_i) in equation (12.23) we get:

$$b = \frac{\displaystyle\sum_{i=1}^{n} x_i y_i}{\displaystyle\sum_{i=1}^{n} x_i^2} \tag{12.30}$$

Note that, for exponential distribution b = 1/MTTF.

Example 12.5

The following failure data were observed on Actuators. Fit the data to an exponential distribution and find the MTTF and the correlation coefficient.

14, 27, 32, 34, 54, 57, 61, 66, 67, 102, 134, 152, 209, 230

SOLUTION:

First we carry out least square regression on $t_i, \ln[\dfrac{1}{1-F(t_i)}]$, various calculations are tabulated in Table 12.8.

Table 12.8. Regression analysis for the data in example 12.5

i	$t_i (= x_i)$	$F(t_i)$	$y_i = \ln[1/(1-F(t_i))]$
1	14	0.0486	0.0498
2	27	0.1180	0.1256
3	32	0.1875	0.2076
4	34	0.2569	0.2969
5	54	0.3263	0.3951
6	57	0.3958	0.5039
7	61	0.4652	0.6260
8	66	0.5347	0.7651
9	67	0.6041	0.9267
10	102	0.6736	1.1196
11	134	0.7430	1.3588
12	152	0.8125	1.6739
13	209	0.8819	2.1366
14	230	0.9513	3.0239

The value of b is given by:

$$b = \frac{\displaystyle\sum_{i=1}^{n} x_i y_i}{\displaystyle\sum_{i=1}^{n} x_i^2} = \frac{\displaystyle\sum_{i=1}^{n} t_i \times \ln[\dfrac{1}{1-F(t_i)}]}{\displaystyle\sum_{i=1}^{n} t_i^2} = 0.01126$$

MTTF is given by $1/b = 1/0.01126 = 88.73$. The corresponding correlation coefficient is 0.9666.

12.5.3   Linear Regression for Weibull Distribution

Cumulative distribution of Weibull distribution is given by:

$$F(t) = 1 - \exp(-(\frac{t}{\eta})^{\beta})$$

That is, $\ln[\ln(\frac{1}{1-F(t)})] = \beta \ln(t) - \beta \ln(\eta)$, which is a linear function. Thus to fit the data to a Weibull distribution, we set:

$$x_i = \ln(t_i) \qquad\qquad (12.31)$$

$$y_i = \ln[\ln(\frac{1}{1-F(t_i)})] \qquad\qquad (12.32)$$

From least square regression, it is evident that the shape and scale parameters of the distribution are given by:

$$\beta = b \qquad\qquad (12.34)$$

$$\eta = \exp(-a/\beta) \qquad\qquad (12.35)$$

Example 12.6

Construct a least square regression for the following failure data:

17, 21, 33, 37, 39, 42, 56, 98, 129, 132, 140

SOLUTION:

Making use of equations (12.31) and (12.32), we construct the least square regression, which are presented in Table 12.9.

Table 12.9. Weibull regression for the data in example 12.6

i	t_i	$F(t_i)$	$x_i = ln(t_i)$	$Y_i = lnln(1/1-F(t_i))$
1	17	0.0614	2.8332	- 2.7581
2	21	0.1491	3.0445	- 1.8233
3	33	0.2368	3.4965	- 1.3082
4	37	0.3245	3.6109	- 0.9354
5	39	0.4122	3.6635	- 0.6320
6	42	0.5	3.7376	- 0.3665
7	56	0.5877	4.0253	- 0.1209
8	98	0.6754	4.5849	0.1180
9	129	0.7631	4.8598	0.3648
10	132	0.8508	4.8828	0.6434
11	140	0.9385	4.9416	1.0261

Using equations (12.34) and (12.35), we get $\beta = 1.4355$, $\eta = 76.54$ and the correlation coefficient $r = 0.9133$.

12.5.4   Linear regression for Normal Distribution

For normal distribution,

$$F(t) = \Phi(\frac{t-\mu}{\sigma}) = \Phi(z)$$

Now z can be written as:

$$z_i = \Phi^{-1}[F(t)] = \frac{t_i - \mu}{\sigma} = \frac{t_i}{\sigma} - \frac{\mu}{\sigma} \tag{12.36}$$

Which is a linear function. Now for regression, we set $x_i = t_i$ and $y_i = z_i = \Phi^{-1}[F(t_i)]$. The value of z can be obtained from standard normal distribution table. One can also use the following expression that gives polynomial approximation for z_i.

$$x_i = t_i \tag{12.37}$$

$$P = \sqrt{\ln[\frac{1}{[1 - F(t_i)]^2}]}$$

$$y_i = P - \frac{C_0 + C_1 P + C_2 P^2}{1 + d_1 P + d_2 P^2 + d_3 P^3} \tag{12.38}$$

where

$C_0 = 2.515517$, $C_1 = 0.802853$, $C_2 = 0.010328$, $d_1 = 1.432788$, $d_2 = 0.189269$, $d_3 = 0.001308$

The estimate for μ and σ are given by

$$\mu = -\frac{a}{b} \text{ and } \sigma = \frac{1}{b}$$

Example 12.7

Fit the following data into a normal distribution

62, 75, 93, 112, 137, 170, 185

SOLUTION:

Table 12.10 gives various computations involved in regression.

Table 12.10. Normal regression for example 12.7

i	t_i	$F(t_i)$	$z_i = P - (c_0 + c_1 P + c_2 P^2 / 1 + d_1 P + d_2 P^2 + d_3 P^3)$
1	62	0.0945	- 1.2693
2	75	0.2297	- 0.7302
3	93	0.3648	- 0.3434
4	112	0.5	0
5	137	0.6351	0.3450
6	170	0.7702	0.7394
7	185	0.9054	1.3132

The estimate for $\mu = 118.71$, $\sigma = 54.05$ and the correlation coefficient r = 0.9701.

12.5.5   Linear Regression for Log-normal Distribution

For log-normal distribution we set:

$$x_i = \ln(t_i) \tag{12.39}$$

$$P = \sqrt{\ln[\frac{1}{[1 - F(t_i)]^2}]}$$

$$y_i = P - \frac{C_0 + C_1 P + C_2 P^2}{1 + d_1 P + d_2 P^2 + d_3 P^3} \tag{12.40}$$

where
$C_0 = 2.515517$, $C_1 = 0.802853$, $C_2 = 0.010328$, $d_1 = 1.432788$, $d_2 = 0.189269$, $d_3 = 0.001308$

12.6.   MAXIMUM LIKELIHOOD ESTIMATION

Although there are many benefits in using regression methods and probability paper, it should be recognised that there are other methods. One of these, which can be useful in many circumstances, is *maximum likelihood estimation (MLE)*.

The main advantages of this method are that it is more mathematically rigorous and less susceptible to individual values as every time-to-failure (repair or support) has equal weighting. Its disadvantages are that it is more difficult to use, it does not provide a visual check and its point estimates of the shape (β) parameter tends to be biased for small samples.

The method can be used with any type of distribution. It can also be used with data that is censored or in which the failures are only known to have occurred at sometime during a given (time) interval.

12.6.1   Complete and Uncensored Data

Consider a scenario in which n identical items are operated until they have all failed or, equivalently, one item is operated until it has failed n times with each repair restoring the item to an "as-good-as-new" condition and its age being reset to zero. Let us further assume that the times at which each of these items failed are known to be t_1, t_2, ...,t_n. If we now assume that these are all independent and identically distributed (*iid*) with probability density function $f(t : \overline{\theta})$ where $\overline{\theta}$ is the set of parameters then we can calculate the likelihood:

$$l = \prod_{i=1}^{n} f(t_i : \vec{\theta}) \tag{12.41}$$

Strictly speaking, the product should be multiplied by n! as the order of the times is unimportant so that we would say the sequence t_n, t_{n-1}, ...,2, 1 is the same as the one given and is, clearly, equally likely. However, because we are not interested in the actual "probability" of the given scenario, only in what values of the parameters maximise the "likelihood", we can safely ignore the n!.

To make mathematics easier and to reduce the problems of dealing with numbers very close to zero, it is normal practice to use a (natural) logarithm transform:

$$-\log_e(l) = -\ln(l) = L = -\sum_{i=1}^{n} \ln(f(t_i : \vec{\theta})) \tag{12.42}$$

Note: that because *f(t)* will always be between 0 and 1, the (natural) log will always be negative so to make it a positive value add the negative sign and turn it into a minimisation.

Of course, in general, we will not know the true values of the parameters. The maximum likelihood estimation method is based on the assumption that if we can find values for the parameters which maximises the likelihood (i.e. minimises the value of L) then these should be the "best" estimates of the true values.

Now, the maxima and minima of any function, L, occur at:

$$\frac{dL}{d\vec{\theta}} = 0 \tag{12.43}$$

So all we have to do is find the derivatives of L with respect to each of the parameters, substitute in the values of the TTF's and solve the simultaneous equations to obtain the "best" estimates of the parameters.

Note that at this stage we have not specified any particular distribution, only that all of the TTF's are iid (i.e. from the same distribution) so we could actually extend this one stage further by choosing the type of distribution which gives the lowest value of L_{mle} (i.e. maximises the maximum likelihood estimates). However, before we do that, let us consider some of the more common distributions.

Exponential Distribution

The exponential distribution only has one parameter which can either be specified as the mean time to failure (MTTF) λ or as its reciprocal, the "failure rate", γ. The probability density function is given by:

$$f(t) = \frac{1}{\lambda} e^{-\frac{t}{\lambda}} = \gamma e^{-\gamma t}$$

The expression for the likelihood function becomes:

$$l = \prod_{i=1}^{n} f(t_i) = \prod_{i=1}^{n} \gamma e^{-\gamma t_i} \tag{12.44}$$

or, if we consider the negative log-likelihood, this becomes:

$$L = \sum_{i=1}^{n} \{ -\ln(f(t_i)) \} = \sum_{i=1}^{n} \{ \gamma t_i - \ln(\gamma) \} \qquad (12.45)$$

giving

$$L = \gamma \sum_{i=1}^{n} t_i - n\ln(\gamma) \qquad (12.46)$$

Now, the minimum value of L will occur when

$$\frac{\partial L}{\partial \gamma} = 0$$

that is:

$$\frac{\partial L}{\partial \gamma} = \frac{\partial}{\partial \gamma} (\gamma \sum_{i=1}^{n} t_i - n\ln(\gamma)) = \sum_{i=1}^{n} t_i - \frac{n}{\gamma} = 0 \qquad (12.47)$$

or

$$\frac{n}{\gamma} = \sum_{i=1}^{n} t_i$$

giving

$$\frac{1}{\gamma} = \frac{1}{n} \sum_{i=1}^{n} t_i = \bar{t} = \lambda \qquad (12.48)$$

where, $\bar{t}$ is the arithmetic mean of times.

Thus the mean time to failure is, in fact, the maximum likelihood estimator of the parameter of the exponential distribution.

Example 12.8

Suppose it has been decided to demonstrate the reliability of a turbine disc by running a number of discs on spin rigs until they burst.   The

following times were recorded as the times-to-failure for each of 5 discs:
10496, 11701, 7137, 7697 and 7720 respectively.

Solution:

Making the assumption that these times are exponentially distributed
then we can find the MLE of the parameter as

$$\lambda_{MLE} = \frac{1}{n}\sum_1^n t_i = \frac{10496 + 11701 + 7137 + 7697 + 7720}{5} = 8950.2$$

We can also determine the value of L from equation (12.46) as:

$$L = \gamma \sum_{i=1}^n t_i - n\ln(\gamma) = 5 - (-45.497) = 50.497$$

Note: this does not mean that the times are exponentially distributed and
it does not say anything about how well the data fits this distribution, for
that we will need to look at interval estimators, later in the chapter.

Weibull Distribution

The Weibull distribution has two parameters: β and η

Its probability density function is given by:

$$f(t) = \frac{\beta t^{\beta-1}}{\eta^\beta} e^{-\left(\frac{t}{\eta}\right)^\beta} = \beta\psi t^{\beta-1} e^{-\psi t^\beta} \text{ where } \psi = \eta^{-\beta}$$

The expression for the likelihood function becomes:

$$l = \prod_{i=1}^n f(t_i) = \prod_{i=1}^n \beta\psi t_i^{\beta-1} e^{-\psi t_i^\beta} \qquad (12.49)$$

and

$$L = -\sum_{i=1}^n \ln(\beta\psi t_i^{\beta-1} e^{-\psi t_i^\beta}) \qquad (12.50)$$

giving,

$$L = -n\ln(\beta) - (\beta - 1)\sum_{i=1}^{n}\ln(t_i) - n\ln(\psi) + \psi\sum_{i=1}^{n}t_i^{\beta} \qquad (12.51)$$

Now, the minimum is given when

$$\frac{\partial L}{\partial\vec{\theta}} = 0 \text{ i.e. } \frac{\partial L}{\partial\beta} = \frac{\partial L}{\partial\psi} = 0$$

giving

$$\frac{\partial L}{\partial\beta} = -\frac{n}{\beta} - \sum_{i=1}^{n}\ln(t_i) + \psi\sum_{i=1}^{n}t_i^{\beta}\ln(t_i) = 0$$

$$\frac{\partial L}{\partial\psi} = -\frac{n}{\psi} + \sum_{i=1}^{n}t_i^{\beta} = 0$$

Taking the second equation first gives:

$$\frac{1}{\psi} = \frac{1}{n}\sum_{i=1}^{n}t_i^{\beta}$$

or

$$\lambda = \left(\frac{1}{n}\sum_{i=1}^{n}t_i^{\beta}\right)^{1/\beta} \qquad (12.52)$$

We can now substitute this expression for ψ into the first differential equation to give:

$$\frac{n}{\beta} + \sum_{i=1}^{n}\ln(t_i) - \frac{n\sum_{i=1}^{n}t_i^{\beta}\ln(t_i)}{\sum_{i=1}^{n}t_i^{\beta}} = 0 \qquad (12.53)$$

This is now independent of ψ so is relatively easy to solve numerically using Newton-Raphson or similar search method. In MicroSoft™ Excel® then you can use Solver® to solve this expression or, one could actually use this on the original likelihood expression, although one has to be careful in setting the precision and conditions, recognising that neither β nor ψ (η) can take negative values. One may need to scale the times (by dividing them all by the mean or 10^k, say) to avoid overflow problems particularly if β may become "large" (> 10, say) which it can easily do if n is small (<10).

Example 12.9

Using the same data as above in the Exponential case, we can now find, using Solver®, the value of β that satisfies equation 12.53

Table 12.11 Calculations for Weibull MLE

	Times	Ln(T)	T^β	$Ln(T)*T^\beta$
T_1	10496	9.25875	3.18E+21	2.9457E+22
T_2	11701	9.36743	5.69E+21	5.32914E+22
T_3	7137	8.873048	4.04E+20	3.58881E+21
T_4	7697	8.948586	6.06E+20	5.4208E+21
T_5	7720	8.95157	6.16E+20	5.50982E+21
Sums	44751	45.39938	1.05E+22	9.72678E+22

In Table 12.11, the value for β is 5.3475...

If we substitute the values for β and those in Table 12.11 Into equation 12.53 we get:

$$\frac{5}{5.3475} + 45.39938 - \frac{5*9.72678*10^{22}}{1.05*10^{22}} = 6.45237*10^{-10}$$

Note: that the right-hand side is not equal to zero but is within the tolerances set in Solver®.

By using the same values we can calculate L = 44.720.

This value is less than that obtained for the Exponential (= 50.497) which suggests that the Weibull is a better fit which suggests the times to failure are age-related. Unfortunately, it does not tell us how good the fit is or, indeed, whether another (theoretical) distribution might fit the data more closely.

Normal Distribution

The normal distribution differs from the previous two distributions in so far as it is defined for both positive and negative values but this does not affect the MLE process. The probability density function for the normal distribution is given by:

$$f(x:\mu,\sigma) = \frac{1}{\sigma\sqrt{2\pi}} e^{-\frac{(x-\mu)^2}{2\sigma^2}}$$

Giving

$$l = \prod_{i=1}^{n} \frac{1}{\sigma\sqrt{2\pi}} e^{-\frac{(x_i-\mu)^2}{2\sigma^2}} \tag{12.54}$$

and

$$L = \sum_{i=1}^{n} \left\{ \frac{1}{2}\ln(2\pi) + \ln(\sigma) + \frac{(x_i-\mu)^2}{2\sigma^2} \right\} \tag{12.55}$$

The maxima and minima of L occur when

$$\frac{\partial L}{\partial \mu} = \frac{\partial L}{\partial \sigma} = 0$$

$$\frac{\partial L}{\partial \mu} = \frac{1}{2\sigma^2} \sum_{i=1}^{n} -2(x_i - \mu) = 0$$

which, can be reduced to

$$\mu = \frac{1}{n}\sum_{i=1}^{n} t_i = \bar{t} \tag{12.56}$$

That is, the maximum likelihood estimator of the mean is simply the sample mean.

$$\frac{\partial L}{\partial \sigma} = \frac{n}{\sigma} - \frac{1}{\sigma^3} \sum_{i=1}^{n} (x_i - \mu)^2$$

Which can be reduced to

$$\sigma^2 = \frac{1}{n} \sum_{i=1}^{n} (x_i - \mu)^2 \qquad\qquad (12.57)$$

Which is the definition of the (population) variance.

Example 12.10

Returning to our 5 discs, we can now apply these two formulae to determine the MLE of the mean and variance.

Solution:

Table 12.12 Calculations for Normal Distribution MLE

	Times	$(x-\mu)^2$
T_1	10496	2389498
T_2	11701	7566901
T_3	7137	3287694
T_4	7697	1570510
T_5	7720	1513392
Mean & Var.	8950.2	3265599
St. Dev.		1807.097

This gives $\mu_{MLE} = 8950.2$ and $\sigma_{MLE} = 1807.1$.

Substituting these point estimates of μ and σ into equation 12.55 Gives the log-likelihood

L = 44.59

This value is slightly lower than that obtained for the Weibull (44.72) which suggests that the normal distribution provides a (marginally) better fit to the data than the Weibull. This, again, indicates that the cause of failure is age-related.

Minimal Repair or 'Bad-as-Old'

In many cases, when components are repaired, rather than replaced, their ages cannot be reset to zero. For example, if a puncture is repaired by inserting a mushroom plug in the hole made by the foreign object, the age of the tyre, relative to its depth of tread, is completely unaffected.

A similar situation can exist when there are multiple occurrences of a part and the repair is achieved by only replacing the one instance that has failed. Typically, compressor and turbine sets contain up to one hundred identical blades. During a repair, it is possible for only one of these blades to be replaced thus leaving the remainder unaffected and hence in no way rejuvenated. This scenario is slightly more complicated than the simply puncture repair case, as the set will contain one new blade and n-1 old ones after the first repair. It is possible that the next blade to fail is the "new" one although, it is much more likely to be one of the original ones. However, as the number of repairs approaches, and certainly after it has exceeded, the number of blades in the set, the probability that the failure will be to one of the replacement blades will increase.

Now, if the times to failure, from new, are t_1, t_2, ..., t_n (such that $t_1 < t_2 < ... < t_n$) then the likelihood function becomes:

$$l = \prod \frac{f(t_i)}{R(t_{i-1})} \text{ where } t_0 = 0 \tag{12.58}$$

For a Weibull distribution, the likelihood function becomes:

$$l = \prod_{i=1}^{n} \frac{\psi \beta t_i^{\beta-1} e^{-\psi t_i^\beta}}{e^{-\psi t_{i-1}^\beta}} \tag{12.59}$$

Thus

$$L = -n\ln(\psi) - n\ln(\beta) - (\beta - 1)\sum_{i=1}^{n}\ln(t_i) + \psi t_n^{\beta} \qquad (12.60)$$

After differentiation, we get:

$$\psi = \frac{n}{t_n^{\beta}} \text{ or } \eta = \frac{t_n}{\sqrt[\beta]{n}} \qquad (12.61)$$

$$\beta = \frac{n}{n\ln(t_n) - \sum_{i=1}^{n}\ln(t_i)} \qquad (12.62)$$

Note: that the original likelihood function is only applicable if one component has failed n times since new with repairs to "same-as-old".

Example 12.11

Suppose that instead of the five discs that have been used in the preceding examples, we had only one disc that was repaired to same-as-old, after each failure and that the times quoted (in random order) were the times since new.

Solution:

Firstly, we recognise that $t_n = \max(t_i) = 11701$.

$$\beta = \frac{5}{5*9.36743 - 45.39938} = 3.478 \text{ and } \eta = \frac{11701}{5^{1/3.478}} = 7366.0$$

Mixed Repairs and Replacement

Consider a component's history has been a sequence of repairs and replacements, or, equivalently, a number of instances of identical components (operating in different systems) each undergo a number of "same-as-old" repairs. An example of the former might be an axe head in which a "repair" is considered to be "re-sharpening" then we might sharpen the blade several times before deciding it is time to replace it. Another example is the booster rockets on the Challenger space shuttle which are recovered after each launch and re-shaped 19 times before they are replaced.

If we consider n new items/replacements and r_I repairs for each item (i = 1, n) then the likelihood function becomes:

$$l = \prod_{i=1}^{n} \prod_{j=1}^{r_i} \frac{f(t_{i,j})}{R(t_{i,j-1})}$$ (12.63)

Thus

$$L = N \ln(\psi) - N \ln(\beta) - (\beta - 1) \sum_{i=1}^{N} \ln(t_i) + \psi \sum_{i=1}^{n} t_{r_i}^{\beta}$$ (12.64)

where N is total number of failures.

Giving, after differentiating

$$\psi = \frac{N}{\sum\limits_{i=1}^{n} t_{r_i}^{\beta}} \quad \text{or} \quad \eta = \left(\frac{\sum\limits_{i=1}^{n} t_{r_i}^{\beta}}{N} \right)^{\frac{1}{\beta}}$$ (12.65)

$$\frac{N}{\beta} + \sum_{i=1}^{N} \ln(t_i) = \frac{N \sum\limits_{i=1}^{n} t_{r_i}^{\beta} \ln(t_{r_i})}{\sum\limits_{i=1}^{n} t_{r_i}^{\beta}}$$ (12.66)

Example 12.12

Suppose a disc was tested until it failed at time 7137 hr, when it was replaced. The second disc failed at 7697 hr was repaired (same-as-old) and failed again at 11701 hr. The third disc was repaired at 7720 hr and finally failed at 10496 hr.

Solution

N = 5, n = 3, t_r = 7137, 11701 and 10496 respectively

Using Solver® gives,   β = 4.57 and η = 9256.1

Censored Data

In most situations, we are unlikely to have complete or uncensored time to failure, repair or support data. Generally, we need to have an idea of how reliable a component or system is from quite early on in the life of these items. This may be so that we can monitor/check the in-service "reliability" against the target, stated, guaranteed or desired "reliability" (maintainability or supportability). It may also be so that we can make better predictions of demands on facilities, equipment, resources and spare parts. We also want to know how well we have designed the various parts of the system so that we can best direct our efforts in future designs.

As we have seen earlier in this chapter, there are various types of censoring but, just as it was not critical for the median rank regression (or use of probability paper) so it is similarly not critical with the MLE method. The major difference between MRR and MLE, however, is that all of the actual ages, whether at the time of failure or the time of censoring, are used explicitly.

If we accept that f(t: θ) is the probability or likelihood that a component will have failed at time t given it has probability density function f(t) with set of parameters θ then we can say that the probability that it has not failed by time t is R(t: θ). This means that the likelihood function (l) for a set of times {t_1, t_2, ..., t_r, t_{r+1}, t_{r+2}, ..., t_{r+s}} where the first r are failures and the last s are suspensions (or censored) is given by:

$$l = \prod_{i=1}^{r} f(t_i) \prod_{i=r+1}^{r+s} R(t_i) \qquad (12.67)$$

and

$$L = -\sum_{i=1}^{r} \ln(f(t_i)) - \sum_{i=r+1}^{N} \ln(R(t_i)) \text{ where N} = r + s \qquad (12.68)$$

Note: the MLE method is not affected by the ordering of the times, provided f(t) is used for failures, (completed repair or support times) and R(t) is used for those which have not yet failed (and unfinished repairs or support tasks).

Weibull distribution for multiply censored samples

For the Weibull distribution with shape β and scale η (= $\psi^{-1/\beta}$) for components whose repairs are to as-good-as-new or the times are to first failure

$$l = \prod_{i=1}^{r} \psi\beta t_i^{\beta-1} e^{-\psi t_i^{\beta}} \prod_{i=r+1}^{r+s} e^{-\psi t_i^{\beta}} \qquad (12.69)$$

Giving

$$L = -r\ln(\psi) - r\ln(\beta) - (\beta-1)\sum_{i=1}^{r}\ln(t_i) + \psi\sum_{i=1}^{r} t_i^{\beta} + \psi\sum_{i=r+1}^{r+s} t_i^{\beta} \qquad (12.70)$$

Which simplifies to

$$L = -r\ln(\psi) - r\ln(\beta) - (\beta-1)\sum_{i=1}^{r}\ln(t_i) + \psi\sum_{i=1}^{N} t_i^{\beta} \qquad (12.71)$$

Giving on differentiating

$$\frac{\partial L}{\partial \psi} = -\frac{r}{\psi} + \sum_{i=1}^{N} t_i^{\beta} = 0$$

or

$$\psi = \frac{r}{\sum_{i=1}^{N} t_i^{\beta}} \text{ or } \eta = \left(\frac{\sum_{i=1}^{N} t_i^{\beta}}{r} \right)^{1/\beta} \qquad (12.72)$$

and

$$\frac{\partial L}{\partial \beta} = -\frac{r}{\beta} - \sum_{i=1}^{r} \ln(t_i) + \psi \sum_{i=1}^{N} t_i^{\beta} \ln(t_i) = 0$$

giving

$$\frac{1}{\beta} + \frac{1}{r} \sum_{i=1}^{r} \ln(t_i) = \frac{\displaystyle\sum_{i=1}^{N} t_i^{\beta} \ln(t_i)}{\displaystyle\sum_{i=1}^{N} t_i^{\beta}} \qquad (12.73)$$

Recognising that the terms on the left-hand-side (LHS) involve only the failures whereas those on RHS involve the whole sample.

Example 12.13

Suppose 10 discs were to be tested to destruction and were put on test at various different starting times. At a particular moment, 5 of the discs had failed at times, 7137, 7697, 7720, 10496 and 11701 and 5 were still running having achieved times of 4222, 4993, 6164, 7440 and 13233.

Solution

Using Solver® we get

$$\beta = \frac{5*7.32148*10^{17}}{5*6.82039*10^{18} - 45.39938*7.32148*10^{17}} + \varepsilon = 4.24$$

and

$$\eta = \left(\frac{7.32148*10^{17}}{5} \right)^{1/4.24} = 11122$$

Multiply Censored, Mixed Repair and Replace Data

In practice, reliability data is likely to contain a mixture of times to first failure, times to second, third, ..., failure, following same-as-old repairs and suspensions which again may be before the first, second, third, ... failure. It is also quite possible that repairs are neither to same-as-old nor to same-as-new but to somewhere in between the two, i.e. the "repair effectiveness" may be between 0 and 1. Unfortunately, it is extremely unlikely that this value will be known and, indeed, it is also likely to be different for ever repair so we will have to decide, based on whatever facts can be obtained, whether to assume same-as-old or same-as-new.

It is quite feasible to use MLE for such a complex scenario. Basically, it is a question of combining some of the above. We shall leave the actual mathematics to the interested reader.

12.6.2   Interval Data

In complex systems, it is often quite possible for a component to fail but go undetected for some time. With systems, such as aircraft or nuclear power plants, every effort is made to minimise the number of "single point failures" – i.e. components whose failure is likely to cause the (catastrophic) failure of the system. This usually involves a level of duplication or redundancy. If these are "hidden" deep inside an engine, airframe, computer or whatever then, unless some warning mechanism is in place, their failure may not become apparent until the parent unit is next in need of repair, (invasive) maintenance or until the remaining (duplicate/redundant) component fails.

In this type of scenario, we would usually have two pieces of information about the time of failure: that the component had not failed at time t_l but had been found in a state of failure at time t_u. The actual time to failure t therefore lies somewhere in the interval $t_l < t \le t_u$.

The probability that a failure will occur at some time in the interval $(t_l, t_u]$ is given by:

$$\Pr\{t_l < t \le t_u\} = F(t_u) - F(t_l) \tag{12.74}$$

However, we can actually be a little more specific because we also know that $t > t_l$ so the probability now becomes:

$$\Pr\{t_l < t \le t_u \mid t > t_l\} = \frac{F(t_u) - F(t_l)}{R(t_l)}$$

The likelihood function is then given by:

$$l = \prod_{i=1}^{n} \frac{F(t_{u,i}) - F(t_{l,i})}{R(t_{l,i})} = \prod_{i=1}^{n} \left\{ 1 - \frac{R(t_{u,i})}{R(t_{l,i})} \right\} \tag{12.75}$$

If we consider the Weibull case we will see that the likelihood function is given by:

$$l = \prod_{i=1}^{n} 1 - e^{-\psi(t_{u,i}^{\beta} - t_{l,i}^{\beta})} \tag{12.76}$$

Unfortunately, this expression is not greatly simplified by taking logarithms. However, we could consider trying to minimise the likelihood that the failures did **not** occur in the given intervals. The probability that a component will not have failed by time t_u given that it had not failed by time t_l (where $t_u > t_l$) is given by:

$$\Pr\{t > t_u \mid t > t_l\} = \frac{R(t_u)}{R(t_l)} \tag{12.77}$$

So, for the Weibull case, this becomes:

$$l = \prod_{i=1}^{n} e^{-\psi(t_{u,i}^{\beta} - t_{l,i}^{\beta})} \tag{12.78}$$

Giving

$$L = \sum_{i=1}^{n} \psi t_{u,i}^{\beta} - \sum_{i=1}^{n} \psi t_{l,i}^{\beta} \tag{12.78}$$

and

$$\frac{\partial L}{\partial \beta} = \psi \left(\sum_{i=1}^{n} t_{u,i}^{\beta} \ln(t_{u,i}) - \sum_{i=1}^{n} t_{l,i}^{\beta} \ln(t_{l,i}) \right) = 0 \tag{12.80}$$

$$\frac{\partial L}{\partial \psi} = \left(\sum_{i=1}^{n} t_{u,i}^{\beta} - \sum_{i=1}^{n} t_{l,i}^{\beta} \right) = 0 \tag{12.81}$$

Which are both independent of ψ and only have solutions if $t_{u,i} = t_{l,i}$ for all i. In other words, the MLE method cannot be used for interval data. One way round this would be to consider the two extreme conditions: that the component failed at $t_l + 1$ or, at t_u. This would set bounds between which we could expect the true parameters to lie. Another possibility is to use a method that has been referred to as *Weibayes*. In this case, however, we need to have an (*a priori*) estimate of the shape (β). We can then use the t_l values as suspensions.

Appendix

1. Standard normal distribution table.
2. Gamma function table
3. t-distribution table

The Φ(z) for Standardised Normal Variable

Z	Φ(z)	Z	Φ(z)
-6	9.9E-10	0	0.5
-5.9	1.82E-09	0.1	0.539827896
-5.8	3.33E-09	0.2	0.579259687
-5.7	6.01E-09	0.3	0.617911357
-5.6	1.07E-08	0.4	0.655421697
-5.5	1.9E-08	0.5	0.691462467
-5.4	3.34E-08	0.6	0.725746935
-5.3	5.8E-08	0.7	0.758036422
-5.2	9.98E-08	0.8	0.788144666
-5.1	1.7E-07	0.9	0.815939908
-5	2.87E-07	1	0.841344740
-4.9	4.8E-07	1.1	0.864333898
-4.8	7.94E-07	1.2	0.884930268
-4.7	1.3E-06	1.3	0.903199451
-4.6	2.11E-06	1.4	0.919243289
-4.5	3.4E-06	1.5	0.933192771
-4.4	5.42E-06	1.6	0.945200711
-4.3	8.55E-06	1.7	0.955434568
-4.2	1.34E-05	1.8	0.964069734
-4.1	2.07E-05	1.9	0.971283507
-4	3.17E-05	2	0.977249938
-3.9	4.81E-05	2.1	0.982135643
-3.8	7.24E-05	2.2	0.986096601
-3.7	0.000108	2.3	0.989275919
-3.6	0.000159	2.4	0.991802471
-3.5	0.000233	2.5	0.993790320
-3.4	0.000337	2.6	0.995338778
-3.3	0.000483	2.7	0.996532977
-3.2	0.000687	2.8	0.997444809
-3.1	0.000968	2.9	0.998134120
-3	0.00135	3	0.998650033
-2.9	0.001866	3.1	0.999032329
-2.8	0.002555	3.2	0.999312798

The Φ(z) for Standardised Normal Variable (Continued)

-2.7	0.003467		3.3	0.999516517
-2.6	0.004661		3.4	0.999663019
-2.5	0.00621		3.5	0.999767327
-2.4	0.008198		3.6	0.999840854
-2.3	0.010724		3.7	0.999892170
-2.2	0.013903		3.8	0.999927628
-2.1	0.017864		3.9	0.999951884
-2	0.02275		4	0.999968314
-1.9	0.028716		4.1	0.999979331
-1.8	0.03593		4.2	0.999986646
-1.7	0.044565		4.3	0.999991454
-1.6	0.054799		4.4	0.999994583
-1.5	0.066807		4.5	0.999996599
-1.4	0.080757		4.6	0.999997885
-1.3	0.096801		4.7	0.999998698
-1.2	0.11507		4.8	0.999999206
-1.1	0.135666		4.9	0.999999520
-1	0.158655		5	0.999999713
-0.9	0.18406		5.1	0.999999830
-0.8	0.211855		5.2	0.999999900
-0.7	0.241964		5.3	0.999999942
-0.6	0.274253		5.4	0.999999967
-0.5	0.308538		5.5	0.999999981
-0.4	0.344578		5.6	0.999999989
-0.3	0.382089		5.7	0.999999994
-0.2	0.42074		5.8	0.999999997
-0.1	0.460172		5.9	0.999999998
0	0.5		6	0.999999999

Gamma Function

x	$\Gamma(x)$	x	$\Gamma(x)$	x	$\Gamma(x)$	x	$\Gamma(x)$
1.01	0.99433	1.51	0.88659	2.01	1.00427	2.51	1.33875
1.02	0.9884	1.52	0.88704	2.02	1.00862	2.52	1.34830
1.03	0.98355	1.53	0.88757	2.03	1.01306	2.53	1.35798
1.04	0.97844	1.54	0.88818	2.04	1.01758	2.54	1.36779
1.05	0.97350	1.55	0.88887	2.05	1.02218	2.55	1.37775
1.06	0.96874	1.56	0.88964	2.06	1.02687	2.56	1.38784
1.07	0.96415	1.57	0.89049	2.07	1.03164	2.57	1.39807
1.08	0.95973	1.58	0.89142	2.08	1.03650	2.58	1.40844
1.09	0.95546	1.59	0.89243	2.09	1.04145	2.59	1.41896
1.10	0.95135	1.60	0.89352	2.10	1.04649	2.60	1.42962
1.11	0.94740	1.61	0.89468	2.11	1.05161	2.61	1.44044
1.12	0.94359	1.62	0.89592	2.12	1.05682	2.62	1.45140
1.13	0.93993	1.63	0.89724	2.13	1.06212	2.63	1.46251
1.14	0.93642	1.64	0.89864	2.14	1.06751	2.64	1.47377
1.15	0.93304	1.65	0.90012	2.15	1.07300	2.65	1.48519
1.16	0.92980	1.66	0.90167	2.16	1.07857	2.66	1.49677
1.17	0.92670	1.67	0.90330	2.17	1.08424	2.67	1.50851
1.18	0.92373	1.68	0.90500	2.18	1.09000	2.68	1.52040
1.19	0.92089	1.69	0.90678	2.19	1.09585	2.69	1.53246
1.20	0.91817	1.70	0.90864	2.20	1.10180	2.70	1.54469
1.21	0.91558	1.71	0.91057	2.21	1.10785	2.71	1.55708
1.22	0.91311	1.72	0.91258	2.22	1.11399	2.72	1.56964
1.23	0.91075	1.73	0.91467	2.23	1.12023	2.73	1.58237
1.24	0.90852	1.74	0.91683	2.24	1.12657	2.74	1.59528
1.25	0.90640	1.75	0.91906	2.25	1.13300	2.75	1.60836
1.26	0.90440	1.76	0.92137	2.26	1.13954	2.76	1.62162
1.27	0.90250	1.77	0.92376	2.27	1.14618	2.77	1.63506
1.28	0.90072	1.78	0.92623	2.28	1.15292	2.78	1.64868

Gamma Function (Continued)

1.29	0.89904	1.79	0.92877	2.29	1.15976	2.79	1.66249
1.30	0.89747	1.80	0.93138	2.30	1.16671	2.80	1.67649
1.31	0.89600	1.81	0.93408	2.31	1.17377	2.81	1.69068
1.32	0.89464	1.82	0.93685	2.32	1.8093	2.82	1.70506
1.33	0.89338	1.83	0.93969	2.33	1.18819	2.83	1.71963
1.34	0.89222	1.84	0.94261	2.34	1.19557	2.84	1.73441
1.35	0.89115	1.85	0.94561	2.35	1.20305	2.85	1.74938
1.36	0.89018	1.86	0.94869	2.36	1.21065	2.86	1.76456
1.37	0.88931	1.87	0.95184	2.37	1.21836	2.87	1.77994
1.38	0.88854	1.88	0.95507	2.38	1.22618	2.88	1.79553
1.39	0.88778	1.89	0.95838	2.39	1.23412	2.89	1.81134
1.40	0.88726	1.90	0.96177	2.40	1.24217	2.90	1.82736
1.41	0.88676	1.91	0.96523	2.41	1.25034	2.91	1.84359
1.42	0.88636	1.92	0.96877	2.42	1.25863	2.92	1.86005
1.43	0.88604	1.93	0.97240	2.43	1.26703	2.93	1.87673
1.44	0.88581	1.94	0.97610	2.44	1.27556	2.94	1.89363
1.45	0.88566	1.95	0.97988	2.45	1.28421	2.95	1.91077
1.46	0.88560	1.96	0.98374	2.46	1.29298	2.96	1.92814
1.47	0.88563	1.97	0.98769	2.47	1.30188	2.97	1.94574
1.48	0.88575	1.98	0.99171	2.48	1.31091	2.98	1.96358
1.49	0.88595	1.99	0.99581	2.49	1.32006	2.99	1.98167
1.50	0.88623	2.00	1.00	2.50	1.32934	3.00	2.00

Critical t values with ν degrees of freedom

ν	α				
	0.100	0.050	0.025	0.010	0.005
1	3.078	6.314	12.706	31.821	63.657
2	1.886	2.920	4.303	6.695	6.625
3	1.639	2.353	3.182	4.541	5.841
4	1.533	2.132	2.776	3.747	4.604
5	1.476	2.015	2.571	3.365	4.032
6	1.440	1.943	2.447	3.143	3.707
7	1.415	1.895	2.365	2.998	3.499
8	1.397	1.860	2.306	2.896	3.355
9	1.383	1.833	2.626	2.821	3.250
10	1.372	1.812	2.228	2.764	3.169
11	1.363	1.796	2.201	2.718	3.106
12	1.356	1.782	2.179	2.681	3.066
13	1.350	1.771	2.160	2.650	3.012
14	1.345	1.761	2.145	2.624	2.977
15	1.341	1.753	2.131	2.602	2.947
16	1.337	1.746	2.120	2.583	2.921
17	1.333	1.740	2.110	2.567	2.898
18	1.330	1.734	2.101	2.552	2.878
19	1.328	1.729	2.093	2.539	2.861
20	1.325	1.725	2.086	2.528	2.845
21	1.323	1.721	2.080	2.518	2.831
22	1.321	1.717	2.074	2.508	2.819
23	1.319	1.714	2.069	2.500	2.807
24	1.318	1.711	2.064	2.492	2.797
25	1.316	1.708	2.060	2.485	2.787
26	1.315	1.706	2.056	2.479	2.799
27	1.314	1.703	2.052	2.473	2.771
28	1.313	1.701	2.048	2.467	2.763
29	1.311	1.699	2.045	2.462	2.756
∞	1.282	1.645	1.960	2.326	2.576

References

Anderson R. T. and Neri L., (1990), *Reliability-Centred Maintenance: Management and Engineering Methods*, Elsevier Science Publishers Ltd, London.

Anon, (1984), 'Electronic Reliability Design Handbook (MIL-HDBK-388),' US Department of Defense.

Asher H.E., and Kobbacy K.A.H., (1995), Modelling Preventive Maintenance for Deteriorating Repairable Systems, *IMA J. Math. APPl. Business & Industry* Vol. 6, pp. 85-99.

Baker R., (1995), 'Airline Operations,' in *The Handbook of Airline Economics,'* Aviation Week Group, 307-320.

Barlow R. E., and Proschan F., (1975), *Statistical Theory of Reliability and Life Testing*, Holt, Rhinehart & Winston, Inc., New York.

Barlow, R. E., Proschan, F., and Hunter, L C., (1967), *'Mathematical Theory of Reliability,'* John Wiley and Sons, New York.

Barros, L, (1996), 'Optimization Model for Level of Repair Analysis,' M.Sc Course Notes, *University of Exeter*, UK

Bateman (1999), *'The Development of Post-Strategic Defence Review (SDR) Business and Inventory Models for RAF Logistics,'* M Sc Dissertation, University of Exeter.

Berman, O and Dinesh Kumar, U., (1999) Optimisation Models for Recovery Block Schemes, *European Journal of Operational Research,* Volume 115, No. 2, 368-379.

Berman, O., and Dinesh Kumar, U., (1999) " Optimisation Models for Complex Recovery Block Schemes," *Computers and Operations Research,* 26 (5), 525-544.

Bazovsky, I, (1961), *'Reliability Theory and Practice*, Prentice Hall, New York

Berg M., Bienvenu M., and Cleroux, (1986), Age Replacement Policy with Age-Dependent Minimal Repair, *Infro.* 24, 26-32.

Birolini, A., (1997), *'Quality and Reliability of Technical Systems,'* Second Edition, Springer-Verlag.

Black W.H., Borhes W.S., and Savits T.H., (1988), A General Age Replacement Model with Minimal Repair, *Naval Research Logistics,* 35, 365-372.

Blanchard, B S., and Fabrycky, W J. (1991), *'Systems Engineering and Analysis,'* Prentice Hall, New York.

Blanchard B. S., Dinesh V., and Peterson E. L., (1995), *Maintainability: A key to Effective Serviceability and Maintenance Management*, John Wiley & Sons, Inc.

Blanchard B.S and Fabrycky W. J., (1999), *'Systems Engineering and Analysis* (Third Edition),' Prentice Hall, New York

Blanchard B.S and Fabrycky W. J., (1991) *Life-Cycle Cost and Economic Analysis*, Prentice Hall.

Brauer D.C. and Brauer G.D., (1987), Reliability-Centered Maintenance, *IEEE Transactions on Reliability*, 36(1), 17-24.

Brain P., (1995), *'Health and Usage Monitoring Systems into Helicopter Support process,'* Project Report, University of Exeter, UK

British Standard BS 4778, (1991), Section 3.2; *Glossary of International Terms*, British Standard Institutes, London.

BS 5760 Reliability of systems, equipment and components, (1997), Part 23, *Guild to life cycle costing*. British Standards Institution, London.

Building Maintenance Information, *Report 254*, 1996.

Bull J. W., (1993), *Life Cycle Costing for Construction*, Chapman & Hall. London

Cini, P F and Griffith, P., (1999), 'Designing for MFOP: Towards the Autonomous Aircraft,' *Journal of Quality in Maintenance Engineering*, 5(4).

Crawford, L.R., (1995), 'World Forecast 1994-2010 of Aircraft Maintenance Capacity, Versus Demand for Independent Maintenance Providers,' in *The Handbook of Airline Economics,' Aviation Week Group*, 385-396.

Chorley E., (1998), *'Field Data – A Life Cycle Management Tool for Sustainable in-Service Support*,' M.Sc. Dissertation, University of Exeter, UK

Crocker, J., Dinesh Kumar, U, and (1999), 'Age-Related Maintenance Versus Reliability Centered Maintenance: A Case Study on Aero-Engines, *Reliability Engineering and System Safety,'* 67, 113-118.

Cox, D R. (1962), *Renewal Theory,* Methuen & Co Ltd.

Dinesh Kumar, U., and J Knezevic, (1998), 'Supportability - Critical Factor on Operational Availability, *International Journal of Quality and Reliability Management,* 15(4), 481-488.

Dinesh Kumar, U., Crocker, J., and Knezevic (1999), 'Maintenance free Operating Period – An Alternative Measure to MTBF and Failure for Specifying Reliability,' *Reliability Engineering and System Safety,* 64, 127-131

Dinesh Kumar, U., (1999), 'New Trends in Aircraft Reliability and Maintenance Measures,' *Journal of Quality in Maintenance Engineering,'* 5(4), 287-295.

Dinesh Kumar, U., J Crocker, J Knezevic (1999), 'Evolutionary Maintenance for Aircraft Engines,' *Proceedings of the Annual Reliability and Maintainability Symposium,* 62-68,

Dinesh Kumar, U., and Birolini, A., (1999), Approximate Expressions for Availability of Repairable Series-Parallel Structures, *International Journal of Reliability, Quality and Safety Engineering.*

Dinesh Kumar, U., (1998). Reliability Analysis of Fault Tolerant Recovery Blocks, *OPSEARCH,* 35(4), 281 – 294.

Dinesh Kumar, U and Gopalan, M N., (1997), Analysis of Consecutive k-out-of-n:F Systems with Single Repair Facility, *Microelectronics and Reliability ,* 37(4), 587-590.

Dinesh Kumar, U., (1999), 'Spares Forecasting and Constant Failure Rate - The magic and the Myth,' *Proceedings of the 15th International Logistics Congress,'* Exeter, UK, 91-96.

Dinesh Kumar, U and J Knezevic, (1998) Availability Based Spare Optimisation Using Renewal Process, *Reliability Engineering and System Safety,* Vol. 59, pp. 217-223

Dinesh Kumar U., and Knezevic, J., (1997), Spare Optimisation Models For Series and Parallel Structures, *Journal of Quality in Maintenance Engineering,* 3(3), 177-188.

Dinesh Kumar U., (1997) Spares Modelling Using Alternating Renewal Process, *Proceedings of the 7-th International M.I.R.C.E Symposium on System Operational Effectiveness, Exeter.*

Dhillon, B S., (1981), *'Engineering Reliability: New techniques and Applications,* John Wiley & Sons, New York.

Dekker P., (1994), *Applications of Maintenance Optimisation Models: A Review and Analysis*, Report 9228/A, Econometric Institute, Erasmus University Rotterdam, The Netherlands.

Drenick, D F., (1960), ''The failure law of complex equipment,' *Journal of the Society for Industrial Applied Mathematics,* 8, pp. 680-690.

Ebeling, C., (1997), *'An Introduction to Reliability and Maintainability Engineering,'* McGraw Hill International Editions, New York

El-Haram, M., and J. Knezevic, (1995), *Indicator and Predictor as Two Distinct Condition Monitoring Parameters, Proc. of the 5th International Logistics symposium,* 179-186, Exeter, UK.

El-Haram M., (1995), *Integration Approach to Condition-Based Reliability Assessment and Maintenance Planning*, Ph D. Thesis, University of Exeter.

El-Haram M. A., Horner R. M., and Munns A., (1996), Application of RCM to Building Maintenance Strategies, *Proc. of the 6th International Logistics Symposium*, Exeter, UK, 133-143.

M. El-Haram and R. M. Horner (1998), Factors affecting Life Cycle Cost in the Construction Industry, *Proc. of the 8th International Logistics Symposium*, 229-240, Exeter, UK.

Gopalan, M N., and Dinesh Kumar, U., (1996) ,'Approximate Analysis of n-Unit Cold-Standby Systems, *Microelectronics and Reliability,* 36(4), 505-509.

Flanagan R. and Norman G., (1983), *Life-Cycle Costing for Construction*, Surveyors Publications Ltd.

Gnedenko, B.V., Belyayev, K., and Solovyev., ' (1969), '*Mathematical Methods of Reliability Theory,'* Academic Press, New York.

Griffith, G., (1998). '*A Study into the Effects of Reliability and Maintainability on Future Combat Aircraft Logistic Support Costs*,' M.Sc. Dissertation, University of Exeter, UK

Hess, J A., (1988), 'Measuring Software for its Reuse Potential,' *Proceedings of the Reliability and Maintainability Symposium,'* 202-207

Hillman, H., (1997), '*Impreciseness and Ambiguity in Logistics Support*,' *M.Sc. Dissertation,* University of Exeter, UK

Hockley, C J and Appleton, D. P., (1997), 'Setting the Requirements for the Royal Air Force's Next Generation Aircraft,' *Annual Reliability and Maintainability Symposium,* 44-49.

Kapur, K C., and Lamberson, (1977), '*Reliability Engineering in Engineering Design,'* John Wiley & Sons, New York

Kelly A. & Harris M. J., (1978), *Management of Industrial Maintenance,* Butterworths Co., UK.

Kelly A. & Harris M. J., (1978), *Management of Industrial Maintenance,* Butterworths Co., UK.

Handlarski J., (1980), Mathematical Analysis of Preventive Maintenance Schemes, *J. Opl. Res. Soc.*, , 33, 227-237.

Kendrick, D., Reynolds, S., Power, R., and Gore, D., (1998) 'Expert provisioner: the development and implementation of a knowledge based decision support tool for Royal Air Force (RAF) reprovisioning,' *Proceedings of the 8th International M.I.R.C.E symposium (Editors: J Knezevic, D Kumar and C Nicholas)*, 89-102

Knezevic J., (1987a), 'Condition Parameter Based Approach to Calculation of Reliability Characteristics, Reliability Engineering,, 19(1), 29-39.

Knezevic J., (1987b), Required reliability level as the optimisation criterion, *Maintenance Management International*, Elsevier, 6(4), 249-256.

Knezevic, J., (1993), '*Reliability, Maintainability and Supportability – A Probabilistic Approach,'* McGraw-Hill Book Company, London

Knezevic J., (1997), *Systems Maintainability; Analysis, Engineering and Management,* Chapman & Hall, London.

Knotts, R., (1996) '*Analysis of the Impact of Reliability, Maintainability and Supportability on the Business and Economics of Civil Air Transport Aircraft Maintenance and Support,* M.Phil. Thesis, University of Exeter, UK

Knotts R., 1999, '*Integrated Logistic Support*,' M.Sc. Course Notes, University of Exeter, UK

Knowles, D I, (1995), 'Should we Move Away from the Acceptable Failure Rate,' *Communications in Reliability, Maintainability and Supportability,* 2(1), 23-28

482

Koss, E., (1988), 'Software Reliability Metrics for Military Systems,' *Proceedings of the Reliability and Maintainability Symposium*, 190-194.

Lam., M., (1995), 'An Introduction to Airline Maintenance,' *Handook of Airline Economics, Avaition Week Group*, 397-406.

Lam, M., (1997), 'Nothing Exceeds Like Excess,' *Aircraft Economics*, Aircraft Economics.

Lavalle J. C., Collantes R., Sanz M. A., and Palacios R., (1993), SEDIMAHE: An Expert System to Help the Maintenance of Machine Tools, *Maintenance*, 8(3), 10-14.

Lewis, C., (1871). *Through the Looking Glass and What Alice Found There*, Macmillan, London

Lloyd, D K and Lipow, M. (1962), '*Reliability: Management, Methods and Mathematics*, Prentice Hall, New York.

Maintaining the Boeing 747, Aircraft Aconomics, 1994.

Mann Jr, L., Saxena A., and Knapp G. M., (1995), Statistical-based or Condition-based Preventive Maintenance, *Quality in Maintenance Engineering*, 1(1), 46-59.

Marshall H. E., (1991), Economic Methods and Risk Analysis Techniques for Evaluation Building Investment - A Survey, *CIB Report 136 for W55*.

Matt, R., (1993) '*The Red Queen: Sex and the Evolution of Human Nature*, Penguin.

MIL-STD 721C, (1966), *Military Standard, Definitions of Effectiveness Terms for Reliability, Maintainability, Human Factors and Safety*, Department of Defense, Washington, DC.

MIL-STD-1629A, *Procedures for Performing a Failure Mode, Effects and Criticality Analysis*, Department of Defense, Washington D.C., 1984.

Misra, K B., (1991), '*Reliability Analysis and Prediction – A Methodology Oriented Treatement,*' Elsevier, Amsterdam.

Mitchell, P., (1998), '*A New Approach to Reliability*, M.Sc. Dissertation, University of Exeter, UK

Mobley R. K., (1990), *An Introduction to Predictive Maintenance*, Van Nostrand Reinhold, New York.

Mobley R. K., (1994), *The Horizons of Maintenance Management*, Maintenance Handbook, 5th edition, McGraw-Hill, New York.

Morgan P., (1999), '*No Fault Found – The Human Factors and the Way Forward ,*' Project Report, University of Exeter, UK

Moubray J., (1997), *Reliability-centred Maintenance - RCM II*, Butterworth and Heinemann, Oxford, UK

Nakagawa T., and Kowada M., (1983), Analysis of a system with minimal repair and its application to replacement policy, *European Journal of Operational Research* 12, 176-182.

Nakagawa T., (1983), Optimal number of failure before Replacement Time, *IEEE Transactions on Reliability* 23, 155-116.

Nakajima S., (1986), TPM - Challenge to the Improvement of the Productivity by Small Group Activities, *Maintenance Management International*, 6, 73-83.

Nakajima S., (1989), *TPM Development Program: Implementing Total Productive Maintenance*, Productivity Press, Inc., Cambridge.

Nowlan F. S. and Heap H. F., (1978), *Reliability Centred Maintenance*, Springfield, Verginia, National Technical Information Services, US Department of Commerce.

Palm, C., (1938), Analysis of Erlang Formulae for Busy Signal Arrangements,' *Ericsson Techniques*, 4, 39-58.

Paulsen J. L. & Lauridsen K., (1991), *Information Flow in a Decision Support System for Maintenance Planning*, 261-270 Operational Reliability and Systematic Maintenance, Elsevier Science Publisher, Ltd., UK.

Pearcy, D (1999), 'A Study into Operational Life Cycle Costs by Investigating the Interaction Between 'Out-of-Service' Time and the Maintenance Concept,' M.Sc Dissertation, University of Exeter, UK

Pierskalla, W. and Voelker, J. (1976), A Survey of Maintenance Models: The Control and Surveillance of Deteriorating System, Naval Research Logistics Quarterly 23, 353-388.

Pironet (1998), 'Multiple provisioning strategy for weapons system used by multiple missions,' Proceedings of the 8th International M.I.R.C.E symposium (Editors: J Knezevic, D Kumar and C Nicholas), 147-160

Robinson, I (1998), 'Investigation of Appropriateness of Statistical Distributions in Spares Scaling,' M.Sc. Dissertation, University of Exeter, UK.

Sherbrooke, C C., (1992), Optimal Inventory Modeling of Systems, John Wiley, New York

Sheu S.H., (1994), Extended Block Replacement Policy with used Item and General Random Minimal Repair Cost, European Journal of Operational Research 79, pp. 405-416.

Sherif Y. S. & Smith M. L., (1981), Optimal Maintenance Models for Systems Subject to Failure- a Review, Naval Research Logistics Quarterly, 28, 47-74,

Smith A. M., (1993), Reliability Centred Maintenance, McGraw-Hill, Inc., New York.

Tijms, H. (1995). Stochastic Models – An Algorithmic Approach, John Wiley.

Turner M. (1999), 'Unmanned Compact Aircraft,' project Report, University of Exeter, UK

US MIL-HDBK-217, 'Reliability Prediction for Electronic Systems,' USAF, Rome

Valdes-Flores, C., and Feldman R., (1989), A Survey of Preventive Maintenance Models for Stochastically Deteriorating Single-Unit Systems, Naval Research Logistics Quarterly 36, 419-446.

Verma D., and Knezevic. J., (1995), 'Conceptual System Design Evaluation: Handling the Uncertainty,' Proceedings of the International MIRCE Symposium, Exeter, 118-130.

Verma, D. (1993), 'A Causal Emphasis During The Failure Mode, Effects and Criticality Analysis,' Proceedings of the International Logistics Symposium, 120-126.

Verma, D. (1999), 'Fault Tree Analysis,' M.Sc. Logistics Engineering Course Notes, University of Exeter, UK

Vanneste S.G., (1992), A Generalized Age-Replacement Model, Probability in the Engineering and Information Sciences, 6, 525-541.

Watson C., (1970), Is preventive Maintenance Worthwhile?, Operational Research in Maintenance, Edt. By Jardin A.K.S., Manchester University Press, Manchester.

Williams J. H., Davies A., and Drake P. R., (1994), Condition-based Maintenance and Machine Diagnostics, Chapman & Hall.

Willmott P., (1989), Maintenance Engineering in Europe-the Scope for Collaborative Technology Transfer and Joint Venture, Maintenance Journal, Vol. 4, No. 4, pp 10-13.

Xie M. (1989). " On the Solution of Renewal-Type Integral Equations,' Communications in Statistics., B., 18, 281-293.

Xie M., (1991), Software Reliability Modelling,' World Scientific Publishing Company, Singapore.

INDEX